Sharks Get Cancer, Mole Rats Don't

Sharks Get Cancer, Mole Rats Don't

How Animals Could Hold the Key to Unlocking Cancer Immunity in Humans

JAMES S. WELSH, MD

59 John Glenn Drive
Amherst, New York 14228

Published 2016 by Prometheus Books

Cover design by Jacqueline Nasso Cooke
Cover images © Media Bakery

Inquiries should be addressed to
Prometheus Books
59 John Glenn Drive
Amherst, New York 14228
VOICE: 716–691–0133
FAX: 716–691–0137
WWW.PROMETHEUSBOOKS.COM

20 19 18 17 16 5 4 3 2 1

Library of Congress Cataloging-in-Publication Data Pending

Printed in the United States of America

CONTENTS

PREFACE

People sometimes arrive at their destinations through the oddest pathways, and this is particularly true in my case. When I was young, I knew that I wanted to be a scientist, veterinarian, or physician, but I was having difficulty deciding, and my path was anything but straight. Looking back, I now realize what a profound influence a tortoise named "Jalopy" had on my life's trajectory.

Many years ago at the Staten Island Zoo in New York City, some lucky children had the pleasure of riding on the back of Jalopy, a giant Galapagos tortoise. With fond memories, I can count myself among those fortunate few. Jalopy arrived at the Staten Island Zoo in 1937 at a svelte 85 pounds. I was four years old when I met him, and by then he had grown to a sturdy 150 pounds. Even with only a little kid on his back, Jalopy was such a slow-moving vehicle that a ride around the tortoise pen could take an hour. But it was a thrilling ride for me.

In later years when I was in college, Jalopy had attained a weight of 175 pounds, and I recall occasionally going to see him during spring break (while most of my friends were off doing the more fashionable things that college students do). One sad day, however, I learned that Jalopy was not there—he was at the University of Arizona undergoing radiation therapy for a sarcoma (a type of cancer) between his neck and his left front leg. I had never heard of a sarcoma before, nor did I even know that tortoises got cancer. I was under the mistaken impression that cancer was exclusively a human disease brought about by smoking or other unhealthy habits or by occupational exposure to certain chemicals. And I certainly did not know anything about radiation therapy. But I was elated to hear that my venerable reptile friend had just about completed his course of treatment and was coming home soon. To keep his neck and left leg elevated and out of the dirt, the lumbering giant had a skateboard strapped under his shell, which helped him get about. For a brief period, I think Jalopy was the world's fastest tortoise!

Unfortunately, Jalopy was not a long-term cancer survivor. He succumbed to his malignancy at the relatively tender age of approximately seventy-five (which is just middle age for a Galapagos tortoise). Nevertheless, upon reflection of his most unusual case, I was inspired to enter the field of cancer medicine, specifically radiation oncology.

We are now in the midst of an exciting revolution in cancer immunotherapy, but when I began writing this book several years ago, there were few credible hints that any big breakthroughs were imminent. It was not immediately obvious that the fascinating tales from the animal kingdom I was learning and writing about would ever prove to be medically relevant at all. But just as Jalopy inspired me decades ago, these observations have inspired me during this writing campaign. While trying to better understand spontaneous remissions of cancer and the abscopal phenomenon, I have encountered some surprising and truly tantalizing clues from the strangest sources: Tasmanian devils, tapeworms, naked mole rats, elephants, sharks, dogs, clams and others play important roles in this amazing journey. Beholding how Nature tangles with cancer brings me full circle—back to the cancer center, where today's exciting new immunotherapy is yielding unprecedented success for certain cancers.

I hope the readers who embark on this exotic journey with me will find that the final product—an exciting and optimistic future for cancer immunotherapy—was strangely predictable. The clues to beating cancer have been right there before us, just waiting for us to find them and logically piece them together. Searching far, wide, and deep, this book brings those pieces together.

Chapter 1

WHAT JUST HAPPENED?

Allow me to present a very distressing case . . .

Daniel—a seemingly healthy, soft-spoken fellow in his early thirties—was about to get news no one ever wants to hear. For years, this athletic, young engineer was eager to soak up some sun and enjoy weekend basketball with his friends. No one can ever know if these habits contributed to the development of an ominous black blotch that began growing on the back of his neck. Like a shark's fin, just beneath the surface was a potential killer.

One day after a pickup game, a friend asked him about a little "freckle" that in some way looked different and angrier. "Hey, Dan, what's that on your neck?" pointing to the dark, discolored patch of skin that slightly resembled tree bark. After seeing it himself with the aid of a couple of mirrors, Daniel decided to have it checked out.

The news was certainly not what he wanted; he had a highly aggressive skin cancer called malignant melanoma. The *wide local excision* surgical procedure appeared to remove all of the cancer, but melanoma has a notorious tendency to spread far and wide. Most unfortunately for Daniel, within a year his cancer did just that. Despite state-of-the-art combination chemotherapy (meaning multiple drugs given at the same time), the cancer progressed unabated. Almost taunting the chemotherapy, the malignancy soon inundated his body, spreading first to his lymph nodes and then to his lungs. The next line of defense was interferon, an early edition of cancer immunotherapy. Clinical studies had suggested that this then relatively new form of cancer immunotherapy could be of benefit to some melanoma patients. Daniel was not one of them. Interferon can be rough on patients, and the harrowing flu-like symptoms of headache, fever, and malaise were just too intense. He capitulated, electing to discontinue treatment. In defiance of the brief interferon treatment, his disease spread, or *metastasized*, unrelentingly, as melanoma is wont to do.

Although advanced cancers can ultimately spread practically anywhere, most have preferred sites of initial migration. For example, prostate cancer normally first metastasizes to bone, whereas colon cancer often first spreads to the liver. Melanoma tends to initially involve the lymph nodes, lungs, and liver—and also the brain. At that time, brain metastases heralded an imminent death sentence. Fortunately for Daniel, his MRI (magnetic resonance imaging) studies never showed brain metastases. But melanoma also has a nasty penchant for attacking bone. And in due course, Daniel's inexorable cancer began assailing his skeleton, causing at first only a dull, throbbing ache in his femur (thighbone). Sadly, this pain magnified unremittingly over time to the point where it was all he could think about.

By the time I first saw Daniel in the radiation oncology clinic he was pale, emaciated, weak, and in a wheelchair, unable to walk because of his now lancinating leg pain. He had that all-too-familiar lifeless look: sallow, wasted, and withered beyond his years. Given his recently discovered liver metastases, his medical oncologist had already answered—quite realistically—that dreaded question that no physician wants to address; Daniel had perhaps four more months to live.

Daniel was given heavy doses of morphine, yet the pain pierced through the drug's defensive layer. I could sense that he was suffering quietly as we spoke about his symptoms. We both hoped that I could target a dose of radiation to the bone tumor, thereby shrinking it somewhat and afford him at least a modicum of relief. My first concern however was his risk of a possible *pathological fracture*—Was his femur in imminent danger of fracturing due to the erosion brought on by the bone-eating tumor? If so, he would require prophylactic surgery to avert a disaster that would likely render him bedridden for the rest of his days. Fortunately the x-rays indicated that the bone had not yet disintegrated to that degree. Nevertheless, cancer was rasping away at this bone, and possibly several others, and needed to be dealt with expeditiously. Before initiating radiotherapy I obtained a *bone scan* (a nuclear medicine study designed to determine if other bones were involved). If one envisions tumors as tiny light bulbs, Daniel's skeleton lit up like a Broadway marquee.

I offered him *palliative* radiation therapy to relieve his excruciating pain and to reduce the risk of future pathological fracture. By definition, although this palliative treatment might reduce his pain and forestall a fracture, it was unable to significantly prolong his life. As far as any long-term prospects, it looked like all hope was lost. This radiotherapy was not going to change his prognosis; it was strictly symptom relieving.

I know Daniel tried to pay attention during our consultation, but his mind was unfocussed, distracted by the agony in his leg. I wanted to have an honest and open discussion with him about his prognosis and what the radiotherapy could and could not do. He brusquely concluded the conversation, "Doctor, I just want this pain to go away. When can we get started?"

At the time, a typical course of palliative radiation therapy was around ten to fifteen treatments spread out over two to three weeks. Given his dismal forecast and knowing that he didn't have much time left, in order to accelerate things I decided to treat him with a *hypofractionated* course of radiation therapy. Hypofractionation is simply the use of fewer, but larger, daily doses of radiation. Although hypofractionation is now recognized as a superior way of treating certain cancers, and today we attempt to "biologically optimize" our radiotherapy courses through hypofractionation, back then in Daniel's case I did this solely to speed the treatment course along.

Daniel's course of radiation therapy consisted of twenty Gray directed exclusively to the rapidly growing tumor gnawing away at his femur. (The Gray is the standard unit of radiation dose. One Gray is one joule of ionizing radiation energy deposited in one kilogram of matter.) It was finished in a few days. At a follow-up only a week after completion of treatment, Daniel was pleased to declare that his pain was already markedly reduced. Not entirely gone by any means, but considerably better than just a week earlier.

Needless to say, I was elated about his newfound pain relief, but I also couldn't help but feel a sense of despair. Why was this previously healthy, young man enduring this tragic early ending? I wished there was something more I, or someone, could do. But there was not. So I did all I could—I renewed his pain medications. I encouraged him to call me if he ever developed any new pain that I might similarly dispel. We scheduled a tentative three-month follow-up with a CT scan (computed axial tomography or CAT scan). I say "tentative" since odds were that he would not be around that long; I most likely would never see him again.

It would be a monumental understatement to say I was surprised when he actually showed up three months later. Seeing him was like being in a waking dream. It was Daniel, but it wasn't. He was now completely pain free; in fact, he said he had discontinued all pain medication shortly after his last visit. And ceasing all those narcotic pain pills (oxycodone and morphine) had another agreeable benefit—he was not constipated for the first time in nearly a year. Mystifyingly, but more satisfying still, he looked somehow healthier. He'd gained a good ten pounds, and more than that, there was a puzzling new vital spark about him.

This was all wonderful, but as to his scheduled follow-up CT scan, I subtly tried to dissuade him. Back then it was customary to obtain a follow-up CT scan after any course of chemotherapy or radiation therapy, and I had reflexively ordered it. But in fact, there was nothing to be gained from the scan—the only thing it could bring was bad news. After all, we had exhausted essentially all available options, so why bother? What would we gain? Nevertheless, despite my gentle protests, Daniel desired the CT scan so I personally walked him over to Radiology. It was a genuine pleasure to be strolling beside this man who only a few months back had been wheelchair-bound.

The Radiology Department was right around the corner, so I invited him to return after his imaging study and we could discuss the results if he was interested. This was way back in the days when everything was still on film rather than on a computer screen. Since I had some time, I strolled over to Radiology early to review the scan with the diagnostic team. At first I was tempted to berate the novice radiology resident for an absolutely amateur error—he must have mixed up the name tags. The scan in front of me was that of a person with no evidence of cancer. Upon closer scrutiny, I could see several small radiographic "scars" or tiny traces of where the tumors once were. On detailed inspection of his recently irradiated thigh bone, I could discern some unmistakable effects of his radiotherapy along with increased *ossification*, or formation of new bone. The young radiologist had made no mistake. This was in fact Daniel's scan.

Just three months earlier, Daniel's body was riddled with cancer. Today he had a "clean" CT scan with no trace of malignancy anywhere. Astonishing!

Returning to my office, I reviewed the scans with Daniel, who naturally was ecstatic about the great news. Being quite bewildered myself, I asked him a series of probing questions. First I asked him if he had had any recent chemotherapy that I wasn't aware of. The answer was no. What about other early variants of cancer immunotherapy such as interleukin-2 or anything like that? Again, no. I began grasping, "Any special vitamins or over-the-counter supplements?" "Nope." Then I really started stretching. "You didn't go to Mexico and try laetrile or some other esoteric alternative therapy by any chance did you?" The reply was again a resounding NO! He emphatically asserted that aside from a lot of people praying for him, the *only* thing he did since the radiation therapy was steadily taper off his oxycodone and morphine pain medications.

By now it had become abundantly clear to him that I, his cancer doctor, was at a complete loss as to how to explain this. And although I thought I made it unambiguous at the time of our consultation, it only now truly dawned on him that the radiation therapy I administered was *not* supposed to do this. Nevertheless, for whatever reason, it apparently had. Daniel didn't seem to care. He was happy—and so was I. A good bit befuddled, but happy.

So what just happened?

Radiation therapy is a *local* treatment, meaning that it should not have body-wide, or *systemic*, effects. Like surgery and other local therapies, radiation therapy normally only affects the specific anatomical region being targeted. This differs from systemic therapies such as chemotherapy, which address the entire body from head to toe. A systemic therapy could, in principle, simultaneously induce remission of lung, liver, lymph node, and bone metastases, as in Daniel's case, but local therapies such as radiation therapy, surgery, cryotherapy (localized freezing), or hyperthermia (localized heating) certainly should not. The radiotherapy should have affected only the single thigh bone addressed. Obviously something far larger and unexpected occurred. I began to ask colleagues and read up on the subject of spontaneous remissions and radiation-therapy-induced body-wide remissions. I came across a then rather obscure phenomenon called the *abscopal effect*.

In an article many decades ahead of its time, the abscopal phenomenon was first described by Dr. R. H. Mole back in 1953.[1] The term is derived from the Greek root "*ab*" and Latin root "*scopus*" for "away from the target" and aptly describes what happened. There was an effect—an astounding effect—far from the target. Although it was rather obscure when I first learned of it, if one scans the Internet or does a PubMed search for "abscopal effect" today, one will find scores of hits and recently published papers. In fact, I first described Daniel's case in *Discover* magazine in 2014, well over a decade after he was treated.[2] Despite the increased awareness and attention, at the time I am writing this, there is still no clear understanding of just how or why this curious and exceedingly rare phenomenon occurs. I, along with most others interested in this phenomenon, are convinced that the abscopal phenomenon represents an extreme example of the immune system's ability to recognize and occasionally overpower even highly advanced cancer.

Daniel came in for routine checkup visits for several more years. Each physical examination and follow-up scan showed no evidence whatsoever of cancer. He gained back all of his lost weight and even proudly packed on some new muscle. A year after his four days of treatment, he was back on the basketball court (albeit with a lot more sunscreen) and again enjoying life. After several years of routine follow-up appointments, Daniel shook my hand firmly and said goodbye, electing not to come back for any more scans or doctor visits. After all, who wants to waste time in the doctor's office when there is so much to be done, so much life to live? For all intents and purposes, he was cured. When I first met him he wasn't expected to live beyond four *months*—yet here he was, thanking me and striding away four *years* later.

To date, cases like Daniel's remain extremely uncommon, and there is no obvious rhyme or reason linking them together. Moreover, Daniel's case was extraordinary on another level. Beyond just the rarity of an abscopal phenomenon, his complete response was *long lasting*. This is a welcomed, but far from guaranteed, outcome with abscopal effects and spontaneous remissions. In fact, the majority are not enduring.

I have only witnessed what appeared to have been durable abscopal effects twice in my clinical career. The other long-lasting response involved a sixty-year-old gentleman with metastatic esophageal cancer. He was diagnosed with adenocarcinoma of the lower esophagus after presenting with progressively worsening *epigastric* pain (upper abdominal pain just at the lower end of his breastbone, or sternum) and *dysphagia* (difficulty swallowing). Esophageal cancer is generally considered one of the worst, even when it is caught early. When diagnosed at a more advanced stage (which is all too often), esophageal cancer is considered incurable. His initial medical evaluation included a CT scan, which showed that the cancer had already metastasized to his liver. Because of this diagnosis of stage IV esophageal cancer, curative surgery and definitive radiation therapy were considered futile, and he began chemotherapy alone. (Cancers are typically assigned stages of I to IV, with stage I being the earliest and stage IV being the most advanced.)

Despite several cycles of chemo, his overall condition continued to deteriorate, and his dysphagia progressed to where he began choking on his own backed-up saliva. Saliva, or any liquid, entering the lungs can lead to infection and lung injury, so he was at risk for *aspiration pneumonia*—a very serious and potentially fatal complication. Although palliative radiation had the potential of relieving his esophageal obstruction and allowing food, liquid, and secretions to pass, it really had no potential to impinge on the numerous metastases still evident in his liver. Or so I thought.

The two-week course of radiation therapy successfully shrank his obstructive esophageal tumor surprisingly quickly, and we were both quite pleased with that outcome. After the first week of treatment, he was no longer choking on his secretions, and by the completion of treatment, he was able to swallow both solids and liquids far easier than he had been able to for many months. Shortly thereafter, his epigastric pain had also fully abated. He was more comfortable than he had been in quite a while. The palliative radiation therapy was a big success!

In this instance, I did not bother him with a CT scan to assess response. Since the treatment was designed strictly for symptom relief, if the treatment worked, we would both know it. On the other hand, if the treatment failed, we would unambiguously know it without any sophisticated medical imaging to enlighten us. Furthermore, since radiation therapy generally cannot be delivered more than once to a specific site (because normal tissues can only tolerate so much radiation before serious long-term consequences might develop), there was little logic in getting a follow up CT. Additionally, by this time insurance companies were beginning to balk at routinely covering such follow-up imaging studies unless a clear benefit could be documented, and in this case it was considered unjustified. He discontinued follow-up in radiation oncology and elected to continue follow-up exclusively with his medical oncology team. This was perfectly reasonable since he was expected to live only a few more months—and if there was to be any breakthrough for his metastatic cancer, it would most certainly come from medical oncology rather than radiation oncology.

To my amazement, we met again in a grocery store nearly *five years* later—but at first I didn't recognize him at all. Fortunately, he recognized me. When I was treating him years before, he was bald from his chemotherapy, pale, skinny, and always hunched over in pain and discomfort. Yet when I saw him in the supermarket that day, he was standing strong and walking tall at six feet six.

We began chatting, and, most unexpectedly, he told me that after he completed the radiation therapy, he never had any additional treatment whatsoever. Yet over the years he did get a series of follow-up imaging studies, including several PET scans (positron emission tomography), a more cancer-specific form of medical imaging, which at that time had only recently come along in the clinic. Regardless, on all his follow-up studies, there was no trace of cancer anywhere. He told me that his medical oncologists attributed the remission to a delayed response to the chemotherapy, but he thought the radiation did the trick, and he thanked me profusely.

Utterly dumbfounded, I certainly didn't think the supermarket was the place to remind him that the radiotherapy I gave him was solely for symptom relief and in no way a definitive attempt to control his advanced cancer. At a loss for words, I told him sincerely how great it was to see him then nerdily wished him something to the effect that he should live long and prosper. Here was another patient whom neither I—nor anyone else—thought would be around for even five months, let alone five years. Yet here he was, robust and tough.

So why then would I start this chapter by describing these miraculous cases as "distressing"? They were certainly not distressing to the cured patients. They were distressing to me—the confused physician! To this day, I remain intrigued and perplexed by the abscopal effect. These extraordinary cases long ago triggered an Ahab-like obsession in me. For quite some time now, I have been fixated on trying to better understand just what happened—and more importantly, to willfully and repeatedly replicate it.

Chapter 2

ACTION AT A DISTANCE

Physics should represent a reality in time and space, free from spooky actions at a distance.

—Albert Einstein

Is it possible that these rare examples of enduring abscopal effects offer tantalizing clues to a real cure for cancer? Personally I believe they could. Unfortunately, the first case was around fifteen years in the past, while the other was now over a decade ago, and I have seen nothing remotely similar since. As mentioned, radiotherapy-related abscopal effects are extremely scarce; it allegedly is unusual for a cancer physician to see anything of the sort over a thirty-year career. Nevertheless, having witnessed it twice, I am beginning to suspect that such abscopal phenomena might not be quite as uncommon as we all once believed. Pinning down exactly what the common denominator is remains challenging. But right now, huge strides are being taken in cancer immunotherapy as scientists slowly but surely unravel more and more of the mysteries of molecular oncology and reveal the various secrets of tumor immunology.

Leading mainstream scientists and physicians are finally taking heed of the abscopal effect. Recently, a highly publicized case study led by researchers from Memorial Sloan Kettering Cancer Center was published in the *New England Journal of Medicine*.[1] This paper detailed the case of a young female with metastatic melanoma who experienced what appeared to be an abscopal effect. Superficially, her case was quite similar to the situation with Daniel. Her disease was diagnosed in 2004 through a biopsy of a mole on her upper back when she was only thirty-three years old. Like Daniel, she underwent a wide local excision that removed all visible cancer. Unlike Daniel, she also had a sentinel lymph node biopsy—the removal of the lymph node or nodes that are anatomically the first to directly drain the site where the cancer is. In this case, the sentinel nodes were in her left axilla, or armpit. There was no cancer in any of the five lymph nodes removed. She showed no signs of cancer until 2008 when a chest x-ray found a suspicious new *pulmonary nodule* just under an inch in diameter in the lower part of her left lung. She underwent combination chemotherapy and then had the tumor surgically resected (removed) from her lung. Unfortunately, a follow-up CT scan revealed new metastatic disease in lymph nodes within her chest and a mass growing near her spine. At the Memorial Sloan Kettering Cancer Center in New York City, she began a clinical trial involving a new form of cancer immunotherapy. The new drug, ipilimumab, was one of the first so-called checkpoint inhibitors to hit the scene. We will explore checkpoint inhibitors in greater depth in a subsequent chapter, but for now, these checkpoint inhibitors represent a new means by which we might be able to overcome cancer's invisibility to our immune systems.

Despite the new immunotherapy, her disease relentlessly marched onward with new lesions appearing in her spleen. In late 2010, she required palliative radiation therapy for an enlarging, pain-producing tumor near her spine. Like Daniel, she received a hypofractionated radiotherapy course, but in this modern era, full of technological advancements, the dose was far more intense: 28.5 Gray spread over three fractions. Additionally, the treatment was administered using modern *intensity-modulated radiation therapy* (IMRT) and *stereotactic* techniques—a testament to the extensive technological progress made in radiation oncology over the prior decade.

Nevertheless, the treatment was strictly considered palliative treatment, with no expectations of anything more than a local effect on the targeted tumor. As hoped for, this local therapy worked; within a few months, the targeted tumor had shrunk. More remarkably however, the lymph nodes in her chest as well as the multiple lesions in her spleen also vanished. Her remission proved to be durable with minimal disease seen on follow up CT scans nearly a year later. The final report came out in the *New England Journal of Medicine* in 2012, over two years after her radiation.[2]

I believe this was a clear manifestation of the abscopal phenomenon since several diseased regions regressed despite being well outside the local field of radiation. But two key aspects set this study apart from most of the other simple anecdotal reports. The first was the treatment combination with the modern cancer immunotherapy checkpoint inhibitor, ipilimumab, and the second was the detailed laboratory measurements recorded as part of her clinical trial at Memorial Sloan Kettering. Here, the abscopal effect was scrutinized with scientific rigor, and the cellular, biochemical, and molecular details were meticulously recorded.

In a follow-up letter to this *New England Journal* article, a team of researchers from Stanford University Medical Center conveyed a very similar case.[3] This time however, the investigators were *prospectively and intentionally* attempting to induce a radiotherapy-related abscopal response in a metastatic melanoma patient. The regimen consisted of two cycles of ipilimumab, followed by radiation therapy, followed by two more cycles of ipilimumab. In the single case reported, it worked. Although the stereotactic radiotherapy was directed at only two liver lesions, all eight of them dissolved along with a mass in the axilla. The Stanford team suggested that immunotherapy combined with radiation might serve as a sort of "tumor vaccine" and this novel strategy might have applications in other cancers beyond melanoma. With this starting point, will oncologists soon be able to regularly induce long-term abscopal effects in other patients?

We will revisit ipilimumab and other checkpoint inhibitors, along with several other exciting new developments in cancer immunotherapy in later chapters. At this juncture, however it might be instructive to review a lesson from the history of physics.

Albert Einstein, one of the founders of modern physics (thanks to his theory of relativity), was loathe to accept some quirky aspects of another pillar of modern physics, quantum mechanics. One particularly weird prediction of the quantum theory was *entanglement*. Technically, entanglement is a permanent linkage between paired particles. In essence, when two particles are entangled, what affects one instantaneously affects the other—even if they are miles apart. They are a single system irrespective of the space between them. The classic example is the effect of someone determining the angular momentum or "spin" of an entangled particle. If one determines that the spin of a particular particle is clockwise, then automatically the spin of the entangled partner will be counter-clockwise.

Ironically, Einstein was one of the first physicists to realize that quantum mechanics predicted the possibility of entanglement—and he was one of the first to denounce it. To Einstein, this all sounded like voodoo. It violated the common sense principle called *locality*. One should be able to affect only things they are next to, things that are *local*. Adamantly refusing to accept the possibility, Einstein disparagingly described entanglement as "spooky action at a distance." In fact, he argued that entanglement's locality-violating implications *proved* that the quantum theory was incomplete.

But science doesn't bend the rules just because someone objects—

even if that someone is Albert Einstein. Despite opposition by Einstein and others, quantum mechanics has withstood the test of time and passed every quantitative challenge thrown its way. Defying common sense, elegant experiments have recently confirmed the reality of entanglement.[4] Something that happens to an object over there can indeed affect an object over here—and vice versa—if they are entangled. What happens to one really does simultaneously affect the other. They are forever intimately intertwined. Irrespective of distance, and regardless of how long even light would take to cover this distance, this effect occurs instantaneously.

Despite derision as "spooky action at a distance," entanglement is fact. And irrespective of whether or not we can understand it, quantum mechanics is fact. It has ushered in a whole new world of modern physics with a wide range of practical applications in computers and electronics. To quote physicist and *Fabric of the Cosmos* author Brian Greene, "Numerous assaults on our conception of reality are emerging from modern physics. But of those that have been experimentally verified, I find none more mind-boggling than the recent realization that our universe is not local."[5] To me, the nonlocal effects seen in the abscopal phenomenon, "radiation oncology's action at a distance," are equally mind-bending. Will the abscopal effect prove to be as paradigm shifting as quantum mechanics? Despite violating all common sense and conventional wisdom, could nonlocal effects usher in a whole new world of clinical cancer immunotherapy?

Presently, I and others believe there is a quiet revolution going on in cancer medicine. Sundry hints, including the abscopal phenomenon are being handed to us and are at this moment illuminating a heretofore obscured path forward. And some of these pointers are coming from the most unexpected places.

During my decades-long quest, I have uncovered several surprising and strange clues that just *might* have bearing on our ultimate goal of curing cancer. For a glimpse at one of those clues, let's take a brief sojourn to the land Down Under. Before plunging in, however, in the spirit of connecting the dots, allow me to extend an explanation for such seemingly disparate stories.

I recall a very direct and unanticipated question I was asked many years back by the wife of one of my prostate cancer patients. She wanted to know if she could "catch" cancer from her husband through intercourse and wind up with cervical or uterine cancer. My reflexive and reassuring answer was, "No, of course not!" Although I believed my answer was correct, the question did start me wondering. Since then I have been asked similar questions by other patients, and I have even asked myself if it is possible to catch cancer from a patient while in the operating room. For almost two decades, my answer had always remained firm and clear: no, cancer is not contagious. Everyone knows that cancer isn't a transmissible disease. One cannot catch cancer the way one can catch a cold, the flu, or Ebola.

That is, of course, unless you are a Tasmanian devil.

Chapter 3

DISAPPEARING DEVILS

It is in moments of illness that we are compelled to recognize that we live not alone but chained to a creature of a different kingdom, whole worlds apart, who has knowledge of us, and by whom it is impossible to make ourselves understood: our body.

—Marcel Proust

Tasmanian devils are vanishing. In just the past two decades, the natural population of this storied creature—*Sarcophilus harrisii*, the largest extant carnivorous marsupial—has been decimated, with nearly 70 percent already lost and counting. In some regions, as much as 90 percent is now gone. At this rate, some scientists predict a complete extinction of Tasmanian devils by the year 2035. In 2008, the International Union for Conservation of Nature officially listed the Tasmanian devil as an endangered species.[1]

What is going on?

Is it because of pollution? Indiscriminate overhunting? Is there an unstoppable epidemic killing them off? Is an exotic invasive species taking over? Well, it appears that the answer to the last question is yes. In fact, the answer is yes to the last *two* questions. Put them together and you get a most peculiar situation—an epidemic of contagious cancer caused by a parasitic tumor.

In a very real sense, this unstoppable ecological attack can be thought of as an invading alien parasitic species that is commandeering the bodies of Tasmanian devils. In a scenario straight out of Hollywood horror and science fiction, the very same demonic tumor is literally being transferred from one animal to the next like some sort of immortal parasite that jumps from host to host. Correction—it is not *like* some sort of immortal parasite, it *is* an immortal parasite. This evil spirit uses the Tasmanian devil's body simply as a means of transportation to the next body. Lamentably, the parasitic cancer rapidly ravages and discards each living corpse within six months—but often not before it has passed itself along to the next unsuspecting victim.

This contagious cancer, known as devil facial tumor disease (DFTD) because of its presentation as a large, disfiguring, and obstructing tumor on the mouth or snout of the Tasmanian devil, is passed from animal to animal through bites. And this little monster does bite! When competing for mates, warding off unwanted suitors, or when simply venting anger over someone cutting in at the dinner table, these pugnacious, pint-sized beasts are not above letting loose with a bone-crushing bite—right on the schnoz! At not much larger than an average house cat, pound-for-pound these little brutes purportedly boast the most powerful bite of any mammal.

Devils become quite animated when jockeying for position while feasting on a large carcass. In fact, disease transmission is facilitated by a particularly brutal form of combat known as "jaw wrestling" in which contestants will lock jaws and turn, tug, and twist like crocodiles until one surrenders. While such behavior is not uncommon in the fish world (as any tropical-fish hobbyist who keeps cichlids can verify), it seems particularly savage when the combatants are mammals! Tumor cells growing on gums are readily transferred from mouth to mouth in this way.

DFTD customarily causes multiple, large, ulcerated tumors on the head, neck, and face (especially in the mouth) of unfortunate victims. In this regard, the Tasmanian devil cancer is most akin to human "head and neck" cancers such as tongue, tonsil, and throat cancers. Like any malignant tumor, DFTD is capable of metastasizing, and it often does. In about 65 percent of cases the cancer shows signs of spread to lymph nodes, the lungs, the spleen, and, on some occasions, the heart.[2] (Incidentally, metastases to the heart are extremely uncommon in human cancers.) However, unlike the situation with most human cancers, in DFTD it is usually *not* the metastatic cancer and accompanying multi-organ failure that brings about death. The devil's demise more often comes within months of onset due to an inability to eat and drink caused by direct tumor obstruction.

Yes, indeed, this new contagious cancer is wreaking havoc on the wild population of Tasmanian devils and is threatening them with extinction. But this is no ordinary cancer—it is a biological nightmare. This tumor is directly passing itself along from animal to animal in a never-ending chain. What will happen when the last Tasmanian devil succumbs? Will the tumor jump to other species and wipe them out as well? Could it possibly infect humans and force the same course as with the devil? Even if the clear and definite answer to such scary, sensationalistic questions is "probably not," what, if anything, can we learn from asking such questions? Perhaps a lot.

Chapter 4

THE DEVIL HIMSELF: SOME DIABOLICAL BIOLOGY

Before we can legitimately begin ascribing any particular attributes to the contagious DFTD tumor itself, we should first do our due diligence and make certain that there is nothing so very different about the Tasmanian devil itself that *allows* the passing along of contagious cancers. For instance, if the Tasmanian devil's immune system is inherently inferior, could this account for the transmission of this transplanted cancer without the need for any special qualities in the tumor itself? Is it the devil's basic biology that enables the transmission of DFTD? Well, let's get acquainted and meet the devil himself.

The Tasmanian devil alluded to here is not the rapidly rotating cartoon character of Looney Tunes fame but rather *Sarcophilus harrisii*, the largest living carnivorous marsupial. Marsupials, as you might recall, are mammals without true placentas, and thus the newborn (barely more than an embryo) must migrate out of the womb on its own and make its way into a pouch, or *marsupium*, for the final phases of fetal development. This is a rough start to life, mind you, as it is accomplished only a month or so into the pregnancy! As I will describe in greater depth later on, the fact that marsupials do not have a complex placenta to protect the embryo from attack by its mother's immune system might play an important role in our overall story on the immunobiology of cancer.

While the Tasmanian devil is currently the largest living carnivorous marsupial, only recently did it attain this dubious distinction. Up until the twentieth century, this title was held by *Thylacinus cynocephalus*, the thylacine (also called the Tasmanian tiger because of its stripes). At about the size of a German shepherd, the thylacine was probably already extinct on the Australian mainland by the time European settlers arrived, but on Tasmania itself, a large island off the southeast corner of the Australian mainland, the coup de grâce likely was intensive hunting (with enticing bounties). Coupled with this direct assault was the introduction of dogs and the encroachment of humans into its natural habitat. By 1930, the last wild thylacine was shot, and in 1936, Benjamin, the last captive specimen, died.[1]

The name *Thylacinus cynocephalus* originates from the Greek, literally meaning "dog-headed pouched one"—a most fitting scientific name. The equally apt scientific name for the Tasmanian devil, *Sarcophilus harrisii* (Greek for "Harris's meat lover") acknowledges its flesh-eating tendencies and honors naturalist George Harris who first described the animal in 1807.[2]

Other marsupials famously include kangaroos (in which the joey-toting pouch is practically trademarked), along with koalas, opossums, and a host of lesser-known species including bandicoots, quolls, and planigales. Indeed, there are over 330 known species of marsupials, with about 70 percent residing in Australia and the surrounding islands. Although modern marsupials are currently quite limited both geographically and in number of species compared to their more advanced placental mammalian cousins (the eutherians), this was not always the case. At one time, marsupials and other "metatherians" (the larger zoological group, which includes not only living true marsupials but also their extinct ancestors and relatives) had a far more powerful presence worldwide. For example, there was the impressive *Thylacosmilus*, a huge lionlike metatherian maybe weighing up to 120 kg (260 lb.) that roamed South America during the Late Miocene and Early Pliocene epochs from about 11 million to 2.5 million years ago. Curiously, *Thylacosmilus* was a sabertooth. In this regard, it was remarkably similar to the more celebrated "saber-toothed tiger" of the Pleistocene, *Smilodon*. The development of the saber-like teeth of these two remotely related animals is an excellent example of *convergent evolution*, in which an anatomical attribute appears and reappears in unrelated animals or plants thanks to some functional advantage. Another classic example of convergent evolution is the fishlike body found in sharks, dolphins, and the extinct reptilian ichthyosaurs. Incidentally, while *Smilodon* was a member of the cat family (Felidae), it was not really a "tiger," and thus the name saber-toothed *cat* is more appropriate. The name *Smilodon* understandably derives from the Greek words for "carving knife" and "tooth."

While extinct marsupials are undoubtedly fascinating, metatherian paleontology has its challenges—it goes without saying that pouches do not frequently fossilize! But the reconstruction of prehistoric marsupials is aided by one interesting bit of odontological trivia: one can confirm metatherian ancestry by the teeth. Metatherians possess four pairs of molar teeth in each jaw (the lower mandible and upper maxilla), whereas true placental mammals never have more than three pairs. Somewhat surprisingly then, humans, with three pairs of molars (when including the not-always-present wisdom teeth) have the same number of molars as the monster-mouthed hippopotamus. Conversely, the tiniest marsupial, the 0.15 ounce (4.25 grams) Ingram's planigale (*Planigale ingrami*) with its teeny 4 mm tall head has more molars than any human (or hippo for that matter).

Marsupials in general have reproductive systems that can safely be described as odd. For example, female marsupials have two vaginas but both open externally through the same orifice. These two vaginas lead to two separate wombs, or uteri. A third canal, the median vagina, is used for birth. Naturally, to accommodate this unusual anatomy males are equipped with a bifurcated penis whose two ends correspond to the female's two vaginas. Moreover, marsupials have a primitive rear end with a common opening for urinary, fecal, and reproductive purposes. This common opening, or *cloaca* (Latin for "sewer"), representing an earlier stage in evolution, is present in amphibians, reptiles, and birds. In contrast, the most advanced anatomical arrangement—separate orifices for feces, urine, and reproduction—is present in female humans (but of course also in all other female placental mammals).

While not an uncommon feat in reptiles, the Tasmanian devil, fat-tailed sheep, and duck-billed platypus are among the precious few mammals that store fat in their tails. Exactly what are mammals anyway? Sure, during school days we all learned about the common characteristics of mammals such as hair, warm-bloodedness (endothermy), milk production, and live birth (vivipary), but there are some glaring exceptions. For example, the duck-billed platypus (another native of Australia) lays eggs. Additionally, unlike the placental mammals who keep their bodies at around 37°C (98.6°F), the platypus prefers a cool 31°C–32°C—about five degrees Celsius or nine degrees Fahrenheit colder than humans. So, is the platypus a mammal or a reptile? It clearly does not easily fit into the class Mammalia or class Reptilia of Linnaean systematics. One could make an argument, for example, that it is an extant mammal-like reptile—a survivor of the ancient lineage known as therapsids who gave rise to true mammals, roughly around the same time the first dinosaurs appeared way back in the Triassic period.

By using modern cladistic analysis, in which a common trait is traced back to its evolutionary point of origin, one can define "mammals" based on middle–ear-bone anatomy. We humans have three tiny middle-ear bones, or *ossicles*, that help transmit vibrations in the air to the cochlea, where the signal is first recognized as sound. These three tiny ossicles (the malleus, incus, and stapes—sometimes called the hammer, anvil, and stirrup respectively) are not found exclusively in humans. *All* mammals *by definition* have them. Thus, one can use these three ossicles as a cladistic defining feature: keeping it simple, if you have three auditory ossicles, you are a mammal, and if you don't, you are not. Based on this definition, marsupials, such as the Tasmanian devil, and placentals, such as the beloved cat, are both mammals. Despite the fact that birds are warm-blooded and some snakes give birth to live young, since they do not possess the requisite three auditory ossicles, they do not qualify as mammals. Meanwhile the egg-laying, nippleless, cool-blooded, venomous (hind-leg spurs in males), duck-billed platypus does make the cut based on its middle ear anatomy.

Genetic analyses indicate that the marsupials split from the placental mammals somewhere around 160 million years ago, during the Jurassic period. But speaking of genetics, as we shall see in the next chapter, it is germane that the ancestral number of chromosomes in marsupials is believed to be seven pairs, or fourteen in total, including the gender-determining X and Y chromosomes. (Incidentally, human have twenty-three pairs of chromosomes for a total of forty-six.) The marsupial contagious cancer DFTD has apparently decided that that number wasn't good enough and came up with its own unique chromosomal composition. The transmissible tumor, rightfully considered a "new species," hardly qualifies as a marsupial or any other kind of mammal as it has no ossicles (or ears or other organs for that matter); it is just an amorphous blob of incoherent cells.

But let's get back to our original question—Is the Tasmanian devil's immune system incompetent, and is this the explanation for the contagious cancer epidemic? As I shall describe in greater detail in the upcoming chapters, the MHC (Major Histocompatibility Complex) system is the key determinant of compatibility when it comes to organ transplantation in humans and all vertebrates. Basically, every cell in the body must display specific molecular markers on their surfaces called the MHC. These molecular markers serve as flags or ID cards that the immune system demands to see upon inspection. When encountering a new cell, the immune system will ask that cell to show its citizenship papers. In effect, the immune system reads the MHC molecular citizenship papers and decides if the card carrier is friend or foe. If new cells display the wrong MHC or no MHC markers at all, they will be viewed as uninvited intruders and viciously attacked. Some scientists have postulated that because the devil's MHC system lacks genetic diversity it cannot reject the transplanted DFTD tumors.[3] In essence, the devils might be too inbred. But experimental evaluation of this hypothesis has proven otherwise. The devil is not weak! Lack of genetic diversity notwithstanding, there are still enough differences between individuals and enough MHC capacity to prompt obvious immune reactions. Researchers used a standard immunological laboratory assay, the mixed lymphocyte reaction, along with skin grafts on animals to ascertain Tasmanian devil immune capabilities.[4] Although the responses were wide ranging, there were clearly measurable mixed lymphocyte reactions in the assay. This proved first that the devils had some degree of immune function in this assay and second, the blood cells from different individuals were different enough from each other to provoke a response. The most intense reactions occurred when lymphocytes from devils residing on the east coast of Tasmania were mixed with lymphocytes from specimens residing on the west coast.

Of more direct relevance was the fact that all skin grafts were uniformly and completely rejected within two weeks of transplantation, irrespective of the apparent degree of MCH compatibility. These animals may be genetically related but they are not *that* closely related! Also of importance was the observation that the immunological rejection of the skin grafts was accompanied by an intense T cell infiltration, documenting the functional capacity of the Tasmanian devil immune system. (T cells, along with B cells, are types of lymphocytes, key players in the immune system, and will be discussed in depth later on). The bottom line is that in contrast to initial expectations, Tasmanian devils are fully capable of rejecting transplanted organs from strangers. Devil immunity is intact.

So, if the devil's immune system is not the reason for the ease of transmission, it must have to do with the weird features of the DFTD tumor, right? Well, before we can properly jump to that logical conclusion, it might be worthwhile to inspect the weirdness of this and other tumors. Is DFTD truly unique in its ability to pass itself along unrecognized and unaccosted by the immune system—or might all cancers have this same ability?

Chapter 5

DEVIL OF A DISEASE

Because the primary anatomical location of most DFTD cases is in the mouth or somewhere on the face, oncologists would classify this as a "head and neck cancer." In humans, these types of cancers are overwhelmingly *squamous cell cancers*. Squamous cell cancers are derived from *squamous epithelium* made up of scale-like cells (Latin, *squama* = scale). Squamous epithelium serves as a bodily lining. These scale-like cells can be found coating the oral cavity, pharynx, esophagus, and trachea and also appear on the skin surface. Historically, most human head and neck cancers have been attributed to smoking and heavy alcohol consumption, particularly of hard liquor.[1] Recent research has revealed that a substantial fraction of cases are virus-associated and such cases might be on the rise. Specifically, certain strains of HPV, or human papilloma virus, appear to be the guilty culprits. Curiously, victims of HPV-associated head and neck cancers tend to be younger people. Frequently they have never smoked or consumed much hard liquor. Fortunately, the prognosis is a bit better for these individuals with virally induced head and neck cancers.

Returning to our situation in Tasmania, the cancers decimating the devils are definitely not virus related.[2] In fact, as with almost everything about this weird tumor, DFTD deviates from textbook descriptions. First, it is not a classic squamous cell head and neck cancer. Laboratory analyses have shown this is really a *neuroendocrine* cancer, and the original tumor was probably a "nerve sheath tumor"—a tumor of cells that line the nerves (also called Schwann cells). No one knows exactly how this original tumor got transformed into a transmissible head and neck cancer, but my own suspicion is that the devil's penchant for cannibalism is the explanation. I can easily imagine the originally affected Tasmanian devil dying with a nerve sheath tumor somewhere on its body and then another devil chowing down on that original cancer-victim's carcass, chomping into the tumor itself. In this manner, some unlucky devil thereby became the first true DFTD victim, or index case, when the cancerous cells got wedged in its gums, thereby initiating the ill-fated chain reaction. Alternatively, the primary nerve sheath tumor might have been bitten into during a big fight and cancerous cells embedded themselves into the mouth of the aggressor. No one knows for certain.

What *is* known is that this peculiar cancer has evolved into a form that is highly contagious. For a couple of decades now, this singular tumor has managed to effectively pass itself along from host to host. In sharp contrast to the normal devil's lifespan of five or six years (maybe up to eight in captivity), this tumor has been alive and well, propagating from devil to devil since at least 1996, when photographer Christo Baars first captured pictures of animals with horribly mangled faces on Mount William in the far Northeast corner of Tasmania.[3] Since then, the disease has relentlessly marched westward across the island. Dr. Menna Jones of the University of Tasmania may have been the first to truly recognize this as a contagious disease when she captured some animals and performed painstaking studies.[4] Appreciating that this was a cancer initially led to the speculation that it was a virally induced malignancy; virus-associated cancer is not uncommon in animals. Alternatively, given the initial geographical distribution of the early cancer victims in the northeast corner of Tasmania, a toxin of some sort was a reasonable explanation. Laboratory analyses however did not identify any toxin nor was a viral etiology confirmed. Rather, it appeared that each and every cancer was oddly identical.

Ordinary human cancers are grossly and microscopically different in appearance from one another depending on which body part they arise from. Additionally, human cancers are genetically different from one person to another; there might even be significant variability from one biopsy site to another in the very same patient. In startling and perplexing contrast, all the Tasmanian devil cancers appeared uncannily similar. Not just grossly, but microscopically they were remarkably similar. Identical, in fact. Could all the animals somehow have the very same tumor? The real clincher came through cytogenetic analysis—characterization of the chromosomes.

Consisting of the genetic molecule DNA plus associated proteins, *chromosomes* are the carriers of hereditary information. Half of our chromosomes are from our fathers, and the other half are maternally derived. During cell division (mitosis and meiosis), the normally invisible genetic material condenses into visible (with a microscope, that is) chromosomes of characteristic sizes and shapes. These characteristic sizes and shapes allow us to identify individual chromosomes and assign them names or numbers. In humans, the normal chromosomal makeup, or *karyotype*, consists of twenty-three pairs for a total of forty-six. There is one pair that determines a person's gender (the X and Y "sex chromosomes") along with twenty-two distinct pairs of *autosomes* that do not influence one's gender. All animals (and plants, for that matter) have different chromosomal constitutions, and in some animals the X and Y pair does *not* determine that animal's gender (birds and butterflies, for example).

As mentioned in the previous chapter, like all other marsupials, the normal Tasmanian devil karyotype contains fourteen total chromosomes: six pairs of autosomes plus an XX or XY sex chromosomal pair. It is this fact that unraveled the contagious cancer mystery. If the mysterious new cancers were arising anew in each animal host, their chromosomes should basically look like their host's chromosomes, save for some subtle variations such as chromosomal deletions or insertions. Instead, the DFTD tumors all demonstrated a very distinct—but consistent—chromosomal make up. Unlike the standard fourteen chromosomes in normal Tasmanian devil cells, *all* DFTD tumors had an absolutely abnormal array of chromosomes. An early study of the disease led by Anne-Marie Pearce demonstrated that the tumors had only thirteen total chromosomes, rather than fourteen. But it was the specific aberrations that were most remarkable: there were *no* X or Y sex chromosomes, no pair of chromosome-2, and only one member of the pair of chromosome-6. Additionally, the long arm (the "q" arm) of chromosome-1 was deleted. (The shorter arm of a chromosome is called the "p" [for petite] arm, while the longer chromosome arm is called the "q" arm). Not only that, there were four additional unpaired new chromosomes that didn't seem to have any obvious counterparts in the normal Tasmanian devil karyotype. This thing looked alien!

The tumor exhibited some major-league chromosomal aberrations and was scarcely similar to its original parent. The fact that each of these tumors had essentially the *same* weird chromosomal count and composition was the clincher. Beyond a doubt, it was now apparent that the DFTD cancers were *not* arising anew in each individual animal like ordinary cancers do. Instead, this tumor was indeed a new life-form that was passing itself along from host to host as a parasite. Its genes, chromosomes, and cells are so radically different from its original host that it can be rightfully considered a new species. Describing it as an alien life-form is not at all unreasonable.

The bizarre biology of the Tasmanian devil cancer could be relevant to human cancer medicine. All cancers in people demonstrate clear differences from their hosts, often including subtle (or not so subtle) chromosomal abnormalities. Nevertheless, despite these differences, cancers somehow gain invisibility to their host's immune surveillance (there will be much more on this later). But DFTD has taken this to an extreme—not only is it invisible to its original host, it remains invisible even when transplanted to another host devil. As we all know, in transplant medicine this is not normally the case. Organ donors and recipients must be a very close "match" to avoid intense immunological rejection of a transplanted organ. This is where the MHC system comes into play in clinical medicine. Only if the organ donor and recipient are close MHC matches will an organ transplant be successful. Even when the match is fairly good, a lifetime of immunosuppressant medication is normally needed to ward off rejection. Not true with DFTD! This contagious cancer is effectively transplanted indiscriminately from one animal to another with no signs of rejection at all! As far as the immune system is concerned, it is completely invisible. But such invisibility could be a huge clue in defeating cancer. If we can understand the mechanism behind this invisibility trick, we might be able to disable it.

Chapter 6

THE PERFECT PARASITE

Cancer is not usually thought of as an alien life-form. Most of us think of cancer as a deadly disease, the medical scourge of modern society. Additionally, cancer is not considered transmissible. One doesn't "catch" cancer the way one can catch a cold; it has not generally gained the ability to jump from one person to another. (Although, as we shall discuss later, many cancers—perhaps up to 15 percent worldwide—are caused by communicable viruses). Cancer is considered by many to be a product of poor lifestyle choices—cigarette smoking, excessive alcohol consumption, overindulgence at the dinner table, and so forth. While modern humans certainly do have more than their fair share of cancer, the disease is not merely a product of access and excess. For example, Egyptian mummies, from a period long before fast-food restaurants and artificial sweeteners, occasionally show evidence of cancer. Going back a bit further, say one hundred million years or so, dinosaur bones have also demonstrated traces of bone cancers.[1] So, scourge of modern society it may be—but it is far from brand new; cancer existed in the Cretaceous!

However, it seems to have gained strength over time. In the United States alone, the National Cancer Institute (NCI) SEER data program (Surveillance, Epidemiology and End Results Program) estimates that for the year 2015 there will be 1,658,370 new cases of cancer (excluding nonmelanoma skin cancers) and 589,430 deaths due to cancer.[2] These are incidence rates, meaning how many per year. Prevalence data in contrast is not cases per year, but instead, the total number of cases. For instance, the present prevalence of cancer in the United States is nearly fourteen million people (again excluding nonmelanoma skin cancer). Based on 2010–12 data there is now about a 40 percent probability of being diagnosed with some form of cancer at some point during our lifetimes, in the United States.

So, has cancer as a disease really grown stronger? It might simply seem so because other diseases that formerly took young lives in earlier eras (such as infectious diseases and cardiac disease) are now being better controlled. But infectious disease, while better controlled, is far from fully controlled. In fact, one might argue that infectious diseases are themselves growing stronger – and staging a comeback.[3]

So, what types of diseases can gain strength over time? Well, certainly not diseases that are too malevolent. For instance, an infectious disease that quickly and invariably kills its host before it can pass on to another host is an infectious disease headed for extinction. As a classic example of just how quickly things change in medicine, when I first started this project, I began writing that Ebola hemorrhagic fever may be one such disease headed for extinction. As anyone who paid any attention to the news in 2014–15 will testify, Ebola certainly did not die out! It reemerged with a vengeance in West Africa during that time span.

Ebola hemorrhagic fever, named for the Ebola River in Africa but often simply referred to as just "Ebola," is a severe, frequently fatal disease caused by a single-stranded RNA virus belonging to the Filoviridae family. Ebola generally has a fatality rate of over 50 percent, and the most lethal strain ("the Zaire virus") can be as high as 90 percent. Clearly, from a virus's perspective, killing everyone you come in contact with is not the ideal way of propagating! If humans were the exclusive host of Ebola, the virus would run a serious risk of extinction in short order. Fortunately for the virus—but most unfortunately for human and other nonhuman primates—there must be a "reservoir" where the viruses can lay low and hide between attacks on its ultravulnerable hosts. One such ultravulnerable nonhuman host is the western lowland gorilla. In 2002–3 and 2003–4, two epidemics (or epizootics as they are referred to in the animal world) struck the Lossi Gorilla Sanctuary in the Odzala-Kokoua National Park in the northwestern Republic of the Congo, which initially homed twenty thousand gorillas.[4] The first, extending from October 2002 to January 2003, killed 91 percent of the known gorillas in the group. A second epizootic, from October 2003 to January 2004, killed 96 percent of individuals in the afflicted troop. All told, these two Ebola epizootics decimated nearly five thousand gorillas and perhaps as much as 25 percent of the world gorilla population. The devastating loss changed their status from "endangered" to "critically endangered" in 2007 on the World Conservation Union's Red List of Threatened Species.

Exactly where Ebola was hiding between the outbreaks was uncertain for a while but after considering plants, various insectivores, rodents, and primates, bats are emerging as the most likely suspects. Of two dozen plant species and nearly twenty animal species tested, only bats have been successfully infected with the virus in the lab. If all the bats had died in short order, they too would be a poor hiding place. But the fact that they became infected and harbored the virus yet exhibited no clinical signs of disease is the hallmark of a reservoir species.

The singular goal of any and all biological entities is simple—to live. So Ebola, with its terrifyingly high death rate might appear to be living dangerously. When it attacks gorillas or humans, it is likely literally encountering a dead end. Odds are high that the virus will die with its host and never pass itself along. A more "successful" pathogen is one that effectively uses its host for its own needs yet also escapes from that host to infect others. A genuinely successful infectious organism will be transmitted readily from host to host but not kill its host in vain before it can reproduce. Several parasites fit this description beautifully.

For instance, *Diphyllobothrium latum*, the fish tapeworm (which is contracted by eating undercooked or raw freshwater fish) is able to reside in a person's gut for nearly two decades but generally causes only moderate health effects. Along with attaining record lengths of over forty feet and producing an astonishing one million eggs a day, which get spread about via feces, the fish tapeworm is quite capable of provoking major disgust when discussed. Infected individuals often never even know they harbor a ten-meter-long egg-laying factory in their guts. Living in a person's digestive tract has its advantages—tapeworms have little need for a digestive system of their own, thus they focus their efforts exclusively on reproduction. Aside from the tiny head, or scolex, the rest of the stringy body is made up of thousands of repeating units, or proglottids, whose sole function is reproduction. For such a creature, landing in the human gut is hitting the jackpot. It can live happily for years with all the food, warmth, and protection it needs to crank out billions of eggs during its stay.

The lifecycle of the fish tapeworm is quite convoluted. If the eggs, passed out via human feces, somehow work their way back to fresh water, they might then develop into several intermediate forms (including one stage curiously called an oncosphere because of its shape, which was reminiscent of tumors to early zoologists). One of the several intermediate forms is the coracidium, which gets gobbled up by certain freshwater crustaceans (copepods). In the copepod, the coracidium develops into a procercoid larva. The copepod then gets eaten by a small fish (e.g., a minnow) where the procercoid larva emerges and migrates into the fish's flesh and develops into another intermediate form called the plerocercoid larva, or sparganum. The minnows are then eaten by larger freshwater fish, such as pike, perch, or trout, where the plerocercoid larva migrates into the flesh of this larger fish. If a person then eats one of these larger fish without thorough cooking, the larvae then gain access to the small intestine where they mature into egg-laying machines.

This has led some to ill-advisedly advocate harboring a twenty-foot fish tapeworm in their gut as a great weight-loss strategy. After all, if the worm is consuming all your food rather than you absorbing it all, might you not shed those unwanted pounds the easy way? One tapeworm legend has it that in the 1950s opera diva Maria Callas exploited a tapeworm to drop a meaningful amount of weight in short order.[5] What is known is that she did somehow manage to lose about sixty pounds in under a year. She was also known to have contracted a tapeworm at some point in her life. In all likelihood the two separate events were conflated into an enduring rumor. The only fact known with certainty is that intentionally consuming tapeworm eggs in an effort to drop a few pounds is a very bad idea![6] (However, thanks to the emergence of "helminth therapy" [worm therapy] for asthma and other autoimmune diseases, roundworms may still make a comeback in the clinic).

The majority of fish tapeworm cases (medically known by the daunting name of diphyllobothriasis) are entirely asymptomatic, although occasionally people will complain of nonspecific gastrointestinal (GI) symptoms such as indigestion, bloating, nausea, and diarrhea. Rarely, the parasite can cause more important symptoms due to vitamin B12 deficiency. With its voracious appetite, the worm can consume more than 80 percent of a person's normally modest intake of vitamin B12.

In addition to causing anemia, vitamin B12 deficiency can lead to neurological complications including subacute combined degeneration of the spinal cord, also known as *Lichtheim disease*. (Just to clear up some common confusion regarding proper use of eponyms - although we often talk about Alzheimer's disease, Parkinson's disease, Hodgkin's disease, etc., the formally proper nomenclature is without the apostrophe "s" when referring to a disease named after the scientist or physician who studied it—Parkinson disease, for instance. If the disease is named for a well-known sufferer, the apostrophe "s" is maintained—Lou Gehrig's disease, for example). Long-standing, uncorrected vitamin B12 deficiency can cause weakness, tingling and numbness, depression, diminished memory and vision, and even psychosis. In the opinion of most professionals, this is a good reason to avoid suspicious weight-loss wonder treatments boasting that "you can eat as much as you want and never gain a pound after just one little pill!" Along with the fact that they generally don't work (yes, people can eat as much as they want—but then they get fat), the possibility of chronic vitamin B12 deficiency is just not worth it. Finally, it might be worthwhile to dispel one lingering tapeworm myth. Allegedly one can rid themselves of a tapeworm by coaxing it out with a bowl of milk and cookies placed near the anus. Not true. Not true at all.

What proof is there that the fish tapeworm is a truly successful parasite? In addition to an estimated twenty million people infected worldwide today, coprolites (fossilized fecal matter) containing fish tapeworm eggs in Chinchorro mummies from Chile date back to maybe ten thousand years. Although those eggs might be another species of fish tapeworm (*D. pacificum*), it is evident that humans and fish tapeworms go back quite a way. Thus the fish tapeworm has both an incredible capacity for spread and has withstood the test of time—attributes of a real winner.

Rounding out the conversation at the other end of the spectrum (i.e., an unsuccessful pathogen), although I initially chose Ebola as the prime example of a big loser headed for extinction since it kills so quickly, recent events might challenge this assertion. Ebola, it seems, is not going away anytime soon.

Since its first recognition in 1975, there have been twenty-two documented outbreaks of Ebola. Prior to the major outbreak in 2014, the largest Ebola epidemic occurred in Zaire in 1995. A total of 315 people were diagnosed of whom 254 died. As to the recent West African Ebola epidemic, records show that it was on Christmas Day, 2013, when young Emile Ouamouno grew ill in the village of Meliandou, Guinea. Many authorities believe that two-year-old Emile was the index case, or first victim, of the recent West African Ebola epidemic. Unbeknownst to Emile, his relatives, and all the villagers, a hollowed-out tree where he and his friends played in the outskirts of Meliandou served as the roosting place for thousands of Angolan free-tailed fruit bats (*Mops condylurus*). Also unrealized at that time was the fact that these fruit bats likely carried the Ebola virus. Up to 40 percent of examined fruit bats show immunological evidence of prior Ebola virus infection. Yet they weather these encounters with ease. In contrast, any brushes between Ebola and *Homo sapiens* are likely to be lethal. As a reservoir species, fruit bats have reached an evolutionary détente with the Ebola virus. They are not killed or even harmed by the virus but instead carry it along with them, quietly awaiting transmission to the next human victim playing too close to a previously sequestered bat colony. At the time I write this in late 2015, there have been over twenty-eight thousand people infected during the 2014–15 Ebola outbreak, and over eleven thousand victims have died. Unlike bats, which have an evolutionary truce with the killer virus, humans and other primates all too often pay with their lives for any contact.

Curiously, the majority of Ebola outbreaks originated in areas of extensive and recent deforestation. As eloquently explained by Robert Dorit in the August 2015 issue of *American Scientist*, Ebola (and other emerging infectious diseases) may be a consequence of ecological disruptions.[7] Deforestation or any major disturbance of the sylvan (i.e. forest) ecology, increases contact between people and animal reservoirs for rarely encountered viruses. Zoonotic pathogens (i.e., germs that normally cause diseases of animals but can affect humans under certain circumstances) do not encroach haphazardly and indiscriminately into the human population and they do not suddenly become supervirulent. Instead, these viruses have probably resided unchanged in their native environment for centuries—what has changed is the breakdown of the former boundaries between humans and viruses. Thus, although I initially picked Ebola as an example of an unsuccessful pathogen, in light of new information gleaned over the past year, I can see this was a bit hasty—this virus is no loser. Even though the 2014–15 outbreak thankfully is waning, Ebola itself undoubtedly has not disappeared—and we now know it will not. It certainly still skulks in the wild population, most likely within a bat reservoir, poised for the next opportunity to reemerge in an unwary or unprepared human population. Analogously, HIV possibly originated in a nonhuman primate and resided in a primate reservoir for countless generations before environmental degradation afforded contact between the virally infected reservoir and humanity, leading to the pandemic that began in the 1980s.

Sadly, the West African Guinea rainforest now covers less than one-fifth of its original area. Overall, Guinea has been losing about 1 percent of its rainforests every year for the past thirty years. Emile's village of Meliandou was once surrounded by lush jungle, but relatively recently it has been transformed by extensive deforestation. Over the past few decades, more than 80 percent of the previously forested region has been razed and replaced with new crop plantations. As to just what impact this pattern will have on future epidemics of Ebola and other diseases, we shall see soon enough.

Ironically (but relevant to our story about cancer), the Ebola virus seems to work by first silencing the cells intended to detect and destroy it. As is the case with cancer, in Ebola, the immune system (which should be on guard and ready to eliminate any malignant cells or invading viruses) is initially hypnotized. Lulled into laziness, the immune system does not recognize the severity of an Ebola attack until it is too late. Only then does the immune reaction finally step up. But by then anything is too late—and too much. The alarmed immune system *over*reacts with a "cytokine storm," which causes more harm than good. Thanks to an inappropriate deluge of cytokines (chemical compounds released by cells of the immune system, which will be described in greater depth in upcoming chapters), blood vessels lose their integrity and leak, leading to the gruesome Ebola trademark of bleeding from all orifices, including the eyes, in moribund victims. In fact, much like the 1918 Spanish flu pandemic, thanks to their more robust immune systems, young and healthy victims are often the ones who suffer the most severe consequences. Almost 60 percent of the Ebola victims in the 2014–15 outbreak were in the fifteen-to-forty-five-year-old age bracket. These are the people who tend to have the strongest immune systems. And when the immune system becomes our enemy, the young and healthy suffer the most.

Returning to the general concept of perfect parasites, the ideal relationship permits parasite propagation indefinitely via its hosts. The fish tapeworm, a successful parasite by any definition, perpetuates itself by laying billions of eggs in the hopes that one of those eggs will fulfill the dream of again landing in a human alimentary canal. But eventually its own life will come to an end, either when it reaches a ripe old age or when its host dies. Even the super-successful fish tapeworm, like all classical parasites, must resort to reproduction in order to survive. Is there a better way of achieving this ultimate aim? Yes. The definitive way of achieving this aim is simply by never dying.

If the goal is everlasting life, cancer's standard strategy fails dismally. Malignant tumors arise in their hosts—but they also die in their hosts, all too often directly causing the death of their creators. By causing illness and shortening a person's lifespan, the tumor fails to live on. From a parasite's perspective, this is clearly not the route to success. Cancer, like a parasite in a dead-end host, will ultimately die in a person's body—unless it can find an escape hatch. There is a newcomer to this survival game—the transmissible tumor—that may have discovered a way out. Rather than dying along with its host, devil facial tumor disease jumps ship. The average lifespan of a Tasmanian devil in the wild is only about five years; the contagious cancer has now been alive for nearly twenty years and counting.

Furthermore, DFTD as a parasitic cancer has found the perfect host in the Tasmanian devil. Other parasites occasionally have to alter their host's behavior to ensure their propagation. For example, the protozoan that causes the disease toxoplasmosis, *Toxoplasma gondii*, induces daredevil behavior in mice—they no longer fear cats. Tom and Jerry's relationship notwithstanding, this newfound bravery is seriously misguided! Mice may think otherwise, but the truth is that toxoplasmosis does not bestow them with kung fu fighting skills. Thus they will approach cats without trepidation—and then are readily eaten! Not good at all for the mouse. But this is an ingenious strategy for the parasite. Through this mind-control mechanism, the parasite successfully gains access to its desired definitive host, the cat.[8] Interestingly, toxoplasmosis in people may subtly alter human behavior and there is some (controversial) data suggesting a link to schizophrenia.[9]

Less controversial is the fact that congenital toxoplasmosis may lead to blindness, deafness, and mental disability in children.[10] Toxoplasmosis can also affect adults who are immunologically compromised. As cerebral toxoplasmosis can mimic brain metastases on CT and MRI imaging studies,[11] it is imperative that we verify that we are truly dealing with what we think we are. In a patient with a recent history of cancer, this usually doesn't pose much of a challenge. For instance, if we are dealing with an individual with small cell lung cancer (which very frequently travels to the brain) whose disease is progressing rapidly despite chemotherapy, a brain MRI that shows new, suspicious lesions is quite sufficient for documenting brain metastases. On the other hand, a completely different situation is presented by an individual with a remote history of prostate cancer (which doesn't usually travel to the brain) who has a new abnormality on a brain MRI; brain metastases would be quite unlikely. During my training at Johns Hopkins, cases such as these had to undergo biopsy to definitively confirm the existence of brain metastases before we could proceed with radiation therapy. In one case during my residency, this policy proved to be most memorable.

My patient was a woman, eighty years old, with new neurological signs and symptoms, so a brain MRI was performed. This revealed a "ring-enhancing lesion" that was consistent with a brain tumor. Because she had a past medical history of cancer, the radiation oncology team was called to provide palliative whole-brain radiation therapy for this presumed metastatic disease. However, things just weren't adding up. We requested a brain biopsy before proceeding, which resulted in some lively debate.

Although the woman did indeed have a past history of cancer, the diagnosis and treatment for her uterine (endometrial) cancer was quite remote at nearly thirty years ago. She was presumed cured of this cancer for decades. Nevertheless, I recall one of the residents arguing, "What else do you think this can be in a patient with a past history of cancer? Take a look at that MRI! Why are you radiation guys wasting time by demanding a biopsy?" At that time I didn't have a better counterargument than, "Well, my attending insists on it."

I did know the brain abnormality could possibly represent something else, although given her medical records, I knew the odds were heavily in favor of the metastatic cancer diagnosis. The neurosurgical team performed the biopsy and the results were most startling: toxoplasmosis.

Symptomatic cerebral toxoplasmosis is quite rare in healthy individuals but does occur in immunocompromised people. Toxoplasmosis of the brain is unfortunately not uncommon in people with full-blown AIDS. The question was: How did she become so immunologically compromised that she developed symptomatic cerebral toxoplasmosis? The answer was quite sad.

The woman was a diabetic and her grandson was abusing IV drugs. It turned out that he was sneaking out with her insulin needles and syringes, shooting up, and then returning the used syringes to her medicine cabinet. Eventually, luck ran out. He came down with HIV infection and via the used syringes, passed it on to his grandmother. Perhaps because of her advanced age and her slight diabetes-related immune debilitation, her HIV infection exploded into full-blown AIDS (before her grandson even knew he was HIV positive), leaving her immune system severely compromised and receptive to the toxoplasmosis. This remains one of the saddest stories I ever encountered in the clinic.

One final comment about toxoplasmosis: While it is true that cats are the only known definitive hosts, the cat is not the villain. For one, indoor cats are rarely infected (unless they catch and eat an infected mouse or other vermin). Many people mentally associate the disease with scooping kitty litter, but unless their cat is infected, there is no danger. Most of the three million exposures annually are due to eating undercooked/raw contaminated food or drinking contaminated water. The pussy cat is not the problem!

Returning to Tasmania, unlike the situation in toxoplasmosis, DFTD doesn't need to alter the behavior of its host—its host is already perfect. The belligerent behavior of Tasmanian devils ensures that through biting and fighting, the tumor will be passed on from host to host and live on (much to the chagrin of the thousands of host carcasses it has passed through along the way). Thus the parasitic DFTD has paired up with an ideal host that will reliably pass it along indefinitely; in this fashion, it has reached biological nirvana. It has achieved immortality.

In a sense, a strange, new alien life-form has surfaced from the depths: transmissible tumors, or parasitic cancers. Since the emergence of one such alien parasite only twenty years ago, the wild population of Tasmanian devils has been decimated and could go extinct within another decade or two. This alien entity could very well pass itself along from animal to animal, in a never-ending chain, until the last Tasmanian devil succumbs. What next? Will it then leap to another host species? Will such a cancer ever appear in humans? While no one knows the answer to that, as we shall discuss later, there is another recently identified and highly virulent contagious cancer currently afflicting, of all creatures . . . clams.

One may legitimately think of a transmissible tumor as an alien being, which, like all forms of life, simply strives to survive—but it has a unique new wrinkle. This is no ordinary tumor. It is the ultimate survivor. It is just what we have been fearing—an immortal, transmissible, parasitic, new life-form.

Huh? You just don't buy it? You think all this talk about immortal tumors is just gibberish? Alright, I certainly could understand someone saying, "Big deal if DFTD has been around for twenty years. Two decades proves nothing! Show me something alive for ten thousand years and maybe we can keep talking about immortality. . . ." Alright, I will. And it happens to be another contagious cancer. But unlike DFTD, which affects a rare, exotic marsupial, this contagious cancer affects a far more familiar host. *Canis familiaris* to be specific.

Chapter 7

A MALIGNANT MALADY IN MAN'S BEST FRIEND

Dogs are better than human beings because they know but do not tell.

—Emily Dickinson

There are indeed a few things dogs have been somewhat reluctant to tell. On rare occasions, they are cursed with a *very* odd tumor involving their genitals. Most patients are free-roaming, sexually active dogs between two and five years old living in tropical and subtropical countries. Because the geographical distribution of the disease also matches the distribution of "intact" (i.e., unneutered or unspayed) dogs, it was initially suspected that the cancer was caused by some unknown sexually transmitted organism. Since the cancer has a higher prevalence in tropical regions, it seemed plausible that a virus or other infectious agent skulking in warm, wet, tropical regions was to blame. Further research revealed no viral, bacterial, or parasitic etiology, but instead something far more fascinating.

While it remains doubtful that an epidemic of contagious cancer will ever arise in humans, it turns out that man's best friend already *is* afflicted with a contagious cancer. Called *canine transmissible venereal tumor* (CTVT), or *Sticker sarcoma*, this is a *sexually transmitted tumor*—a sure-bet method of being passed from dog to dog in regions where spaying and neutering are less common. But here is the kicker with Sticker—in stark contrast to the uniformly fatal Tasmanian devil disease, the contagious cancer in dogs *often undergoes spontaneous and permanent remission.*[1]

When I asked some of my veterinary colleagues about how they deal with cases of CTVT, many have joked that the best thing to do is quickly treat the dog with chemotherapy or radiation and take credit for curing it—before it disappears!

This would appear to present another tantalizing clue and another mystery to unravel. And here are two crucial parts of the puzzle: The majority of advanced human cancer cases end badly—they are like Tasmanian devils in this regard. However, what happens in the dog is somewhat reminiscent of Daniel's abscopal effect. Our challenge is to find a way to make all cancers behave like those in dogs rather than those in devils—the "devil to dog transformation."

Although the canine communicable cancer most often involves the penis or labia, on occasion it may also spread to the nose or mouth through licking, biting, and sniffing tumor-bearing areas on cancer-carrying dogs (including themselves!). CTVT is more prevalent where dogs are not commonly neutered, and in India, for instance, this cancer constitutes approximately 29 percent of all dog tumors.[2] The eponym for CTVT, Sticker sarcoma, honors German veterinarian and physician Anton Sticker (1861–1944), who performed detailed studies on this tumor in Frankfurt, Germany, between 1902 and 1905.[3] Other early work was done by Russian veterinarian M. A. Novinsky (1841–1914), who in 1876 first showed that the tumor could be transplanted from one dog to another by infecting them with tumor cells in the laboratory.[4] This was the first instance of a tumor being experimentally transplanted and marked the dawn of research in tumor transplantation.

Whether in the lab or in the wild, CTVT undergoes a very predictable cycle in healthy adult dogs with a normal immune system: an initial phase of progressive tumor growth over two to six months (the P phase), followed next by a phase of stability (S phase), and finally by a phase in which the tumor typically regresses in adult dogs (R phase). On the other hand, in immunosuppressed animals, the tumor frequently continues to grow and can metastasize. Laboratory studies have shown that the tumor can be transplanted by subcutaneous injection of the cancerous cells and that there is a curious temporal window of susceptibility. After the P phase and S phase, the tumor regresses spontaneously in adults but metastasizes in immunosuppressed hosts and neonates.[5]

Both the natural and the experimental forms of the tumor are exquisitely sensitive to radiation therapy. For instance, four doses of five Gray each for a total of twenty Gray will usually cause complete regression of this neoplasm and such irradiated animals are permanently resistant to further transplantation. (Incidentally, this is nearly the same dose-fractionation regimen I used for Daniel, as described in chapter 1). Regardless of whether the cancer is cured by radiation therapy, chemotherapy, or regresses on its own, once the cancer has gone into remission, that dog will not again get this type of cancer.

Another fascinating aspect of CTVT is that this cancer can not only pass from dog to dog but also to a few other species. Other members of the canine family such as coyotes, foxes, and jackals have also been known to host this transmissible cancer.[6] While the tumor regresses spontaneously in these animals, the fact that these other animals can contract the disease at all is quite surprising and frankly disconcerting. It is alarming enough that a tumor can be relocated from one dog to another but to be capable of surviving transplantation across species is most unanticipated. This is because the so-called *major histocompatibility complex* (MHC) barriers are expected to impede indiscriminate tumor transplantation between any two unrelated dogs, but the MHC barriers between different species should be *absolutely* insurmountable.

In very simple terms, all cells must constantly carry "ID cards" with them that the immune system can inspect at any time to verify identification. This identification is displayed on the surfaces of cells, as a means of making their allegiance very visible. In this manner, cells display molecular "flags" on their surfaces, which the immune system recognizes and uses to decide if the cells are friend or foe. One key set of molecular flags displayed by essentially all cells (in vertebrate animals) is the major histocompatibility complex or MHC. If foreign cells display the wrong flag they will be tagged as unwelcomed intruders by a certain branch of the immune system (the adaptive immune system) and mercilessly attacked. Similarly, if the cells display no molecular flags at all, another branch of the immune system (the innate immune system) will identify them as invaders and eliminate them. Thus there are two major roadblocks to any cell trying to parasitize a vertebrate animal—the adaptive immune system and the innate immune system that team up in a double defense.

Based on the specific molecular flags displayed, the immune system decides whether or not the cells in question are "self" or "non-self." It is the MHC molecular identification system that allows the immune system to recognize and tolerate self or conversely recognize and reject non-self. An appreciation of the MHC system was crucial to a better understanding of organ-transplantation biology.

Depending on the degree of MHC similarity between an organ donor and a recipient, the recipient's immune system may identify a donated organ as foreign and attempt to eliminate it. The risk of such transplant rejection can be estimated ahead of time by the degree of MHC compatibility between the potential donor and the recipient. Compatibility is determined by the similarity of the molecular flags (more precisely called MHC *alleles*) in question.

Just in the way of terminology, an *autograft* is when an organ is transplanted from one person back into that same person. This is sometimes done, for instance, when blood vessels are blocked in the heart and "bypass surgery" is needed. In this case, veins from the leg might be used to bypass clogged coronary arteries. An *allograft* is when an organ transplant is done between two genetically nonidentical members of the same species, as is normally the case in human organ transplants. Finally, a *xenograft* is a transplanted organ from one species to another. Ordinarily this is unequivocally forbidden thanks to the MHC system.

An understanding of the MHC (also called the *HLA* for *human leukocyte antigen* system in humans) ushered in the era of successful organ transplantation in human medicine. In human organ transplantation, MHC compatibility is of paramount importance, and unless the degree of compatibility is quite close, organ rejection is inevitable. For organ donor/recipient pairs to be an adequate match, multiple MHC alleles must be closely matched for success. Organs transplanted between MHC identical individuals can be accepted, whereas organs transplanted between MHC-mismatched individuals are vigorously rejected even in the face of intensive immunosuppressive therapy. However, it is important to keep in mind that although the MHC alleles are the most crucial markers to match up when considering organ transplantations, there are several other molecular markers that also need to be considered such as minor histocompatibility antigens, ABO blood group antigens, and monocyte/endothelial cell antigens.

Returning to CTVT, given the formidable MCH barriers in place, the observation of interspecies transplantation of this tumor is perplexing to say the least. It is possible that the MHC and other barriers might have evolved *expressly* for the purposes of *preventing* tumors from being transplantable; were it not for the antigenic diversity found in most species and the existence of mechanisms to react against these antigens, contagious tumors could perhaps be relatively common.[7] The recent discovery of contagious cancers in Tasmanian devils as well as dogs (and as we shall see soon, in some surprising other examples as well) has bolstered the credibility of this idea.

Definitive confirmation that CTVT of dogs, like DFTD of devils, is a communicable cancer (rather than a cancer that arises anew in each animal) came through chromosomal analyses.[8] Normal dog cells have seventy-eight chromosomes—thirty-eight pairs of autosomes and one gender-determining X/Y pair. (Parenthetically, the number of chromosomes in a given species has little to do with where they might stand on the fictitious "evolutionary ladder," with several species putting the human forty-six to shame. For instance, several rat species boast over ninety chromosomes, a few fish exceed one hundred, some butterflies surpass two hundred, and certain plants thoroughly trounce most multicellular animals with figures beyond one thousand.) However, rather than the expected seventy-eight chromosomes of normal dogs, CTVT cells contain a highly abnormal number of chromosomes, averaging fifty-nine. Intriguingly, there are some slight differences between tumors, especially when they are from dogs on different continents. For example although the malignant cells average fifty-nine chromosomes, the number ranges from fifty-seven to sixty-four. This subtle geographic variability hints that the tumor might be quite old and has had ample time for evolution, and the minor chromosomal changes reflect this evolution. Regardless of these minor differences, the *similarities* shared by the various tumors vastly outstrip their differences—and any similarities they have with normal dog cells. They are quite distinct from ordinary dog cells and are remarkably similar to each other, corroborating their common origin. As additional confirmation, a DNA molecular marker in the genome called LINE-1 is consistently found in a different location in tumor DNA compared to the LINE-1 location in normal dog cell DNA.[9] This offers further proof that the tumors do not arise anew as separate cancers in each individual animal (since if that were the case, they would have the LINE-1 element in the typical dog DNA location rather than in this weird other location common to all CTVT tumors). It therefore appears that all CTVT tumors are the same clone of cancer cells that have been passed along from one animal to another for years. How many years is a most interesting question.

In an earlier chapter we observed that DFTD has been decimating devils for nearly two decades, but just how long has CTVT been beleaguering man's best friend? Detailed studies of the chromosomes and DNA of these tumors have provided an answer—and it's a doozy!

In order to answer this question, one must recall that all living things undergo spontaneous *mutations* or changes in their DNA compositions (i.e., their genomes). Many of these spontaneous mutations neither help nor hinder the cells they are in; they are inherited and just accumulate in the genome over time. Thanks to this "background" mutation rate, if one starts out with a tumor having a certain DNA sequence, over a given period of time some of the descendant cells will evolve slightly different DNA compositions from the parent. And if one knows the rate at which these random mutations occur, one might be able to estimate just how many years have passed between the original parent tumor and the descendants. In other words, one can create a "molecular clock" based on DNA changes.

Researchers have compiled and compared the myriad mutations (about five hundred thousand of them) in the CTVT DNA from various dogs and created a molecular clock to approximate the tumor's age.[10] Simply collating an assortment of mutations is not enough to generate a real clock however. The various CTVT mutations must be compared both with each other and with some "standard" in order to calibrate the clock and subsequently estimate the age of the tumor.

Thus a "raw" CTVT molecular clock was calibrated by comparison with *medulloblastoma* (a human malignant brain tumor that most often afflicts children), which has a known mutation rate of about 43.3 mutations per genome per year. (Medulloblastoma was chosen for this since it has a proven, strong correlation between age and the number of mutations and is thus relatively reliable for this purpose.) By using the medulloblastoma mutation rate as a molecular-clock calibrator, investigators calculated the age of CTVT to be . . . drum roll please . . . 11,368 years!

In other words, this contagious cancer, this alien parasite, arose in a single dog or wolf *over ten thousand years ago* and has been passing itself along from animal to animal ever since. Instead of just reproducing its own kind, this *creature* has found an even better way to survive—by living forever and transferring itself from one host to another. It has fulfilled its dream of true immortality. And if 113 centuries isn't, for all intents and purposes immortal, I don't know what is.

What has allowed CTVT to survive for so many centuries? What is its secret? Since the tumor is ultimately rejected by the host dog, one might have logically anticipated that it would have gone extinct early on. However, before the canine immune system awakens and kills it, the tumor manages to lie low just long enough to leap off and hitch a ride on its next victim through sexual contact. While thousands of generations of dogs and wolves have come and gone, this cloned cancer has persevered for millennia. There are several peculiar properties that have facilitated its transmission and permitted its persistence for so long. As mentioned earlier, if foreign cells display the "wrong flag"—incompatible MHC molecules—they will be singled out as unwelcomed intruders by the adaptive immune system and confronted in a hostile manner. Conversely, if the foreign cells display *no* molecular flags at all (i.e., an absence of MHC molecules on their surfaces), the other main branch of the immune system (the innate immune system) will classify them as trespassers and eradicate them.

The first thing CTVT very cleverly does upon entering a new host is *reduce* its MHC class I expression, thereby minimizing its initial visibility to the host's adaptive immune system. Its molecular flags fly at "half-mast" in essence. This trick is elegant. Reduction—but not complete elimination—of class I MHC molecules on its surface allows CTVT to escape adaptive *T cell*–mediated immunity (which would occur if its MHC class I antigens were fully expressed). But this ingenious half-staff maneuver allows the cancer cells to also evade the innate immune system's *natural killer cells* (which would eradicate the cells if they were completely devoid of *all* their MHC class I antigens). This approach—hoisting the flags to half-mast, confers a perfectly balanced means of concealing itself from both major branches of immunity. For further insurance, the tumor also stimulates the release of certain *cytokines* (chemical substances derived from immune cells that stimulate or repress the activity of other immune cells). In the case of CTVT, the cytokines in question serve to suppress the activity of the natural killer cells just enough to ensure that the innate immune system doesn't see through the ruse.

As mentioned, in most healthy adult dogs, CTVT spontaneously regresses after a few months of sustained growth. It is as though the host's immune system eventually awakens, realizes that there is some sort of scam going on, and recognizes the foreign cancer for what it is—an uncalled for transplanted organ from another animal. It then promptly rejects this uninvited parasite. What actually happens is a bit more sublime. It is not so much that the dog's immune system truly grows wise but rather, the tumor seems to tire of its own treachery and surrenders. Rather than continually struggling to sustain its flags at half-mast the exhausted cancer reverts to its original form with fully hoisted MHC flags. With dispatch, the adaptive immune system's T cells get to work, and, before long, the tumor is purged, never to return.

Irrespective of the details, the remission provides life-long freedom from reoccurrence of the initial CTVT tumor and, very interestingly, also confers immunity that protects that dog against any future re-inoculations. Thus, in addition to being cured of its cancer, the once-afflicted dog cannot get a new case of CVTV from another infected dog—no matter how promiscuously he or she might behave. Abiding by the old adage "fool me once, shame on you; fool me twice, shame on me," the immune system may have been humiliated once—but won't be fooled again.

Given the drastically different natural histories of the uniformly fatal DFTD of devils versus the normally nonlethal CTVT of dogs, it seems obvious that a comparison between the two could be most instructive—and relevant to human cancer management. As alluded to before, human cancers act more like devils, but we all want human cancers to behave like dogs. One gets the distinct feeling that there is an important and tantalizingly close clue just hovering before us, but it is somehow still hard to grasp. How can we actualize the desired devil-into-dog transformation? First, let us ask just why is DFTD invariably fatal? Why doesn't the devil immune system reawaken and eliminate the cancer just the way dog's does? Is it because, unlike CTVT, DFTD doesn't eventually wave its white MHC flags and surrender to the devil's immune system? Or is the Tasmanian devil's immune system simply too inept to recognize and overcome the deception?

The most common explanation for why DFTD is so easily passed from devil to devil is based on genetic diversity—or the lack thereof. The Tasmanian devil population is highly inbred and relatively bereft of genetic diversity. Thanks to this genetic uniformity, many researchers have argued that it should be no surprise that the transplanted tumors cells are not rejected.[11] These researchers assert that the DFTD tumors are akin to organ transplants donated from "perfect matches." In essence, what this is implying is that the entire population of Tasmanian devils is entirely composed of identical twin brothers and sisters.

According to this hypothesis, thanks to their greater genetic diversity, dogs, on the contrary, will eventually evict any transplanted cancer, just as a human organ transplant from a mismatched donor would be rejected. In fact, based on this premise, one group of prominent researchers has boldly proffered that MHC diversity in vertebrate animals arose for exactly this purpose—to prevent cancers from being communicable. While this speculation regarding the evolutionary origin of the MHC is most fascinating, as to the idea that low genetic diversity alone is the explanation for the absence of rejection in transplanted cancers in Tasmanian devils, I for one, vehemently reject this hypothesis.

First, if lack of genetic diversity were the sole explanation, *any* organ (not just a tumor) should be capable of being transplanted from one Tasmanian devil to another, unmolested by the immune system. This does *not* appear to be the case in Tasmanian devils. Indiscriminately transplanted organs (e.g., skin grafts) from one Tasmanian devil to another are immunologically rejected, as expected. The devil population is not *that* closely related to one another. Thus, it appears unlikely that the dearth of genetic diversity can be the complete answer to how and why DFTD is so readily transmitted.

Furthermore, and more importantly from my clinical perspective, the following observation demands explanation: In humans, as we are all aware, *normal organs* are summarily rejected upon transplantation unless extreme immunosuppression is implemented even when there is an excellent MHC match. In confusing contrast, transplanted *cancers* that tag along for a ride in organ transplants occasionally go unharassed! Albeit rare, there are numerous documented cases where a transplanted cancer has grown, unimpeded by the new host's immune system, and eventually killed that new host.[12] What the devil is going on here?! Well, the immediate logical explanation would be that the organ recipient's immunity was severely compromised by the required immune-suppressing transplantation drugs, and therefore that person was unable to mount an adequate response to the foreign cancer. While this is true, it can't be the complete answer. We shall go into this in greater depth in a later chapter, but in cases where a cancer-infested transplanted organ was not absolutely critical for life (e.g., a kidney transplant where dialysis might be possible), it might be possible to stop the immunosuppressive drugs and allow the organ recipient to regain normal immune function. In this scenario, the patient should reject the transplanted kidney—along with its cargo of cancer. What actually happens however is alarming and ominous. The transplanted kidney is rejected as expected, but the cancer continues to progress, unfettered and unmolested by the host's newly awakened immune system! It is as though the malignant tissue, unlike the normal tissue, possesses some cloak of invisibility.

We shall revisit this bewildering and portentous paradox in greater detail in upcoming chapters, but for now let's end on one final related note. In stark contrast to the typical pattern in immunologically healthy dogs, in immunologically *incompetent* dogs the transplanted CTVT tumor will grow, begin to fester, bleed and ulcerate, and eventually metastasize far and wide. Ultimately, just as in the Tasmanian devil scenario, death ensues in these unlucky animals. (Incidentally, the disease, even when advanced, appears relatively responsive to both radiation therapy and chemotherapy. A chemotherapy drug often selected for treatment of advanced cases is vincristine, one of the so-called vinca alkaloids.)

The fate of immunologically compromised dogs with CTVT is not dissimilar to that of Tasmanian devils with DFTD—or ordinary humans with typical cancers for that matter. Noting that under normal conditions, CTVT is immunologically rejected, beginning about a century ago some veterinarians attempted to treat advanced CTVT cases with an early iteration of cancer immunotherapy. [13] The idea was to insistently prod and awaken the slumbering immune system by administering living or killed bacteria or bacterial toxins to the dogs. In some cases, good results were obtained by administering preparations of the bacterium *Chromobacterium prodigiosum*, which instigates a furious, but unfocused, immune reaction. In its effort to fight the bacteria, this ferocious immune response took no prisoners. In the antibacterial assault, anything and everything in its wake (including an unwary tumor) was attacked—and occasionally destroyed. Other early veterinary researchers observed some benefit by injecting isolated bacterial toxins, which similarly, but more safely, provoked a vigorous immune reaction. However, given that CTVT usually doesn't progress to an advanced stage in immuno-*competent* dogs in the first place, it isn't surprising that in these immuno-*compromised* animals, their feeble immune systems often simply couldn't cure them. Nevertheless, the idea of administering bacteria or bacterial toxins to awaken the immune system and fight cancer was not lost on one individual—a human physician named Dr. William Coley. We will delve into his most incredible—but largely forgotten—story in the next chapter.

Chapter 8

THE CURIOUS CASE OF COLEY'S TOXINS (OR SOMETIMES THE TREATMENT WORKED)

I recall hearing very briefly about "Coley's toxins" during medical school and my oncology training. It was all but dismissed. It was *not* considered something to take seriously. In my readings however, I recalled stumbling upon a particularly perplexing phrase: "Sometimes the treatment worked."

What?! Would you please repeat that? To my mind, when it comes to cancer treatment anything that "worked"—even if only in rare cases—deserves a bit more press than that. At the very least, please define the word "worked" in this context. I found that it was not easy to tease out the details back then, and seeking information regarding Coley's toxins meant pouring through, not the medical journals, but rather the history books. Today, however, there is a resurgence of interest in Coley's toxins and in his work in general. Before diving in, let's set the stage a bit . . .

On May 1, 2012, Aimee Copeland, a vivacious twenty-four-year-old graduate student, was kayaking with friends on the Little Tallapoosa River in Carrollton, Georgia.[1] It was a sunbaked day with a cloudless sky overhead. The ambitious University of West Georgia psychology student was enjoying a day on the river to decompress when the group stopped to have some fun soaring across the river on a homemade zip line. When it was Aimee's turn, the jerry-rigged apparatus snapped, plunging her into the river and leaving her with a nasty gash in her thigh. She was rushed to the emergency room where she was sutured up and given an antibiotic. Three days later Aimee was on life support. When she regained consciousness her world had been changed forever. Her left leg had been amputated and massive amounts of abdominal tissue were surgically removed. Within days her right foot and both hands were also amputated.

Necrotizing fasciitis—a virulent bacterial infection that destroys skin and the subcutaneous tissues between the skin and the underlying muscle—was ravaging Aimee's body. Not unlike firefighters cutting a firebreak in the terrain to contain a wildfire, Aimee's surgeons were amputating parts of her body to stymie the raging infection's progress. Remarkably, she survived—necrotizing fasciitis is often fatal. In recent years, there have been several dramatic accounts of the "flesh-eating disease" in the popular media. In most cases, they get their start through an open wound, but on rare occasions an insect bite serves as the portal. Although there are now several different, highly resistant flesh-eaters out there (such as *Staphylococcus aureus*, *Vibrio vulnificus*, and *Aeromonas hydrophila*, among others), in most cases the flesh-eating bacteria is an especially vicious strain of *Streptococcus pyogenes*.

But why can't our immune system stop the flesh-eating bacterial infections from spreading? While in many cases of necrotizing fasciitis it is the host's immune system that is the cause (e.g., when it is enfeebled by diabetes), in some cases it is due to the bacteria themselves. Basically, the human immune system, as powerful as it is, represents no match for these devastating adversaries. In simple terms, our immune system attacks and destroys pathogens such as bacteria through two broad components—the innate and the adaptive responses. The innate immune system is our first line of defense, marshaling white blood cells called phagocytes (such as macrophages and neutrophils) to engulf and eliminate harmful bacteria. But no one's immune system, even boosted by antibiotics, stands a chance against this "strep on steroids" once it really takes hold in the subcutaneous tissues.

Now contrast our relatively impotent immune systems to that of crocodiles. For hundreds of millions of years, crocodiles have lived in boggy, bacteria-laden bodies of water—far filthier than the water Amiee fell into. Much of their time is spent trying to kill insubmissive prey and fighting with other crocs. It's a rough life, and they frequently get wounded. Open wounds in the marshy muck should be a fast track to deadly infection. Yet crocodiles as a rule don't get fatal infections. In fact, they rarely get sick. Their innate immune systems are so supercharged that the same devastating flesh-eating bacteria that wreak havoc on humans would perish instantly in crocodile blood—along with dozens of other pathogens lethal to humans, including HIV.

At this point you might be wondering what this has to do with cancer. Surprisingly, there is more than initially meets the eye. To explain, let's introduce Dr. William B. Coley, who in the 1890s began directly injecting streptococcal organisms into patients with seemingly hopeless advanced cancer.

And sometimes this treatment worked.

Coley, a Yale graduate who received his medical degree from Harvard Medical School, certainly had the credentials. He became an eminent New York City surgeon and researcher at Memorial Sloan Kettering Cancer Center (which back then was simply Memorial Hospital). In fact, he served as head of the Bone Tumor Service. With his keen clinical insights, compassion for patients, scientific discipline, and unparalleled drive, he was the model clinician-scientist. Coley was deeply moved by the deaths of his early patients who had succumbed to a type of cancer known as sarcoma—including a close friend of John D. Rockefeller, Elizabeth Dashiell.[2] (It was Dashiell's death that inspired Rockefeller to fund cancer research). Coley scoured the scientific literature looking for a clue that might lead to an effective treatment. After finding hints of a possible connection between severe infection and tumor regression, Coley pressed on with his research on "fever therapy" with maniacal enthusiasm.

Although Coley was not the first to consider and explore fever therapy, he was the first to do so in a methodical, scientific fashion. In fact, the observation of fever-induced cancer remissions may date back to the time of the Egyptian pharaohs as historical writings intimate that the Egyptian physician Imhotep might have used a primitive "incise-and-infect" technique to treat tumors. Another ancient account describes how the thirteenth-century Italian saint, Peregrine Laziosi (1265–1345), became the "patron saint of cancer patients."[3] Consumed by his never-ending obligations of preaching and proselytizing, he failed to notice the progressively enlarging lump on his leg. By the time he brought it to the attention of his physicians, it was unambiguously diagnosed as a malignant cancer. The only known option was amputation. Before that, however, the gargantuan tumor had erupted, breaching the skin and resulting in a horrific infection. It was said that the infection exuded "such a horrible stench . . . that it could be endured by no one sitting by him."[4] In danger of losing his life to infection even sooner than to cancer, his physicians planned to amputate. Miraculously, he fought off the infection, and by the time he was due for his operation, his physicians were stupefied to find no evidence of the tumor. And his cancer never returned. Several centuries later, he was canonized as Saint Peregrine and named the patron saint of cancer patients.

Coley might have been particularly inspired by a relatively contemporaneous 1868 account by German physician Wilhelm Busch. Busch described how he had deliberately infected a neck sarcoma patient with pathogenic bacteria. Naturally, in this pre-antibiotic era, the infection almost killed her, but her huge tumor subsequently softened and shrank.[5] Coley was probably most directly influenced by the account of a poor German immigrant named Fred Stein, who by 1885 had had surgery for his neck sarcoma on four separate occasions.[6] Following each operation, the tumor returned with renewed fury; it seemed to grow larger, meaner, and angrier each time it was disturbed. As if this weren't enough, following his last operation, he developed a serious bacterial disease called *erysipelas*—an infection caused by none other than the potential flesh-eater, *Streptococcus pyogenes*.

After his wound got infected, Mr. Stein predictably suffered through numerous bouts of high fevers, bone-rattling chills, and drenching sweats. To everyone's amazement, however, with each round of fever, the tumor mysteriously shrank. Not surprisingly, Stein nearly died from the infection. Nevertheless, against all odds and a four-and-a-half-month stay, he eventually recovered and was discharged from the hospital—without his tumor. The sarcoma had undergone a complete spontaneous regression. The only remnant was a large scar below his left ear.

After hearing of Stein's miraculous spontaneous remission, Coley became a man on a mission. He doggedly searched for Stein until he finally found him six years later in one of the Lower East Side ghettos of Manhattan. To Coley's stunned surprise, Stein was fine; he had no evidence of cancer. Aside from the scar below his left ear, there was no trace whatsoever that this man had ever endured this harrowing experience years ago.

Having now seen it with his own eyes, Coley became a believer, more motivated than ever to pursue this line of clinical research. Could he somehow routinely replicate what had happened to Stein? Was it possible to intentionally incite the immune system to viciously massacre a malignant bystander during its battle with a bacterial infection?

The Stein case and other anecdotes convinced Coley that a severe bacterial infection could inspire an immunological reaction powerful enough to kill cancerous tumors. In a (very) bold move, he began injecting live streptococcal bacteria into some of his cancer patients, intentionally inducing potent infections in the hope of stimulating anticancer immune responses. His first patient was Mr. John Ficken, a bedridden young man with a monstrous abdominal sarcoma invading his abdominal wall, bladder, and pelvis. Coley injected his bacterial concoction right into the tumor every few days beginning in January 1891. As anticipated, this yielded dramatic fevers and shaking chills. But in addition, Ficken's tumor gradually shrank. By May it was down to one-fifth its original volume; in August (eight months after Coley started treatment) it was gone. Just as Coley had predicted, the tumor had vanished, a casualty in an epic battle between bacteria and a forceful, if undirected, immunological attack. And this previously extensive and unresectable tumor, considered a terminal case, remained in remission for nearly another three decades when Mr. Ficken passed away from a heart attack.

As one might have predicted however, Coley's subsequent results were mixed. Not infrequently, Coley was unable to successfully give his patients erysipelas (the bacterial infection caused by *Streptococcus pyogenes*). On these occasions the streptococcal infection failed to "take," and the cancer progressed unimpeded. Conversely, but not unexpectedly, in other cases the infection did more than just take—two of his early patients died of uncontrolled infection. Given that Alexander Fleming's 1928 discovery of penicillin was still several decades away, Coley's perilous approach of administering live bacteria seemed seriously at odds with the "First, do no harm" principle of the Hippocratic Oath.

Undaunted, Coley persisted. In an effort to mitigate the dangers, rather than using live bacteria, he began combining heat-killed streptococcal organisms with another bacterium, *Serratia marcescens*. This "pasteurized" version, subsequently known as "Coley's toxin," was used for several decades. Even in this watered down version, Coley's toxins were no picnic. Following the injections of living bacteria, or even heat-killed bacterial preparations, patients would sweat profusely, spike fevers to as high as 105°F, and suffer though intermittent, violent shivering chills. Finally, after a day of this torment, they were ready for round two. And the regular injections of Coley's toxins went on for months. This was not a treatment for the faint of heart. While there were some startling success stories, inconsistencies in his techniques and deficiencies in his data collection invited harsh criticism from his peers, especially from those in the burgeoning field of radiation therapy.

One such critic was Dr. James Ewing. Ewing, a prominent pathologist credited with the first detailed description of the bone sarcoma that bears his name ("Ewing sarcoma," a principally pediatric malignant bone tumor). Ewing was one of the early proponents of radiation therapy—

and a staunch opponent of Coley's toxins. It was particularly problematic that Ewing was such an inflexible critic of Coley since Ewing eventually became the medical director of Memorial Hospital—and Coley's boss.[7] Apparently there was a deep personal animosity between the two, which was most ironic since at the time, Coley was the country's most experienced surgeon in treating Ewing sarcoma.

Further insult came through Ewing's role in the world's first "cancer registry"—the Bone Sarcoma Registry. By collecting cases of bone sarcomas from all over the country, carefully scrutinizing them, and evaluating their treatment, it was hoped that the registry would someday lead to standardization of the diagnosis and treatment of this cancer. Submitted cases were to be reviewed by a team that included Dr. Ewing. Despite being the leading bone cancer surgeon in the country, Coley had a tough time getting many of his cases into the registry. In a catch-22, some members of the registry criticized Coley's claimed results on the basis that he simply misdiagnosed the cancers. Eventually, Ewing revoked Coley's privileges to use his toxins at Memorial Hospital. In the end, Ewing won his battle with William Coley, and Coley's toxins all but vanished save for a few lines in old textbooks, while radiation therapy advanced into one of the three present pillars of modern cancer treatment, along with surgery and chemotherapy.

Still, Coley, who can rightfully be considered the "father of cancer immunotherapy," pursued his work in this nascent field until his death in 1936. In 1953, Coley's daughter, Helen Coley-Nauts, dusted off the clinical cases described by her father. This was no small feat. As forewarned, she discovered that the records were often incomplete, and his bacterial extracts were not prepared in a consistent fashion. In fact, Coley-Nauts found that her father had used at least fifteen different formulas with widely varying potency. Nevertheless, the occasional success stories stand out even by today's standards. We now know Coley's intuitions were at least partly correct: stimulating the immune system can be an effective way to treat cancer.

Although Coley apparently achieved some spectacular cures, the treatment failed FDA approval standards.[8] While chemotherapy and radiation evolved into mainstream treatments, Coley's theories and immunological methods were relegated to the margin fringes of medicine; his work consigned to the graveyard of the alternative-therapy world.

Meanwhile, as alluded to in the previous chapter, variants of Coley's

Toxin survived for quite a while in the veterinary world. Beginning in 1907, a la William Coley, veterinary researchers occasionally obtained reasonable results from killed preparations of *Chromobacterium prodigiosum*, either alone or in combination with other bacteria for intractable CTVT. Even as recently as 1975, veterinary scientists were observing and reporting apparently positive effects from injecting bacterial toxins.

Exactly how Coley's toxins and its veterinary counterparts truly work remain a bit obscure even today. Many now believe Coley's toxins induced the host's white blood cells to elaborate certain cancer-fighting chemical compounds called cytokines. Specifically, the bacterial cell walls contain something called *lipopolysaccharide*, and this bacterial lipopolysaccharide induces the secretion of certain cytokines such as *tumor necrosis factor* from host white blood cells. As its name implies, tumor necrosis factor, or TNF, can really do a number on tumors. This raises the larger question: Could we simply administer TNF to patients and cure them? Alternatively, can we make vaccines against cancer like we have for influenza and hepatitis B? Where are we today in the fields of tumor immunobiology and clinical cancer immunotherapy? Well, some might rightly argue that we are in the midst of a real renaissance and about to enter the golden age.

Professor Uwe Hobohm of the University of Applied Sciences in Giessen, Germany, and author of *Healing Heat: An Essay on Cancer Immune Defense*, is a modern explorer of spontaneous and infection-induced cancer remissions. He has been investigating the effects of certain chemical "danger signals" associated with infectious bacteria, viruses and fungi called *pathogen-associated molecular patterns*, or PAMPs.[9] In addition to pathogen-associated molecular patterns, there are clear *damage-associated molecular patterns*, or DAMPs, that our immune system recognizes. PAMPs and DAMPs are identified by *pattern recognition receptors* (also called *pathogen recognition receptors* or *primitive pattern recognition receptors*). Recognition of (and reaction to) basic danger signals is a primitive aspect of our immune systems, and this component of *innate immunity* predates the more sophisticated *adaptive* immune system. It appears that exposure to PAMPs and DAMPs gear up our innate immune systems and prepare them for battle. With tumors, however, our immune system appears too weak to recognize and kill the cancer. As an explanation for Coley's toxins and similar phenomena, Hobohm proposes that exposures to these immune-stimulating, infection-associated patterns elevates our baseline, feeble immune response beyond the threshold required for cancer recognition and elimination. The stimulated immune system is now armed and ready to fight anything—including the cancer that was sitting right beneath its nose all this time.

Dendritic cells are the first immune cells to notice and react to these various pathogen-associated patterns. In turn, the activated dendritic cells trigger *T lymphocytes*, or *T cells*, to then identify and attack cancers they previously could not see. In this fashion, the immune response to a pathogen such as *Streptococcus pyogenes* could trigger an immune response against cancer cells. Could administration of pharmaceutical preparations of these microbe-associated molecules prove to be a safer and more effective modern counterpart to Coley's nineteenth-century concoction? Stay tuned.

Elucidation of the cellular, biochemical, and molecular mechanisms of Coley's toxins and the like are in the realm of the *basic* (i.e., laboratory-based) science of *tumor immunology*. The *clinical* science counterpart that involves patients is the field of *cancer immunotherapy*. William Coley's particular brand of bacteria-based immunotherapy historically represents the earliest methodical venture into the field of clinical cancer immunotherapy. The basic science of tumor immunology was very far behind him, and it would be many decades before an understanding of the relevant immunobiology and molecular biology of cancer was to emerge. However, over the past few decades, the field of molecular oncology has been exploding, becoming one of the most rapidly evolving disciplines in all of biology—arguably in all of science.

Chapter 9

THE DOG KNOWS

For better or worse, I have been a cancer physician long enough to have witnessed the dawn of PET (positron emission tomography) scanning in oncology. I will explain more about PET in the next chapter, but for now I would like to focus on a most unusual form of "pet" scan I encountered years back.[1]

A forty-four-year-old patient told me that for about a month, every night when she would sit down on the couch to enjoy some evening television, her dachshund puppy would persistently sniff and poke around in her left armpit and breast area. This annoying habit was ignored for weeks until one day she decided to feel around in her own armpit to find out just what her puppy was so obsessed with. This was how she discovered the lump in her breast. Alarmed, she contacted her primary physician who confirmed the presence of a breast mass and ordered a mammogram. The ensuing workup led to the discovery of breast cancer.

This was not the first time a cancer was detected by "pet scan"—meaning by a pet dog. In 1989, there was a well-publicized case of a young woman in her forties, whose border collie/Doberman mix sniffed and nipped at a specific mole on her leg while ignoring other moles and freckles. This spurred her on to a clinic visit, and she was eventually diagnosed with malignant melanoma. Given that melanoma can all too often prove fatal, her dog, which detected this cancer at an early stage, may very well have saved her life. In 2001, another case was reported wherein a Labrador retriever doggedly sniffed at a lesion on a person's left thigh, eventually leading to a diagnosis of another, far less dangerous form of skin cancer, basal cell carcinoma. In both cases, the dogs apparently detected the lesions right through the patients' clothing. Probably of significance was the fact that their curiosity ceased once the cancerous lesions were fully excised.

But these previous cases were skin cancers—visible to anyone and right there under each dog's nose. To my knowledge, the above described breast cancer case was the first instance in which an internal malignancy was discovered by "pet scan," so, in 2004, I decided to formally write the case up along with my colleagues Drs. Harish Ahuja and Darryl Barton of Wausau, Wisconsin.

Since then there have been numerous additional cases of lung cancers and bladder cancers that have been detected serendipitously by pet dogs, as well as by specially trained dogs. Dogs purportedly possess approximately three hundred million olfactory receptors compared to a human's five million—a sixty-fold enhancement, which confers a sense of smell perhaps one hundred thousand times more sensitive than people. Given such olfactory prowess, dogs understandably have found important roles in police work. For example, specially trained dogs are often used for detecting hidden explosives and drugs, tracking criminals, and finding cadavers or missing persons. It therefore doesn't seem too far-fetched to think that certain dogs could detect subtle odors emitted by human cancers, even if they were internal.

This burgeoning field—called "dognoseis" by some—evolved into a legitimate scientific discipline when researchers from Tallahassee, Florida, began conducting rigorous experiments involving a standard schnauzer and a golden retriever who were specifically taught to sense melanoma. The dogs were first trained to find melanoma tissue samples that were hidden somewhere on healthy human volunteers. After this preliminary training, they were tested on seven real patients suspected of having melanoma. One dog "reported" melanoma in five patients, all of whom were confirmed by biopsy to actually have melanoma. Amazingly, a sixth individual with suspected, but not biopsy-proven, disease was also reported as "positive" by the dog. Initially recorded as a false positive, additional testing on this patient confirmed the diagnosis of early melanoma, thereby vindicating the dog. The second dog examined four individuals and agreed with the first dog in every case. The calculated odds of these results being obtained purely by chance were less than one in ten million for the first dog and less than one in one thousand for the second dog.

Other early dognoseis investigators examined the ability of dogs to detect patients with bladder cancer through urine odor. Dogs were first trained to identify urine from bladder cancer patients. After this education, they were then tested for their ability to specifically select cancer patients through their urine. The dogs correctly identified patients 41 percent of the time—substantially better than the 14 percent odds that chance alone would predict. Curiously, some patients were consistently identified or missed by the dogs. This might indicate that the pungency of the cancer varied between patients. Interestingly, several dogs consistently identified one of the controls as a cancer case—and as in the melanoma study, further evaluation of this particular patient revealed an undiscovered cancer.

Exactly what the dog was able to smell in our breast cancer patient remains uncertain. All odors, whether the pleasing scent of fresh flowers or the fetid stench of feces, are due to volatile molecules diffusing through the air. These airborne molecules are detected by olfactory receptors in the noses of humans and animals, sending signals to the brain that then interprets the smell. There have been some interesting reports of volatile organic substances, or odorants, found in the breath of lung and breast cancer patients, and detection of such odorants either by dogs or an instrument could lead to novel means of diagnosing these cancers early. Odorants have also been associated with ovarian cancer, and researchers at the University of Pennsylvania have been exploring the potential of specially trained dogs to sniff out early-stage ovarian cancer. One can now comb the Internet and come across countless extraordinary cases and entire websites devoted to the topic.

For example, researchers at the University of Arkansas for Medical Sciences have worked with a German shepherd mix named Frankie, who was found as a stray pup and rescued from a busy street in Little Rock. After his intensive training, Frankie could predict—with an impressive 88 percent accuracy—which patients had thyroid cancer and which had a benign disease. Another inspiring study took place in Japan in 2011. There, specifically trained dogs detected colorectal cancer with an astounding 98 percent accuracy by sniffing breath samples. This beats the accuracy of even the most sophisticated diagnostic medical tests. Italian researchers announced at the American Urological Association annual meeting in 2014 that based on 320 men with prostate cancer and 357 without it, two highly trained dogs had a sensitivity of 98 percent in detecting prostate cancer through urine samples.

One article by Amanda Cable for the *Daily Mail* especially caught my attention last year because it mentioned animal behavioral psychologist Dr. Claire Guest, a co-investigator in the above-mentioned study on bladder cancer detection by dogs. She has apparently continued her research and even formed a charity, Medical Detection Dogs, devoted to trained dogs capable of detecting human cancers from breath and urine samples. One superstar sniffer, a Labrador named Daisy, has sniffed over six thousand samples of urine and detected over 550 cases of cancer with an impressive diagnostic accuracy of 93 percent. But Daisy did more for Dr. Guest than just further her scientific career. Daisy might have saved her life.

Daisy was not always a superstar. Daisy started out simply as Dr. Guest's own beloved pet dog, and four years previously, she seemed intent on insistently pawing at the doctor's chest, eventually even bruising her. "I felt the tender area where she'd pushed me, and over the next few days I detected the tiniest lump."[2] Her physicians believed it was a cyst, but obtained a mammogram to be sure. There was indeed a cyst, but further in the breast tissue was a deep-seated cancer. This cancer was fortunately caught very early and was addressed with a lumpectomy, the removal of some lymph nodes, and a course of radiotherapy. Dr. Guest was quoted, "I was forty-six, and the specialist told me that by the time a lump had become noticeable, this cancer would already have spread and my prognosis could have been very different. Just as I was doubting the future of dogs being used to detect cancer, my own pet Labrador saved my life."[3] It thus appears that our initial case of breast cancer detection by a pet dog at the start of this chapter was not a one-of-a-kind freak occurrence but instead did actually have a scientific base.

Going back to the Emily Dickenson quote I used to open chapter 7 ("Dogs are better than human beings because they know but do not tell.") one might note that sometimes dogs do know and they *do* tell—thereby saving people's lives.

Sadly, this is not always the case. Shortly after our case study reached the library, numerous newspapers and television crews wished to speak with me, interview the patient, and, most important, photograph and video the hero dachshund. Regrettably, these media folks didn't read all the sobering details. My patient did not have a good outcome. Most curious, and perhaps tellingly, her dog continued sniffing and poking around after the surgery—even though the lumpectomy appeared to have cleared out all the cancer and there was no proof of metastases to the removed lymph nodes. Despite our best efforts at containing it, her cancer began to spread quickly, and she succumbed approximately a year later. Upon hearing this, the news dropped the story like a hot potato.

The fact that her cancer progressed so quickly made me suspect that her disease was more advanced than we thought it was upon diagnosis. As mentioned, it may have been revealing that her dog continued sniffing and poking around even after her surgery. It remains quite possible that her dog was detecting an odor from her cancer that was more extensive than anything that modern medicine was capable of detecting at that time. (Genuine PET scanning was not routinely available at that time but possibly could have helped us determine if her disease was indeed more advanced.)

It should come as no surprise that some critics have pointed out that having a clinic full of cancer-sniffing canines might not be exactly practical. Unlike conventional CAT and PET scan machines, which don't need to be walked, washed, and fed regularly, dogs need constant care and attention. I don't know what the comparative costs of maintaining radiology equipment versus walking and feeding a kennel full of tumor-detecting dogs might be—but one suspects the latter might be a lot more fun. And patient friendly.

So what is the relevance of our little aside on this unique brand of "pet scanners"? How does it tie back in with our main theme—that immunotherapy might someday be capable of conquering cancer?

Well, if one thinks about it, the fact that dogs can literally smell cancer is consistent with the premise that cancers are indeed vastly different from normal cells, tissues, and organs. So much so that they actually give off a distinct odor that can be detected by highly sensitive sniffers or maybe sensitive electronic instruments. This confirms the concept that cancers are so immensely different from normal tissues that they should indeed be very visible (or smellable, so to speak) to the immune system. The fact that they most often are not is most curious—and concerning. In subsequent chapters we shall explore the reasons why, despite obvious differences from normal tissues, cancers are often unseen by our own immune systems.

Chapter 10

MALES NEED NOT APPLY

Many would view a "Robin's pincushion" as a thing of natural beauty.

A botanist might consider it a distortion in the axillary meristem.

Entomologists might see it as an ingenious wasp's nest.

As an oncologist, I view it in utter amazement and fascination. Robin's pincushion is a plant tumor.

Amazing things about cancer are learned not only from the animal kingdom; plants teach us much and help us a great deal as well. Many highly effective and popular cancer chemotherapy drugs trace their roots to the plant kingdom. For example, the *vinca alkaloids* such as vincristine and vinblastine are alkaloids (a catchall name for any large, naturally occurring, nitrogen-containing, organic molecule from a plant source) and, as the name indicates, are derived from Vinca (periwinkle) plants, specifically *Vinca rosea* (now *Cantharanthus roseus*). Vinca alkaloids interfere with microtubule assembly. Since microtubules are essential in pulling chromosomes away from the middle of dividing cells and into the incipient daughter cells during mitosis, vinca alkaloids inhibit cell division. Cancer cells that can't divide can't cause trouble. Vincristine is frequently used (in combination with other drugs) to treat certain non-Hodgkin lymphomas and leukemias whereas vinblastine is a popular drug (again in combination with other chemotherapy agents) for treating Hodgkin lymphoma, among some other cancers.

A botanical origin is not at all unusual for chemotherapy agents. Taxol (paclitaxel) is another plant-derived chemotherapy drug that interferes with microtubule function during mitosis. This time however, the microtubules used to pull chromosomes apart from one another and into the newly forming daughter cells are not prevented from forming but rather are prevented from disassembling. In this manner, dividing cells are "frozen" in place during late mitosis and their chromosomes might not get to their desired daughter cell destinations. Taxol, which is often used against breast, lung, and ovarian cancers, originates from the bark of Pacific yew trees (*Taxus brevifolia*).

Etoposide (an agent often used in lung and testicular cancers) and teniposide (a drug sometimes used in pediatric acute lymphocytic leukemia) belong to the class of chemotherapy agents called *epipodophyllotoxins*, a daunting name for drugs based on extracts from roots of the American Mayapple (*Podophyllum peltatum*). These drugs work via inhibition of the enzyme *DNA topoisomerase II*, an enzyme that relaxes DNA supercoils (or more simply stated, untangles bunched-up DNA strands). The inhibition of this enzyme introduces breaks in the DNA of cancer cells.

Topotecan and irinotecan (used in advanced colon and rectum cancers) are *camptothecins*, meaning they are semisynthetic agents derived from *Camptotheca acuminata*, the Chinese Xi Shu ("happy tree"). Camptothecins interfere with another enzyme that affects DNA topology, namely DNA topoisomerase I. This enzyme relieves torsional strain in the coiled up DNA while it is undergoing replication. Inhibition of topoisomerase I interferes with the unwinding of DNA and thereby halts DNA replication in cancer cells.

An interesting aside about all these plant-derived chemotherapy agents and the curious way that cancer cells remain one step ahead of them concerns *p-glycoprotein*, which is short for "permeability glycoprotein," and also goes by the more descriptive name *multidrug-resistance protein 1*. One important physiological role of this membrane-bound cellular protein is to prevent certain toxic substances from being absorbed in the intestines. Some drugs and toxins may enter intestinal cells only to immediately be rerouted back out of these intestinal cells by p-glycoprotein. Once dumped back from whence they came from, they are expelled with the feces. Similarly, cells in the liver and kidneys take toxic substances from the blood and promote their excretion into the bile or urine, respectively via p-glycoprotein. Many toxins and drugs are unable to enter the brain, in part thanks to the role of p-glycoprotein. In cancer chemotherapy, p-glycoprotein plays a key role in drug resistance. Many cancer cells will, upon taking up certain chemotherapy drugs, immediately pump them back out where they can cause no harm (no harm to the cancer cells that is). This p-glycoprotein mediated resistance appears to be quite pronounced for plant-derived chemotherapy agents.

Another drug known all too well by oncologists and cancer patients is *senna*. Senna is derived from the leaves and pods of *Cassia acutifolia* and *Cassia angustifolia*. Senna is a laxative that is often used to relieve the severe constipation brought on by pain-relieving *opioid* medications. The opioids include morphine and codeine, which are naturally present in the opium poppy, *Papaver somniferum*, as well as the semisynthetic morphine derivatives hydromorphone, hydrocodone, oxycodone, and heroin (and fully synthetic analogues such as fentanyl).

Not only do plants provide us with lifesaving chemotherapy agents and other drugs used in the cancer clinic—they can teach us a great deal about oncology as well. For they, too, can develop cancer.

Yes, plants do get tumors.

Luckily for plants, their tumors are typically less able to do serious damage compared to what havoc they wreak in animals and people. For one thing, vegetable tumors don't metastasize. Plant cells are locked in place by a high-fiber matrix of rigid cell walls and do not roam about. Thus, even if a rogue plant cell begins proliferating with abandon, the tumor it generates remains stuck in one place. Unlike metastasizing cancer cells in people and animals, a plant tumor is little more than, for lack of a better term, a bump on a log. Furthermore, in contrast to people and animals, which have vital organs such as the brain or liver, plants do not have critical organs that would quickly cause death if compromised. If a stem, leaf, root, or branch goes bad, a plant can usually just make a new one. Nevertheless plant tumors sometimes do cause significant damage—at a great economic expense.

While plant tumors can be caused by spontaneous mutations in the same genes associated with animal tumors, more often they are induced by infections with bacteria, fungi, or viruses. For instance, certain fungal diseases such as black knot disease (caused by *Dibotryon morbosum*) and azalea knot (caused by *Exobasidium vaccinii*) can lead to plant tumors. Two bacterial species, *Agrobacterium tumefaciens* and *Agrobacterium vitums*, are the causes of crown gall disease. And crown gall disease is a major economic problem of walnuts, grape vines, fruit trees, horseradish, sugar beets, and rhubarb.

Agrobacterium tumefaciens contains a *plasmid* (a virus-like segment of DNA that is independent of the main bacterial chromosome). The plasmid DNA can leave the *A. tumefaciens* bacterium and become integrated into the genome of the infected host plant cell. Certain genes on the *A. tumefaciens* plasmid are then expressed, causing the plant cell to proliferate. Such unbridled proliferation ultimately leads to a tumor, or *gall*. Some of the *A. tumefaciens* plasmid genes encode for plant hormones, which stimulate plant cell growth and disrupt the plant's control over cell division in the infected area. This natural, horizontal interkingdom DNA transfer (from kingdom Bacteria to kingdom Plantae) is an astonishing biological phenomenon and has proved to be an invaluable tool in genetic engineering. (Just as an aside to clarify what is meant by "horizontal" DNA or gene transfer: Most genes are passed from generation to generation, from parent to daughter; this is described as *vertical* gene transfer. On the other hand, when genetic material (i.e., DNA) is passed along from daughter to daughter by passing around plasmids, it is called *horizontal* gene transfer. During horizontal gene transfer, pieces of DNA, sometimes with valuable genetic information on them, are actually passed from one organism to another via plasmids. Horizontal gene transfer happens all the time with bacteria—and is a major problem medically. For instance, an antibiotic such as methacillin might initially work wonderfully against a strain of *Staphylococcus*. But in a hospital environment where there are many bacterial strains roaming the halls, these Staph bacteria might come in contact with other bacteria that possess resistance to methicillin. During such encounters, the Staph bacteria acquire methicillin resistance in a quick and dirty fashion. Through horizontal gene transfer, the other bacteria just lend the Strep a plasmid with the methacillin resistance copy on it. They didn't buy it, didn't earn it . . . they were just handed a "get out of jail free card" from a friend. An example of evolution, the easy way. While such unbridled gene transfer between bacteria is the bane of infectious disease control, horizontal DNA transfer has proved highly invaluable in recombinant DNA technology as a means of transplanting certain genes from one species to another in the lab.)

Another common cause of plant cancer is insect infestation. Many plants wind up with tumors that double as homes for insect larvae. Rather than build an elaborate nest for their eggs and developing larvae, some insects simply make the plants do all the work for them. For instance, the tiny gall wasps *Callirhytis cornigera* and *C. quercuspunctata* primarily induce galls in oaks. But these wasps are not alone. The so-called oak apple, a round, spongy, fruitlike object about one to two inches in diameter, is caused by the larvae of another gall wasp, *Biorhiza pallida*. About thirty wasp larvae may develop in a single "apple." The marble gall, a green growth, typically about one inch in diameter, is inspired by the tiny wasp *Andricus kollari*.

Yet another gall wasp, the cypinid rose gall wasp, *Diplolepis rosae*, produces strange but often beautiful tumors on roses known as the Robin's pincushion, the moss gall, or the rose bedeguar gall depending on the shape and appearance. Such a gall may contain fifty or more developing wasp larvae.

Exactly how gall wasps provoke plants to form tumors is unclear. Perhaps the wasp stings the plant or maybe the egg itself in some way deregulates plant cell division and induces cellular proliferation. In any case, in response to the stimulus, the plants produce excess tissue around the wasp egg, which serves as a safe haven for the egg and later feeds the insect larvae.

One fascinating bit of biological trivia about gall wasps: they are *parthenogenic*. Parthenogenesis is the ability to reproduce without males. The eggs laid by females are already fully fertilized. In some of these parthenogenic gall wasps there is an *alternation of generations* such that one generation may have males and the following generation may be completely devoid of males. The generation with males may reproduce sexually whereas the next generation, with no males, will reproduce asexually through parthenogenesis. Remarkably, the galls induced by the different generations are themselves slightly different in appearance and can be distinguished by the careful observer.

The amazing biology of these wasps doesn't end there. Some gall wasp species are gall *inquilines*. These wasps are unable to induce the formation of galls themselves but instead make good use of galls created by other insects. While a gall provides a generally safe refuge for the gall wasp's eggs and the developing larvae, certain other wasps have found a way to breech this sanctuary and parasitize the gall wasp larvae within. Known as inquilines, or parasitoids, these wasps use their specialized egg-laying ovipositors to drill into the gall and lay an egg—right on the hapless gall wasp larva! In this fashion, a robin's pincushion or other gall may not yield the original gall-making wasp in the springtime, but instead a different species of parasitoid wasp. These parasitoids, such as *Eurytoma rosae*, are often beautiful, bright, metallically-shimmering little wasps with long ovipositors. Parasitoid wasps may, in turn, be victimized by other wasps—hyperparasitoids.

In any case, we do know that plants get tumors. And (if one were so inclined to scan a tree), these masses would certainly be seen on CAT scan. What I do not know (and I doubt anyone does!) is whether these tumors would be positive on PET scan!

Chapter 11

COULD BROWN FAT BE THE SECRET TO WEIGHT LOSS?

By now, most folks are familiar with CAT (computed axial tomography) scans, more often just called CT (computed tomography) scans. These medical imaging tools of the trade came to the forefront in the 1980s and were later joined by MRI (magnetic resonance imaging) in the 1990s. More recently, PET scans (positron emission tomography) have joined the lineup as invaluable tools, especially for cancer detection and "staging" (i.e., determining the stage of a given patient's cancer). PET scans are useful for cancer detection because they visualize things that are metabolically active—and malignant tumors are usually quite metabolically active. I recall during the early days of PET being fooled by one of the unusual and then very surprising causes of a false-positive PET scan—*brown fat*. A "false-positive" implies that whatever it is, it is metabolically dynamic. In this way it behaves like a cancer—but it isn't. Being metabolically active implies that it consumes calories—lots of calories. So, could brown fat be the cure for the obesity epidemic? Before we get to that controversial concept, let's examine some of the basics.

PET scanning exploits two fascinating scientific principles, one from physics and one from cancer biology. The fascinating physics concept is *mutual annihilation* when matter meets antimatter, and the intriguing biological principle is the still not fully understood *Warburg effect*.[1]

On a variety of levels there are a host of obvious differences between cancer cells and normal cells. Among some of the clear dissimilarities are the very rapid growth and reproduction rates typically exhibited by malignant cells. However there are some less perceptible distinctions as well. One such dissimilarity, observed nearly a century ago by Otto Warburg, is in their metabolism of the universal cellular fuel, glucose. Warburg, the son of an eminent physicist, earned doctorates in chemistry and in medicine and was awarded the 1931 Nobel Prize in Medicine or Physiology for his work on respiratory enzymes. Today, Warburg is again in the limelight for another important observation he made regarding malignant cells.

Cancer cells, with their accelerated growth and prodigious reproductive rates, require exorbitant amounts of cellular fuel. For reasons still poorly understood, cancer cells seem to prefer the inefficient method of *fermentation*, or *glycolysis*, to extract energy from glucose. This seems quite odd since there is a far more efficient method available to them: *aerobic respiration*.

On occasion (for example when muscles are called on for extreme and rapid effort), the faster, but far less thorough, *an*aerobic glycolysis pathway may be preferentially exploited over the more complete but slower process of aerobic respiration. This more primitive method of metabolism can very rapidly convert the raw fuel, glucose, into the required energy commerce molecules, *ATP* (adenosine triphosphate). Thus, in a pinch, glycolysis can produce ATP in short order without the need for oxygen. However, through the use of oxygen, the more evolutionarily advanced form of ATP production, aerobic respiration, can yield far more ATP molecules per glucose than glycolysis alone can. For example, a sprinter might be far more out of breath after an all-out two hundred meter dash than an athlete who has just finished a forty-five-minute jog. The sprinter, through glycolysis, has incurred an "oxygen debt" in order to complete that maximal effort, and will be huffing and puffing for a few minutes to rid himself of the excess lactic acid buildup. In contrast, the distance runner has efficiently used oxygen via aerobic respiration during the prolonged workout and might not be breathing hard at all at the end. The point is, ATP production is far more effective via aerobic respiration than through anaerobic fermentation. So why then would cancer cells, with their relatively high energy demands, stick with the inefficient, more primitive method of burning glucose even when ample oxygen is available?

On the surface, this predilection for anaerobic glycolysis over aerobic respiration seems quite strange, given that cancer cells grow and reproduce so frantically and should want to maximize their energetic efficiency. In fact, many tumors grow so rapidly that they literally outgrow their blood (and oxygen) supply and are inherently oxygen-depleted. Thus, given this overgrowth tendency, it might make sense that in order for them to survive, they must be capable of tolerating low-oxygen conditions and remain able to readily switch over to *an*aerobic glycolysis. Nevertheless, even with such low-oxygen tolerance, one would expect that when oxygen abounds, cancer cells should switch back to the more efficient means of ATP production.

It was therefore most baffling to observe that even in an oxygen-replete environment, malignant cells stubbornly stick to the inefficient, anaerobic glycolytic fermentation pathway. This paradoxical phenomenon, "*aerobic glycolysis*," or the preference of glycolysis alone even when oxygen is abundant, is now known as the Warburg effect. In order to satisfy their inordinate energy needs through the inefficient glycolytic fermentation method, malignant cells are exceptionally ravenous for glucose. Fortunately, we can diagnostically exploit this hunger via PET scans through the use of glucose analogues (glucose counterfeits) such as the positron-emitting radiopharmaceutical, *fluorodeoxyglucose*, or *FDG*. The radioactive FDG molecule looks sufficiently like the real thing to be avidly taken up by most malignant cells. Once inside a cancer cell however, the FDG molecule is modified during the first step of glycolysis (by the addition of a phosphate group) making it impossible for that modified FDG molecule to ever escape. In this fashion, the Warburg effect facilitates the accumulation of a radioactive tracer molecule into cancerous tissues. It would be quite hard to conceive of a better way of getting a radiotracer into cancer cells than this one that nature has handed us.

It is worth mentioning that the significance of the Warburg effect might go far beyond simply allowing effective detection and staging of cancer via PET. Given the peculiar penchant of cancer for using fermentation, cancer investigators are looking into ways to "force" cancer cells to do what they don't want to do—namely oxidative respiration.[2] Hopefully by driving cancer cell metabolism down this unwanted pathway, the cells will become metabolically confused or disrupted and unable to carry out their nefarious mission of spreading and wreaking havoc throughout the body. Inhibition of some of the molecular mechanisms behind the Warburg effect might enhance the efficacy of radiation therapy for instance.

Alternatively, cancer cells might be thwarted through inhibition of their preferred pathways. Recall that unlike normal cells, which use glucose efficiently via aerobic respiration, malignant cells must guzzle down gross quantities of glucose to sustain themselves through glycolysis. Therefore, agents that inhibit glycolysis might selectively stymie cancerous cells. One experimental anticancer approach works through interference with an enzyme called *pyruvate dehydrogenase kinase*, which is essential for the continuous breakdown of glucose through glycolysis. If this enzyme is blocked, glycolysis comes to a halt, and the cell can no longer produce the ATP it needs. Based on this hypothesis, investigators from the University of Alberta have been exploring the use of a pyruvate dehydrogenase kinase blocker called *dichloroacetate* (DCA) as a novel therapeutic agent based on the Warburg phenomenon.[3] Another approach being explored is the application of *monocarboxylate transporters*, or MCTs. These molecules prevent an end product of glycolysis (lactic acid) from building up in tumors and thereby could cripple their metabolic mechanisms. Some sophisticated MCTs are being designed as anticancer drugs and are currently undergoing clinical testing.

The chemistry and physics behind PET imaging is just as fascinating as the biology. As mentioned, the current most popular chemical agent to bring a positron-emitting radioisotope into a tumor is FDG. Chemically, FDG is simply a glucose molecule where one of the hydroxyl groups (specifically the 2' hydroxyl group) is replaced by a fluorine atom. If the fluorine atom in the FDG is radioactive flourine-18, the positrons emanating from the fluorine-18 will mark the position of the cancer within the body. But this will only work if two criteria are met: first, the carrier molecule must selectively accumulate in the tumor, and second, it must stay there long enough to be practical for PET scanning. The first criterion is readily met by FDG thanks to the Warburg effect. The glucose-ravenous cancer cells gobble up the glucose impersonators en masse. The second criteria, staying put long enough to be imaged, implies that not all positron emitters are equally suited. A positron-emitting radioisotope with a half-life of only a few seconds would not prove practical in the clinic. For instance, labeling a glucose molecule with positron-emitting oxygen-15 might initially appear even more logical than adding a foreign-appearing fluorine-18 atom. But oxygen-15 has a half-life of only 122 seconds, making PET imaging with this isotope more technically challenging. Thus, although one might predict that, based on biochemistry, the oxygen isotope would be the top choice, for practical reasons the fluorine isotope has proven to the preferred choice.

FDG, carrying positron-emitting fluorine-18 atoms with their 1.83-hour half-lives, meets the two criteria. Beyond the sufficiently long half-life of fluorine-18, the fact that when FDG is biochemically modified during the first step of glycolysis by the addition of a phosphate group and cannot subsequently sneak out of a cancer cell it has entered is an additional bonus.

Radioactive isotopes release their energy through decay by alpha, beta, and gamma emission in order of both their historical discovery and in order of their depths of penetration. Alpha radiation is nothing more than the emission of a helium nucleus. This provides a quick way for an overweight atom to shed some excess pounds and gain nuclear stability. On the other hand, gamma radiation, being a form of electromagnetic radiation, is completely massless; emission of gamma rays has no impact on the mass of the nucleus that emitted it (of course, this is not entirely true since energy and mass are equivalent by the Einstein equation, and the energy carried away by the gamma ray must be subtracted from the overall mass-energy of the parent nucleus). Gamma rays and x-rays are high-energy photons. They are the most energetic forms of electromagnetic radiation and are regularly used in diagnostic imaging and radiation therapy.

Beta radiation, the third of the basic type of radioactive decay, comes in a variety of forms. "Standard" beta decay is simply the emission of an electron (plus, for physics fans, an electron antineutrino for conservation of matter/antimatter, momentum, and lepton number). A second form of beta decay is the capture of an orbiting electron by a proton-rich nucleus. As electrons are negatively charged, this *electron capture* neutralizes the excess positive charge of proton-rich nuclei. Competing with electron capture is another form of beta decay that also alleviates surplus positive nuclear charge in proton-rich nuclei: "beta plus" decay, or *positron emission*.

Positrons are the *antimatter* equivalent of electrons. As such, they possess the exact same mass and other physical properties as regular electrons, but have the opposite electrical charge. While the electron has a charge of negative one, the positron has a charge of exactly plus one. Thus, the alternative name for positron emission, "beta plus" decay. In any case, whenever matter meets its antimatter counterpart . . . KABOOM! All of the mass is instantly converted into energy via Einstein's famous equation, $E=mc^2$.

That energy is in the form of two (rarely three) high-energy gamma rays shooting off in opposite directions (or toward the apices of an equilateral triangle in the rare instances of three photon emissions). Through the use of computerized 3-D analysis, the origin of these photons can be calculated in space, and images can be reconstructed. If a patient is injected with a positron-emitting isotope, and if that radioisotope accumulates in tumors, physicians can determine the location and number of malignant masses within a patient. This is the physical basis of PET imaging.

In this way, a newly diagnosed cancer patient can be "staged"—the cancer can be assigned a stage (which typically ranges from stage I, the earliest, to stage IV, the most advanced). Incidentally, there is a stage 0 for some cancers (for instance, in-situ breast cancer), and there is a stage V designation for the pediatric kidney cancer, Wilm's tumor, when the tumor involves both kidneys at the same time. PET scans have emerged as an invaluable tool in oncology and can help determine if a cancer is in an early stage and can be surgically or radiotherapeutically cured—or if such heroic efforts would be in vain because the stage is too advanced. Additionally, for a patient who has already undergone curative treatment in the past, follow-up PET scans are sometimes used to ascertain whether or not the disease remains in remission.

Returning to that strange cause of false-positive PET scans mentioned at the start of the chapter, namely *brown fat*, it is somewhat shocking that this tissue was not even known to exist in adults until about six years ago.[4] As a source of false positive PET results, brown fat must be a metabolically active tissue that takes up FDG—and by association, real glucose. So what is this metabolically active tissue, and what is it doing with all that glucose? And what is becoming of the calories contained within? The answer is that it is burning those calories like crazy to produce heat.

Calorie-consuming, heat-generating brown fat (also called *brown adipose tissue*, or *BAT*) is found in mammals such as seals who live in chilly environments and is also abundant in hibernating mammals. In fact, the brown fat pad between the shoulder blades of some rodents was once inaccurately called the "hibernation gland." Recently, a team from Japan discovered that certain dolphins and porpoises also possess brown fat, distributed around their bodies in a fashion that serves as an electric blanket to keep them warm in cold waters.[5] Therefore, the old concept that cetaceans must constantly swim about in order to maintain their core body temperatures needs revision. It has been known for a while that heat-producing brown fat is also present in human infants, but until recently it was assumed that adults do not harbor any brown fat. In 2009, however, it was discovered that some adults do indeed possess brown fat, predominately in the base of their necks, along their spine, and on their backs between their shoulder blades. Since cancer likes to spread to lymph nodes, and since there are abundant lymph nodes in the base of the neck, it is obvious why this recently discovered phenomenon was initially so confusing when PET scans first hit the scene.

Until very recently it was believed that brown fat stores were genetically determined and some lucky people were endowed with ample amounts to stay warm and burn calories for the rest of their lives whereas others were relatively devoid of brown fat deposits and destined to grow fat and cold with age. In fact, obese people do tend to have particularly small stores of brown fat, and we all tend to deplete what little we have as we age. For these reasons, it was of great interest to learn of an intermediate form of fat known as *beige* fat.[6] Unlike the calorie-burning, heat-generating but anatomically limited brown fat—and also unlike the anatomically abundant, energy-storing but weight-gaining white fat—beige fat seems to serve both functions.

In the simplest terms, brown fat expends calories by consuming the cellular fuel glucose *without* converting it into ATP. Instead of transforming glucose into the universal biochemical commodity, ATP, brown fat cells squander the chemical energy stored in glucose and release heat. (More technically, *oxidative phosphorylation* is "uncoupled." That is, instead of the normal biochemical process of using the chemical energy in glucose to generate an electrochemical gradient across mitochondrial membrane, and then converting this electrical potential into chemical energy in the form of ATP molecules, brown fat mitochondria simply let the electrochemical gradient discharge, dissipating the energy in the form of heat.)

Most encouragingly, there might be a common precursor cell shared by both white fat and beige fat—and under certain circumstances it appears that a white fat cell can transform itself into an energy-producing, weight-losing beige fat cell. The circumstances that seem to favor the transition of white fat into brown fat are the same conditions that activate brown fat—namely exposure to cold. By this reasoning, one silver lining to the grippingly cold winters that have plagued the mid-western United States over the past few years could be that white fat cells *might* transition into calorie-consuming beige fat cells. However, before anyone jumps to the conclusion that "beige is the new brown," at the time I write this in 2015, all conclusive research on the white-to-beige fat transition has been confined only to animal studies. It remains uncertain whether humans really have this beige fat, and if so, to what degree we can transform our excess white fat into calorie-burning beige.

In a fascinating story in *New Scientist*, author Chloe Lambert recounted her own experience in which she subjected herself to cold walks and swims for five days in an effort to engage her brown fat.[7] I read with bated breath for the outcome. The monitors showed that she succeeded—her brown fat was activated. But much to my disappointment (and probably to many others who read this), she did not lose any weight! In fact, she gained about two pounds. Apparently there is a lot more to the story that must yet be unraveled. More research is needed before one can simply take a stroll out in the cold wearing shorts and a tee shirt, build a snowman, or have snowball fight and shed all their unwanted weight. Quite possibly, the activation of brown fat also increases one's appetite, and the increased calorie intake outweighs the calorie expenditure of the brown fat.

There is one final less sanguine aspect to brown fat that relates to cancer. In people who have advanced cancers, a condition called *cachexia* is common. The name aptly derives from the Greek for "bad condition." Perhaps 50 percent of all people with advanced cancer will eventually develop cachexia, which results in major muscle wasting and extreme weight loss. It is well known that cancer patients with cachexia may heed their physician's advice and eat and drink abundantly—yet as if suffering some sort of curse, they alarmingly continue to lose weight. Approximately one-fifth of all cancer deaths may ultimately be attributable to the malnourishment associated with cachexia.[8] Recent research in mice indicates that cancer converts white fat into calorie-consuming beige fat. If the same phenomenon is occurring in people, it could explain why advanced cancer patients inevitably lose weight and fail to thrive no matter how much they earnestly attempt to eat.

Once again, readers might be wondering what, if anything, this digression on PET scans and variegated fats has to do with the immune system and cancer. It turns out that there is one curious link between the Warburg effect and the immune system. *Naïve* T cells, that is, T cells that have not yet been "trained" to identify and eliminate specific enemy targets, use oxidative respiration to generate their ATP. No surprise there. Most cells in the human body that are able to, strongly prefer oxidative respiration over glycolysis. (There is one notable exception to this rule: Red blood cells are devoid of mitochondria and therefore unable to carry out aerobic respiration. This should come as no real surprise since the primary function of red blood cells is oxygen transport and delivery; if they themselves consumed oxygen, it would defeat their purpose.) Brain cells are 100 percent dependent on aerobic respiration and cannot fall back on glycolytic fermentation as a sole means of meeting their ATP needs. This is why only a few short minutes of oxygen deprivation can cause permanent brain damage or even death. Strangely however, *activated* T cells which have gone through their training and are now ready and able to destroy unwanted invaders, perplexingly prefer glycolysis. In essence, they exhibit the Warburg effect. To date, the exact reasons for this discrepancy are unclear. But T cells are not alone in this oddity. Neutrophils, another key cellular component of our immune systems and the most abundant of our white blood cells, also demonstrate the Warburg effect. Fascinatingly, neutrophils consume large amounts of oxygen, but instead of using that oxygen for respiration and production of ATP, they use the oxygen to generate germ-killing oxidizing molecules such as hydrogen peroxide plus superoxide and hydroxyl radicals. Finally, macrophages, yet another key player in immunity, also demonstrate the Warburg effect upon their activation. In fact, there is active research ongoing about the possible role of macrophage-cancer cell fusions. Such hybridization is proposed as a means of gaining the ability to metastasize. Macrophage-cancer cell fusions have been proffered as an explanation for a phenomenon called EMT, or *epithelial-to-mesenchymal transition*; EMT is commonly seen in cancers capable of metastasizing.[9] Could it be that cancer cells switch over to aerobic glycolysis (i.e., the Warburg phenomenon) *because* they have fused with macrophages? Alternatively, could cancer cells that exhibit the Warburg effect be shielded from immune attack because they are masquerading as ordinary activated T cells, macrophages, and neutrophils? Finally, could the fusion between a cancer cell and a macrophage grant it invisibility to other cells in the immune system and shield it from attack? Presently no one knows for certain, but such questions, tangentially arising from our digression on PET scans, certainly pique one's curiosity. If our quest is to unveil cancer and allow the immune system to "see" malignant disease for what it is, such questions could prompt studies that open new doors indeed.

Chapter 12

GAMMA RAYS AND DINOSAUR CANCER

Nevertheless so profound is our ignorance, and so high our presumption, that we marvel when we hear of the extinction of an organic being; and as we do not see the cause, we invoke cataclysms to desolate the world, or invent laws on the duration of the forms of life!

—Charles Darwin, *On the Origin of Species*

Since we just reviewed some of the interesting physics of PET scans, including various forms of radioactivity, allow me to indulge in one of my personal favorite subjects and embark on a brief aside on dinosaur oncology.

Throughout Earth's history there have been periodic major and minor mass extinctions. The five major mass extinctions that have carved their indelible fingerprints in the rocky record include the Ordovician extinction, the Devonian extinctions(s), the Permian extinction, the Triassic extinction, and the Cretaceous-Tertiary extinction, in chronological order.[1]

The most recent mass extinction, the extinction event at the close of the Cretaceous period about sixty-five million years ago was likely triggered by a massive asteroid impact striking the region where the northern end of the Yucatán Peninsula now rests. This extinction ranks third among the "big five," behind the Ordovician extinction (which was circa 444 million years ago and exterminated nearly 85 percent of marine species) and the mother of all extinctions, the Permian extinction (which occurred somewhere between 266 and 251 million years ago and eliminated 95 percent of all marine species and about 70 percent of land species—including insects, which were relatively unscathed by most other extinction events). Regardless, it is the Cretaceous-Tertiary (K-T) extinction that captivates most folks since it is the one that extinguished the dinosaurs sixty-five million years ago.

There is solid supportive evidence for the asteroid theory of extinction, including the "iridium anomaly" (a thin layer of iridium, which is generally rare on Earth's surface, residing precisely at the K-T boundary), shocked quartz (deformed quartz crystals caused by monstrous explosions or impacts), tektites (bits of natural glass created by natural impacts that are scattered about in a distribution of increasing concentration as one nears the Yucatán Peninsula), evidence of massive flooding (consistent with a huge tsunami centered on the region that now corresponds to the northern Yucatán), and the distribution of cenotes (natural sinkholes) scattered about in the Yucatán Peninsula. In fact, the last of these, the distribution of cenotes, while generously distributed across the entire Yucatán, are especially concentrated in a semi-circular distribution on the very northernmost part of the peninsula. When one extrapolates the pattern into the Gulf of Mexico, the full circle corresponds well to what is known as the Chicxulub crater (named after the town near the center of the crater) which was created around sixty-five million years ago.

Collectively, the data for an immense impact caused by an asteroid about 6.2 miles (~10 km) in diameter is quite compelling. The resultant energy released is estimated at about one hundred teratons, or the equivalent of nearly seven trillion Hiroshima atomic bombs. The local explosive devastation and tsunami certainly would have wiped out anything within hundreds, if not thousands, of miles, and the induced dramatic climatic shifts could have served as the coup de grâce for the dinosaurs and countless other species that vanished at this extinction event. One would have been hard pressed to come up with a better location for a killer impact since that region on the Yucatán Peninsula was especially well endowed with sulfur-containing rocks, which upon incineration led to sulfuric acid rain and sun-blocking aerosols, intensifying the already disastrous situation.

Over the years, alternative hypotheses—some reasonable, others outlandish—have come and gone with little supporting evidence leaving the impact theory standing as the most plausible model. Then again, as a radiation oncologist who uses high-energy gamma rays on a daily basis to treat cancer patients, how could I not be intrigued by the wild proposition that a widespread epidemic of cancer induced by a gamma ray burst was what really killed the dinosaurs?!

Just what are gamma ray bursts? The history of their discovery is most intriguing. Back in 1963, the nuclear weapons Test Ban Treaty outlawed aboveground nuclear weapons testing. In order to confirm compliance with the treaty, the United States launched the *Vela satellites* in the 1960s.[2] These satellites were designed to detect the gamma ray signatures of any atomic bomb explosions. Curiously, the Vela satellites began detecting multiple gamma ray events early on. Was there rampant violation of the ban? Well, it turned out that the gamma rays were emanating from outer space, and while the treaty explicitly prohibited nuclear bomb testing on the surface or in the atmosphere, it said little about outer space. Initially this was most perplexing and led some to speculate that the Soviets were conducting nuclear weapons tests on the moon or elsewhere in the solar system.

With time it became clear that these were not the result of Soviet weapons testing, but it wasn't until 1973 that the data became declassified and news regarding an astrophysical phenomenon was released. It seemed that these gamma ray bursts were not even originating from our solar system. If so, they must have been enormously intense to be detectable through the relatively limited Vela satellites. To investigate further, a dedicated set of experiments was designed to specifically explore gamma ray bursts. The Compton Gamma Ray Observatory was launched by NASA in 1991 and hosted the Burst and Transient Source Experiment (BATSE), which was ten times more sensitive to gamma rays than previous missions.[3] The greater sensitivity of BATSE on the Compton Observatory led to the astounding discovery of nearly one gamma ray burst per day.

The next question was, exactly where were these gamma ray bursts emanating from? Our Milky Way galaxy is an enormous collection of stars once estimated to contain about one hundred billion—roughly the same as the number of neurons in the human brain. (More recent estimates have upped the number of stars in our galaxy to between two hundred billion and four hundred billion, while the estimated number of neurons in the brain has dropped to a mere eighty-six billion. Oh well. For those interested in such comparisons, if you include the number of neurons and *glial brain cells*, the figures are again close.) Viewed from above, our galaxy would appear as a spiral galaxy. Viewed from within, however, it looks like a flat disc with a central bulge. Given this distribution of stars, one would expect that gamma ray bursts should lie along this flat disk if they were of Milky Way origin. Most curious, however, it appeared that the 2,700 gamma ray bursts detected by BATSE over its nine-year life were isotropically distributed, meaning they were scattered uniformly in all directions. This implied that the gamma ray bursts were not coming from the Milky Way itself but instead were coming from *beyond* our galaxy. But an extragalactic origin implied energies far beyond anything previously imagined.

BATSE was able to establish that there were two different general types of gamma ray bursts. The majority were "long duration," while a lower number were short duration. (The term "long" must be taken in context here because the cutoff is a mere two seconds). Further exploration of gamma ray bursts was performed by an Italian and Dutch x-ray satellite observatory called BeppoSax.[4] This satellite enabled the localization of gamma ray bursts with unprecedented precision. Pinpointing the precise location of a gamma ray burst allowed other telescopes to train their eyes in that direction and detect visible afterglows. Analysis of the *redshift* associated with some of these gamma ray bursts confirmed that they were truly cosmologically distant—on the order of several billion light-years in many instances! (In astronomy, "redshift" refers to the Doppler Effect on light emanating from a receding source. The greater the redshift, the greater the speed of recession and the greater the distance from us. This was first discovered by Edwin Hubble and is thus known as Hubble's Law.)

So how could gamma ray bursts be so intense that their gamma rays could reach Earth from billions of light-years away? The answer is beaming. Essentially all of the energy is focused into two extremely narrow, laser-like beams less than 1 degree wide. The current most widely accepted theory for long-duration gamma ray bursts is the "hypernova," or "collapsar," model of extreme supernova explosions. After exhausting their nuclear fuel, the cores of some colossal stars will collapse upon themselves and terminate in a black hole. In other words, a gamma ray burst signifies the birth of a black hole. Importantly however, the laser-like gamma rays beam out only along the axis of rotation (that is only from the north and south axial poles).

In this *fireball model* of gamma ray production, an intense stream of charged particles blasts from the poles, and as these particles violently jostle and collide into each other, they generate the beams of gamma rays. Next, the charged particles crash into the gas in the interstellar medium and generate the visible light afterglow. The result is that only two intense, but narrow, jets of gamma rays burst out of the burgeoning black hole, 180 degrees in opposite directions. Only supermassive stars with very little hydrogen or helium can become gamma ray bursts. This is perhaps because only without a thick gas envelope around the star can the particles readily punch through at the poles and create the gamma rays and other radiation.

Given the extremely limited angles of emission, one can reasonably conclude that there must be many, many more gamma ray bursts than we can detect, since we can only identify such bursts when we are directly in our line of sight.

Upon hearing all this, anyone would wonder what would happen if a superintense and nearby gamma ray burst happened to be pointed our way? What if, instead of starting in a distant galaxy, a gamma ray burst came from within our Milky Way galaxy itself? Luckily, it appears that most long-duration gamma ray bursts are associated with dwarf galaxies of low metal content (note—in astronomy "metal" refers to all elements other than hydrogen and helium). Our Milky Way galaxy, relatively rich in metals, is therefore unlikely to be harboring many gamma-ray-burst wannabes. Nevertheless, there are some individual stars within our galaxy that could terminate themselves in such a spectacular fashion. There is an ultramassive star named Beta Carinae that many astrophysicists predict will ultimately explode as a supernova and perhaps as a hypernova, suggesting that it could create a gamma ray burst. At only 7,500 light-years away, the optical afterglow from such a gamma ray burst would be as bright as the daytime sun. While the optical flash from such an event would not be harmful, the associated gamma rays from such a nearby event could have a drastic effect on Earth and on life. But to do any damage, the intense, narrow beam has to be pointing right at us, and, fortunately, it appears that Beta Carinae's rotational axis is not aimed in our direction.

On the other hand, there is one giant whose rotational axis may be pointed too close for comfort. Approximately eight thousand light-years away and discovered in 1998, WR 104 is a Wolf-Rayet star, the most massive type of star. While most estimates put the inclination of WR 104 at a safe 30 to 40 degrees, other calculations have suggested that WR 104's rotational axis is aligned to within 16° of Earth. While most astronomers are confident that there is no real danger and WR 104's aim is off, it is somewhat disquieting that, of all constellations, WR 104 is located in Sagittarius. It would be the ultimate irony (but the revenge of all astrologers!) if humanity ultimately was wiped out by a well-aimed arrow from "the Archer."

But what if? First off, I doubt that anyone would get transformed into the Hulk à la Dr. David Banner by a cosmological gamma ray dose! Nevertheless, as a medical physics researcher interested in gamma rays and a doctor with a fondness for photons, I remain curious. Some calculations have suggested that supernovae and gamma ray bursts produce sea-level radiation exposures of about one Gray every five million years or so despite the enormous distances.[5] If a gamma ray burst were relatively nearby, within our own galaxy, one would expect the radiation dose to be far greater, and such doses can have definite biological impacts. Nevertheless, if a nearby gamma ray burst were to let loose and blast us, while there would be *some* gamma radiation that filters through to the surface (on only one side of the planet of course), the intense, but very brief, pulse of gamma radiation would be unlikely to affect most land animals, let alone deeper sea life. Instead, most of the gamma rays would be absorbed by the atmosphere. But that doesn't imply that we would be safe—such an atmospheric dose could obliterate our stratospheric ozone layer, which filters out harmful ultraviolet radiation.[6] Additionally, the gamma rays might interact with the most abundant gas in our atmosphere, nitrogen, to produce nitrogen dioxide, generating a climate-altering red-brown smog along with acid rain. With the elimination of our ozone layer (actually perhaps only 30 percent of it), the associated increased ultraviolet radiation might spell doom for many microorganisms, which in turn would lead to the demise of many organisms higher in the food chain, possibly culminating in a mass extinction. On the other hand, a recent hypothesis holds that in the year 774 CE a gamma ray burst struck us, accounting for the twenty-fold increase in carbon-14 found in tree rings from that year. If this is true, humanity (and life on Earth in general) appears none the worse for wear. Just how devastating to life gamma ray bursts truly are remains unclear.

In 2004, Adrian Melott and colleagues from the University of Kansas suggested that, based on the fact that superficial sea life was hit harder than deep sea life, the *Ordovician extinction* around 444 million years back was caused by a gamma ray burst. In addition to this differential biological extermination, the global cooling due to the atmospheric effects of dark nitrogen dioxide, the injurious acid rain, and the later global warming would have seriously disrupted ecosystems leading to the widespread loss of life.

As of now, this hypothesis remains highly speculative and while not entirely implausible, it lacks rigorous supportive evidence. As mentioned earlier, another even wilder, purely speculative hypothesis held that a *cancer epidemic* caused the extinction of the dinosaurs. Could a gamma ray burst, leading to transiently increased gamma radiation and prolonged increased ultraviolet radiation, have started a malignant pandemic among dinosaurs? Well, first of all, did dinosaurs even get cancer? Isn't cancer a disease due to smoking, drinking, imprudent eating, pollution, and other ravages associated with modern society? Apparently not.

One of the first reports I encountered was aptly titled, "Metastatic Cancer in the Jurassic" in the highly rated medical journal the *Lancet* in 1999.[7] Not sure what to expect, and admittedly quite skeptical, I said to myself before reading the article, "I hope this isn't some sloppy, superficial treatment of an interesting subject. I hope these guys have done their homework!" Were they really capable of demonstrating that a big bump on a fossilized bone from the Jurassic was metastatic cancer? Come on! The Jurassic period was 201–145 million years ago. Please. What traces of malignancy could possibly persist that long and on a *fragment* of bone, mind you. How would one even prove that this petrified chunk was originally from a dinosaur?

The authors mentioned that the presence of "dense Haversian bone" was characteristic of dinosaur bone and that dinosaurs were the only animals large enough to have left this large bone fragment in that particular time and location. Okay, so it's a dinosaur bone. But how can you convince me that it was really a cancer-related metastatic tumor rather than just a healed fracture, an infected bone, or a zillion other things?

I was encouraged to read that the piece of bone, originally thought to be an osteosarcoma by another scientist was "easily distinguished from the lesions of myeloma, which have a 'punched out' appearance " and that the "preservation of a residual cortical shell also helps to distinguish metastatic cancer from multiple myeloma."[8] Although I still had my doubts, maybe the authors had indeed done their homework.

They went on to say that the stony, bony bump was "distinguished from superficial solitary and coalescing (1–3 mm) pits of leukaemia; sclerotic-rimmed lesions of gout; zones of resorption characteristic of tuberculosis; 'fronts of resorption' of fungal granulomas; sclerotic features of gummatous lesions of treponemal disease; the expansile, soap bubble appearance of aneurysmal bone cysts; sharply defined unicameral bone cysts; enchondromas; osteoblastomas and chondromyoxoid fibromas; the radiolucent nidus of osteoid osteoma; the epiphyseal 'popcorn' calcifications characteristic of chrondroblastomas; the 'ground glass' appearance of fibrous dysplasia, the onion-skin periosteal reaction and ill-defined margins of Ewing sarcoma and osteosarcoma; and the space occupying-mass appearance of eosinophilic granuloma."

Oh.

This silenced at least one critic. Yes, I was indeed impressed with the effective narrowing of what initially seemed an impossibly long differential diagnosis—which finally offered closure to a two-hundred-million-year-old cold case.

But I was also fascinated by the idea that malignancy, far from being an exclusively contemporary disease of alcohol-drinking, cigarette-smoking, fast-food-overeating modern humans, extended back as far as the mid-Mesozoic!

I read more on the subject, including George Johnson's fascinating book *The Cancer Chronicles*.[9] There was one report about a (benign) bone tumor in a mosasaur (incidentally, mosasaurs were not dinosaurs but rather gigantic marine lizards) and the 2003 discovery of a massive, probably fatal brain tumor in the skull of a young *Gorgosaurus*. *Gorgosaurus* was a thirty foot long, relatively slender relative of *Tyrannosaurus rex* that lived in the late Cretaceous, about seventy-two million years ago. Possibly an extraskeletal osteosarcoma (a type of bone-producing malignant tumor that doesn't actually originate from the skeleton), the tumor occupied nearly the entire region of the skull formerly filled by the cerebellum and brainstem. This large, space-occupying mass probably directly impaired function of these two essential brain regions affecting balance and coordination and likely also adversely impacted the cerebrum, the seat of memory and cognitive functions. The myriad healed fractures and major injuries suffered by this probable female is consistent with the idea that she might have taken numerous rough tumbles thanks to balance and movement challenges. As the tumor grew, it might have caused problems with breathing, blood pressure, and heart rate since these "autonomic" functions are controlled by the brainstem. Not surprisingly, the putative example is controversial with a range of differing opinions and alternative possibilities.

One large "epidemiological" study aimed to explore the incidence of tumors among dinosaurs by using x-rays to screen over ten thousand fossilized vertebrae (spine) specimens from 708 dinosaur specimens and then obtaining CT scans to follow up on those suspicious for disease.[10] The researchers found various lumps and bumps consistent with benign desmoplastic fibromas, hemangiomas, and osteoblastomas as well as malignant metastatic cancer. An interesting finding was that, although the researchers x-rayed a wide variety of well-known dinosaurs including *Triceratops*, *Stegosaurus*, *Diplodocus*, *Spinosaurus*, and *Tyrannosaurus* among many others, tumors were restricted to members of the family hadrosauridae, the duck-billed dinosaurs.[11] The relative overrepresentation of tumors in the examined hadrosaurs was substantiated irrespective of whether they were of the crested or flat-headed subtypes. The scientists speculated that this familial concentration of bone tumors might reflect either a genetic propensity (as seen in families with *BRCA1* or *BRCA2* mutations who are at risk for breast, ovarian, and other cancers), an unidentified environmental effect, or a consequence of their diet. "Mummified" hadrosaur stomachs disclosed a diet rich in coniferous foliage that contained mutation-inducing resins, tannins, and phenols. Conifers are rebuffed by most animals around today thanks to their high resin content and repugnant alkaloids. Some paleontologists think this carcinogenic coniferous concoction might have been restricted to hadrosaurs, selectively making them vulnerable. Another proposed possibility is that based on some details of their bone structures, hadrosaurs might have been warm-blooded, increasing their risks of developing cancer. One other proposition was that hadrosaurs lived exceptionally long lives, allowing them to eventually develop tumors, but other studies have not confirmed extremely long lifespans for hadrosaurs.

Alternatively, since the investigation was restricted only to vertebrae, it remains possible that other dinosaur species also had tumors but they were distributed elsewhere in their skeletons. Another curiosity of this study was that, for reasons unknown, in hadrosaurs only tail vertebrae appeared to harbor tumors. While this study showed that hadrosaur bones seemed to have a high rate (perhaps 2–3%) of benign tumors (hemangiomas), true metastatic cancer was actually rare, being found in only a single specimen of *Edmontosaurus*. *Edmontosaurus* seemed to be especially susceptible to tumors—in addition to harboring a genuine example of metastatic cancer, about a full 3 percent of *Edmontosaurus* specimens sported tumors in their tails bones.[12]

Again, the odds of detecting a fossilized neoplasm depended on multiple improbable variables: the dinosaur must have developed a tumor, the tumor must have started in or spread to bone, the affected bone must be among the one-in-a-billion preserved and discovered bones, and the tumor must have lodged in one of the examined vertebral bones. Nevertheless, the bottom line is that that dinosaurs did indeed grow tumors—and not all dinosaur species seemed to be equally susceptible.

The specimen of Edmontosaurus with metastatic cancer was from the late Cretaceous, near the time of the great dinosaur extinction. Intriguingly, many of the hadrosaurs with bone tumors were from this general time period raising suspicions of a widespread epidemic of cancer around this time that could have contributed to the K-T extinction. Following up on the epidemiological study, researchers from the University of Kansas took another step. They performed a statistical analysis to determine whether bone tumors in the late Cretaceous were more or less frequent than the rate observed in modern reptiles and birds. Using the results of the 2003 epidemiological study, which showed one metastatic vertebral tumor in 708 dinosaurs, they compared this 1-in-708 frequency with the incidence in modern birds and reptiles. The statistical analysis showed no significant difference in frequencies of metastatic cancer between dinosaurs and modern birds and reptiles, eliminating an abnormally high rate of malignancy in dinosaurs. The researchers were thus able to rule out the hypothesis that there was an elevated rate of cancer among dinosaurs due to high levels of ionizing or ultraviolet radiation following an atmosphere-crippling gamma ray burst. Whew! I am glad we can put that one to rest!

This digression on dinosaurs might seem a bit off topic, but in fact there is direct relevance to modern cancer biology. Although some dinosaurs were far larger, present day elephants are by no means petite. And as big beasts, one would expect elephants to suffer cancer at rather high rates. This is because of their trillions of cells and the prodigious number of cellular divisions during their lifetimes. According to the current theory (which will be explained in greater depth in a later chapter), cancer arises due to DNA mutations, and although carcinogen exposure can accelerate mutations, there is an unremitting baseline mutation rate. In other words, to a large degree, the development of cancer is just a matter of chance.[13] And those chances should be directly related to an animal's size and lifespan. For each time a cell divides, there is a small but non-zero probability that its DNA will not be replicated with perfect fidelity. If there are a lot of cells dividing over a long period of time, the odds are that somewhere, someday, a deleterious mutation will crop up. Chances are high that among the countless unavoidable mutations in a huge animal, at least one is likely to land in a bad spot. And if that mutation in a bad spot happens to trigger profligate cellular proliferation, that animal is in for trouble.

Well, at least that's what the theory predicts. So, is it true? At least as it relates to today's largest land animal, the elephant, the answer is no![14]

According to Dr. Joshua D. Schiffman of the Huntsman Cancer Institute at the University of Utah, "Every baby elephant should be dropping dead of colon cancer at age three." The fact that they are not raises some interesting questions. For one, it could be that the theory is in need of revision (and this is explained in greater depth in chapter 32). Another explanation is that elephants might have some intrinsic means of fending off cancer. To test this idea, Lisa M. Abegglen and colleagues (including senior scientist Schiffman) reviewed zoo records on the deaths of 644 elephants. They observed that fewer than 5 percent had died of cancer.[15] Considering that full-grown elephants weigh nearly a hundred times as much as people do (and that they live roughly to age forty, and sometimes considerably longer),[16] this is an astonishing observation. According to current thinking, they should have a cancer incidence far greater than our roughly one in four, yet they exhibit a rate of roughly one in twenty.

Abegglen and colleagues examined the *TP53* gene of elephants, since that gene is known to serve crucial anticancer functions in humans and other animals. The team of scientists found that elephants had multiple copies of the *TP53* gene. In contrast to the single pair of *TP53* genes found in humans, elephants possess an impressive twenty pairs of this cancer-fighting gene. We will review the importance of the *TP53* gene in greater depth in subsequent chapters but for now, this endowment of *TP53* appears to be one of the methods elephants use to ward off cancer despite their incredible bulk.

In another study, Michael Sulak and colleagues (including senior researcher Vincent Lynch) attempted to trace the evolutionary history of the TP53 gene in the elephant lineage by examining DNA found in fossils of long-extinct relatives like mastodons and mammoths.[17] Most fascinating—and consistent with theory—the smaller extinct proboscideans possessed only a single pair of this gene whereas the gargantuan antecedents had multiple TP53 copies. Specifically, American mastodons, which were smaller than elephants, had only three to eight copies of TP53, whereas woolly mammoths and Columbian mammoths (which were bigger than today's elephants) had about fourteen copies of the gene.

To ascertain whether or not these extra copies really made any difference, both teams of researchers conducted experiments on elephant cells *in vitro* (i.e., in the lab rather than in a living animals, or *in vivo*). Schiffman's team exposed the elephant cells to high-energy ionizing radiation and carcinogenic chemical compounds whereas Lynch's team used ultraviolet radiation and chemical carcinogens. Either way, upon insult to their DNA, the elephant cells committed suicide (i.e., underwent apoptosis), which is an effective means of taking damaged cells out of circulation before they can turn malignant and propagate as a cancer. Since one way that the p53 protein (the product of the amplified *TP53* genes) acts to stymie cancer is by inducing cells with badly damaged DNA to commit suicide through apoptosis, this could be the mechanism by which elephants defy the odds and exhibit lower rates of cancer than theory would predict.

But as will be described in depth later, there is another possibility.

A competing model of cancer as a dangerous disease (rather than simply the biological phenomenon of uncontrolled growth of cells) is the immune suppression model. In this theory, the immune system is a more important player than the mutations themselves. Mutations may be necessary for the creation of a mass of uncontrollably dividing cells, but they are not sufficient—they need a debilitated immune system that will allow those cells to do the nasty things they potentially can do (i.e., become a clinically significant cancer). According to this concept, an animal would exhibit fewer cancers if they have an exceptionally capable immune system that stays capable over time. In other words, unlike man, where immunity wanes with age, if elephants demonstrated intact immune responses with age, they should have fewer cancers—despite their mammoth proportions. And it appears that this is the case.

For at least some immune responses, elephants do indeed demonstrate superior immune systems.[18] When it comes to their ability to respond to tetanus vaccines, elephants leave humans in the dust, especially in their old age. Unlike in man, dogs, horses, and mice, where antibody-producing capacity declines over the years, geriatric elephants (beyond age forty) actually show strengthened immune capacity with increased age. This could be the real reason why elephants have less cancers compared to humans. This hypothesis has yet to be rigorously tested and of course the response to a tetanus vaccine isn't a full reflection of immune capacity. In all likelihood, the multiple copies of *TP53* probably do play an important role, as their multiplicity seems too much to be a mere coincidence. But maybe that multiplicity plus an exceptional immune system work synergistically to fight off cancer. This fascinating observation demands further exploration.

In any case it is clear that "the gamma ray burst-induced cancer epidemic hypothesis" for the extinction of the dinosaurs can be definitely relegated to the realm of science fiction. Our next story—about a clone of killer cancer cells in a New England setting—is reminiscent of a Steven King horror tale. But this one is real. And it is happening now.

Chapter 13

CANCER OF THE CLAM!

In the study of cancer biology many useful clues and correlations can be gleaned from examining tumors elsewhere in the plant and animal kingdoms. Although plant tumors have been known about for over a century and are a fascinating topic unto themselves, I must admit that until quite recently I hadn't given the concept of cancer in clams much thought.

Can clams get cancer? How about flatworms like planarians and other platyhelminthes? What about arthropods such as the European edible crab, *Cancer pagurus*? Is there such thing as "cancer of *Cancer*"?

Yes, clams, mussels, oysters, cockles, and other bivalve mollusks can and do get cancer. Actually, much to my own surprise, cancer is not as uncommon among invertebrates as I once believed.

I was under the impression that cancer was primarily a disease of vertebrate animals and was not very prevalent amongst "lower organisms" like clams and their kin. Thus, it came as quite a shock to say the least when I learned about the clam cancer epidemic that is presently ravaging soft-shell-clam beds from Prince Edward Island in Canada down to Port Jefferson, New York, in the Long Island Sound, not far from where I once lived when I was a Stony Brook medical student. In fact, as I write this chapter in mid-2015, the cancer epidemic is seriously depleting the number of soft-shell clams along an expanding length of the Atlantic coast.[1] It may now have reached as far south as the Chesapeake Bay in Maryland—a 1,500 km or 940 mile length of shoreline.

While it is primarily the soft-shell clam, *Mya arenarea*, that is being decimated by this strange leukemia-like cancer, similar cancers have been seen in other edible or commercially valuable mollusks including mussels, oysters, and cockles.

The most amazing, bizarre, and frightening aspect of this whole ordeal is that the clam cancer is contagious and is apparently being spread through seawater.[2]

Known as *disseminated neoplasia*, or *hemic neoplasia*, this clam cancer calamity was first discovered in the 1970s. The cancer appears to be most akin to a leukemia, which causes rapid proliferation of certain blood cells in their *hemolymph*—the molluscan equivalent to human blood. Unlike humans, who keep their blood neatly contained in a closed circulatory system of blood vessels, mollusks such as bivalves, snails, and squid allow their blood to roam around freely within a large cavity, the *hemocoel*. Furthermore, mollusks use a different oxygen transport molecule from us. Instead of the iron-containing hemoglobin molecule that gives our blood a deep-red color, mollusks employ copper as the key element in their *hemocyanin* oxygen transport molecule, conferring their blood with a bluish tinge when exposed to air. Regardless of the minutiae of molluscan hematology, the hemic neoplasia clam cancer is typically fatal, and countless clams have perished during the epidemic, accounting for a serious clam population decline. Will the beach clambake of summertime become a thing of the past?

Initially it was suspected that some sort of pollution-related chemical carcinogen was to blame for the outbreak of leukemia. (It is known that leukemia can be experimentally induced in clams by injection of the carcinogen, 5-bromodeoxyuridine.) Alternatively, since many animal cancers were known to be virally induced, maybe a virus was the cause. To test this idea, investigators searched for the presence of *reverse transcriptase*, an enzyme associated with certain viruses known as *retroviruses* (since several retroviruses are known to cause cancer in animals and in humans). Readers may be familiar with retroviruses since HIV, or *human immunodeficiency virus*, the cause of AIDS (*acquired immunodeficiency syndrome*), is a retrovirus. Another retrovirus, *feline leukemia virus* (FeLV), is a retrovirus that causes leukemia in cats. In humans, HTLV-1 (*human T-lymphotropic virus 1*) is known to cause a form of adult leukemia and lymphoma (as well as an unusual neurological condition called *tropical spastic paraparesis*). Another retrovirus, XMRV (*xenotropic murine leukemia virus-related virus*) hit the headlines for a while in 2006 when it was allegedly linked to prostate cancer, but numerous follow-up studies have failed to confirm any definite link.[3] In any case, since retroviruses are ubiquitous and occasionally linked to cancer, a retroviral etiology seemed like a good bet.

Retroviruses are so named because they run in reverse of the so-called *central dogma of molecular biology*. Basically, the central dogma holds that DNA begets RNA begets protein. This concept can be abbreviated as: DNA → RNA → protein. Recall that the sequence of nucleotides in the DNA molecule ultimately holds the genetic code that gets translated into the proteins that carry out all the cellular functions and provide the structural integrity of the human body (or any other cellular life-form). The process wherein the DNA genetic code (i.e., the genome) is transmitted into an RNA molecule is called *transcription*. The resultant messenger RNA molecule inherits the code and then creates polypeptides (the chains of amino acids that constitute proteins). The process of making polypeptides out of RNA is called *translation*. The polypeptide chains are then gathered together into a proper protein molecule. Thus the full sequence of the central dogma might better be written: DNA → RNA → polypeptide → protein.[4]

Retroviruses pose a bit of a problem to the central dogma—their genomes are not composed of DNA but rather are made of RNA, and part of their process of information transmission runs in reverse.

Whether viruses are truly "alive" remains debatable. They are not composed of cells and therefore are inherently different from all unicellular or multicellular life-forms. In fact, they are simply strands of nucleic acid (i.e., DNA or RNA) associated with a protective protein *capsid* with or without a third component—an outside envelope of membrane-like lipids. It was Sir Peter Medawar (who we shall revisit in a later chapter) who succinctly described viruses as a piece of bad news wrapped up in protein. Although many viruses have RNA genomes, retroviruses are unique because of their peculiar violation of the central dogma of molecular biology. In retroviruses, which are composed of single stands of RNA, the central dogma sequence is partly reversed. The single strand of RNA is "reverse transcribed" into DNA by the enzyme *reverse transcriptase*. Once a segment of retrovirus RNA is reversed transcribed into DNA, that newly formed DNA segment is then integrated into the host's chromosome (in a human cell, for example). From there, the host cell treats this new DNA as if it were its own. The integrated retroviral DNA then produces RNA using the host cell machinery, and that RNA in turn is converted into proteins in the normal fashion. Thus the entire modification to the central dogma sequence can be written as: RNA → DNA → RNA → polypeptide → protein.

Returning to our clam saga, simply put, the DNA from clam cancer cells does not match that of the host clam's other tissues, but instead, the cancerous cells genetically match each other—irrespective of which clam they were taken from. This latter point raises the specter of yet another lethal contagious cancer epidemic in the wild. Rather than being virally induced, similar to the case in the Tasmanian devils, the cancer is apparently being directly transmitted from clam to clam. The details of how scientists were able to definitively show this are fascinating. The team of Michael Metzger, Carol Reinisch, James Sherry, and Stephen Goff and colleagues examined clam cancer cells and observed that the samples *did* show ample amounts of reverse transcriptase enzyme, which would be consistent with a viral cause for the cancer.[5] Further experiments, however, most curiously and confusingly demonstrated that the reverse transcriptase did *not* originate from a retrovirus. Rather, the enzyme appeared to come *from the cancer cells themselves*.

The reverse transcriptase, to everyone's astonishment, appeared to be emanating from *transposable elements* (also known as *transposons*) within the clam cancer cell's genome. Transposons are segments of DNA that can figuratively jump around from place to place within the genome and are for this reason sometimes called "jumping genes."[6] Some transposons jump from place to place by simply "cutting and pasting" their DNA from place to place within the genome. Certain others jump about through a more complicated route: they first transcribe themselves into RNA and then reverse transcribe themselves back into DNA before finally settling into a new home somewhere else in the genome. The crazy clam transposon was of this latter sort. It goes through the unusual process of starting out as a segment of clam DNA, copying itself into RNA, reconverting itself back into DNA, and finally reinserting itself again into new locations within the host genome. This type of transposon is what scientists call a *retro-transposon* (or Class I transposable element). The killer leukemia-associated retro-transposon was given the name "Steamer" in honor of the clam's common name.[7] It turns out that the number of copies of the Steamer retro-transposon within the cancer cell's genome was a smoking gun in terms of solving the case. Regular clam cells usually have only two copies of *Steamer* whereas malignant clam cells have hundreds . . .

Curiously, leukemic cells from soft-shell clams ranging from Maine to Long Island all have roughly the *same* highly amplified number of copies (up to three hundred) of the *Steamer* retro-transposon. This seems strikingly suspicious. One would have expected the various clam cancer cells to all possess vastly different numbers of *Steamer* copies, and these numbers in turn should have reflected the number of copies found in the normal clam cells. Oddly and unexpectedly, instead of reflecting the small number (two to ten) of *Steamer* copies found in normal clam cells (which would be the case if they were derived from ordinary clam cells), all the cancerous cells, regardless of which clam they came from, had nearly the same and very high number of copies per cell. As was the case with the Tasmanian devil and the dog communicable cancers, these hemic neoplasia clam cells did not look like their hosts but instead all looked remarkably like each other.

Even more convincingly, the *distribution* of *Steamer* within the clam DNA genome was eerily similar from one clam cancer to another. As mentioned, the number of *Steamer* transposons within a given clam's healthy tissues is small (two to ten), but in addition, they are far apart from each other with a random distribution. In sharp contrast, all the malignant cells tested, irrespective of which clam they were taken from, showed nearly the same number of *Steamer* copies with a *nearly identical distribution* within the genome. This data alone could be considered irrefutable proof that the cancer cells were all related to each other but not to the parent clam they were killing. However, the team of scientists was not yet finished. To further corroborate their findings, the team analyzed two additional molecular markers (for those interested in technical details: mitochondrial DNA single nucleotide polymorphisms [SNPs] and polymorphic microsatellite repeat loci variations). Both of these methods confirmed the conclusions of the *Steamer* integration site analysis: the DNA of the cancer cells do not match the DNA of the clams they affect but instead matched each other. This was no coincidence. The team of scientists had undeniably proved that this was yet another example of contagious cancer in nature.[8]

The inevitable conclusion was that instead of being caused by a virus or because of pollution-produced environmental carcinogens, like the Tasmanian devil contagious cancer, this fatal clam leukemia is similarly spreading *directly* from clam to clam. Instead of simply *affecting* a clam, the cancer is actively *infecting* that clam. Yet another example of a parasitic, transmissible, clone of cancer cells that passes itself along from animal to animal has been discovered.

As in the cases of canine transmissible venereal tumor and the Tasmanian devil facial tumor disease, it appears that the infectious clam cancer cells in hemic neoplasia all originated from one cell gone bad. And this clone of rogue cells has been spreading from clam to clam ever since. Like the immortal canine transmissible venereal tumor, the contemptuous clam cancer cell scoffed at death and opted to live forever rather than go down with its dying host. In this case, the scoundrel cell literally jumped ship to roam the high seas in hopes of finding a new world on its own. Just how long ago this first malignant cell became unbound, decided to go its own way, and landed on another clam is presently unknown. There are some subtle genetic differences between the malignant cells in clams from the northern range compared to those in the southern range. This suggests that the contagious clam cancer might have arisen sufficiently long ago for some evolutionary divergence. Because the disease was first spotted in the 1970s (although only recently has it been recognized as a contagious cancer), we know it has to be at least forty years old.

Could it be as ancient as the eleven-thousand-year-old CTVT cell line of dogs? Most folks feel the answer is probably not, given the current geographic distribution and observed rate of spread—although no one currently knows for sure just how fast that rate of spread truly is. Clams are not known for their ability to get around!

There is a chance that the clam cancer may be quite ancient, very sluggishly inching its way along the north Atlantic coastline at a snail's pace (or even slower, at a clam's pace). Or it may be rather young and getting some unintentional human assistance in creeping along the Atlantic coast. For instance if people have been unwittingly stocking clam beds with infected clams, the speed of spread could be far faster than nature alone would have allowed. This might explain the geographical "skip metastases" currently observed along the Eastern seaboard. Instead of a continuous geographical distribution along the Atlantic shoreline, the cancer is mysteriously cropping up in distinct pockets separated by hundreds of miles.

In any case, that highly determined cellular survivor that first defied fate must have been most delighted to discover that in the new world it could flourish unharassed by a different host clam's immune system. To those initial iniquitous, immortality-seeking cells that broke free after killing their host, the new world was no Botany Bay prison camp; they had reached nirvana. Immortality was theirs.

Just exactly how the cancer spreads itself from one animal to another is presently unknown.[9] It is possible that affected clams are releasing the cancer cells upon their death and healthy clams are sucking these malignant cells in as they filter-feed. Alternatively, the leukemic cells may be released upon spawning. Perhaps they are being jettisoned upon some form of injury or during attack by predators. Maybe they are continuously being spewed out as part of the disease process. No one knows. What is known is that clams and mussels are filter feeders *par excellence*, and anything floating around in their vicinity is likely to be sucked into their systems. Thanks to this, a leukemia cell, weary and bored with its old, damaged or dying home, can boldly venture out on an ocean journey with a fresh start on life—and successfully set up shop in a new shell so to speak.

It is not believed that eating clams with the cancer is in any way harmful to people (but I, for one, would not dine on raw clams, mussels, or oysters). Many mysteries remain. For instance, it remains unknown how these malignant cells, which after all, are foreign tissues, can evade the recipient clam's immune systems. In this regard, the contagious clam cancer is genuinely analogous to the Tasmanian devil pandemic. At this time it is completely unclear if there is *any* immune counterattack whatsoever to the transplanted cancer cells. And if there is any immune reaction, the extent of that immunity and what variability exists from animal to animal also remains unknown.

Although they have some sort of primitive system to help distinguish self from nonself, clams and other invertebrates, unlike (jawed) vertebrates, do not possess the complex major histocompatibility complex (MHC) tissue recognition system described in earlier chapters. In this regard, their immune systems have a very major weakness. Without this histocompatibility system, invertebrate organisms theoretically might have trouble distinguishing *self* from *nonself*, leaving them potentially vulnerable to natural transplantation of tissues from one organism to another. This actually has been observed in certain urochordates (marine invertebrates commonly called tunicates) such as sea squirts or ascidians.

Now and then, tissues from one individual of the ascidian species, *Botryllus schlosseri* can graft onto another individual and take hold unimpeded by any immunity. This particular species is of great scientific interest since on occasion, rather than acceptance of the grafted cells, there can be an inflammatory response leading to *rejection*. In other words, sometimes the grafted tissues are accepted whereas at other times the grafted tissues are rejected. It thus appears that in this group of lower organisms, the earliest beginnings of a primitive tissue compatibility network or histocompatibility system can be discerned. Because it either enables transplantation (i.e., fusion) between different organisms or it fights off transplantation as if there were some form of histocompatibility problem, this primitive immune network is called the *fusion/histocompatibility (Fu/HC) system*. Nevertheless, without the advanced MHC of jawed vertebrates, clams (and other mollusks) might be at risk for uninvited tissue transplants—even malignant ones. As mentioned earlier, one very interesting hypothesis regarding the evolution of the MHC system is that this aspect of the immune system evolved not just to improve immunity against pathogenic bacteria, viruses, fungi, and parasites but also explicitly to reduce the chances of contagious cancer. If the MHC did indeed evolve to minimize the odds of transmissible cancers, since clams don't have the MHC, it bodes badly for their future. On the other hand, the cancer, if viewed as an independent new species in and of itself, has a bright future indeed.

As stated, the cancer, not content to die a natural death along with the host it has inflicted mayhem on, moves on to its next victim. In *sessile* (i.e., stationary) organisms such as clams, it appears that this clever contagious cancer is unsatisfied with simply spreading from tissue to tissue or organ to organ like a metastatic cancer in man. Spreading from organ to organ in a single body will not slake its wanderlust. The renegade cancer refuses to go down and instead jumps overboard, pirating the next organism. Instead of spreading *from organ to organ* like an ordinary cancer in a person, this "super cancer" is spreading *from animal to animal* along the Atlantic seaboard. Where it will stop nobody knows.

One frightening thing that is known is that soft-shelled clams are not the only species afflicted with a weird form of leukemia. Mussels, cockles, oysters, and other clams species are as well. But the details of their disease have not yet been fully explored. Could it be that they, too, are examples of contagious cancer? Preliminary data in the form of chromosome counts suggests this could be the case. In the clam contagious cancer, the number of chromosomes is abnormal (that is, they are *aneuploid*). Our normal human cells are described as *diploid* meaning we have two copies of each chromosome, one from the mother and one from the father. Whereas normal clam cells are also diploid, the contagious malignant cells have twice this number. Instead of just two copies of each chromosome, they have four (in technical terms, rather than being diploid they are *tetraploid*). In mussels with hemic neoplasia leukemia there also appears to be a consistent pattern of abnormal chromosome counts. In fact, there seems to be at least two distinct forms of hemic neoplasia in mussels—one with four sets of chromosomes (tetraploid) and another with five sets of chromosomes (*pentaploid*). Similar chromosomal aberrations have been seen in hemic neoplasia of cockles, and this evidence is being backed up by more conclusive mitochondrial DNA sequencing data. Whether these are additional examples of contagious cancers in nature remains to be seem, but at this point I wouldn't be at all surprised.

An even more chilling question is: Could all of these bivalves be suffering from the *same* contagious cancer? In other words, has some form of cancer learned not only how to jump from clam to clam but from species to species? Without the intricate MHC tissue recognition system to ward off contagious cancer, this unsettling possibility remains. Even if bivalves possess some sort of primitive version of a compatibility check (such as the Fu/HC system of the colonial ascidians mentioned earlier), the fact is that it is failing miserably.

I am now convinced that without the MHC network, contagious cancer could indeed be a very serious threat to the survival of clams and other bivalves at this time. The problem will prove especially vexing if it turns out that the current clam leukemia truly is able to jump from clams to mussels, cockles, and oysters. One wonders what might have gone on throughout time. Could contagious cancers have wiped out entire species of invertebrates in the past? In the previous chapter, we alluded to the major and minor extinctions throughout Earth's history. Although contagious neoplasia has never been proposed as one of the causes of mass extinctions, given what is now known, one has to wonder.

Geneticist Elizabeth Murchison, an expert in contagious cancers among Tasmanian devils and dogs, was quoted in the journal *Science*, "It could be that this type of disease might not be so rare as we thought."[10]

Chapter 14

SHARKS DO GET CANCER (OR HOW SHARK CARTILAGE CAN KILL YOU)

> ***It has not changed in over 400 million years. It never sleeps or rests. It is said to be "the perfect living machine." And, within the last decade, it has been found to hold the key to reversing cancer as well as numerous other major diseases.* Sharks Don't Get Cancer *is the story of this amazing breakthrough.***
>
> **—Avery Publishing Group regarding the 1992 book, *Sharks Don't Get Cancer: How Shark Cartilage Could Save Your Life* by I. William Lane and Linda Comac**

I remember the 1992 tall tale *Sharks Don't Get Cancer: How Shark Cartilage Could Save Your Life* quite well thanks in part to the television special on *60 Minutes* a year later. And who could forget the 1996 sequel, *Sharks Still Don't Get Cancer: The Continuing Story of Shark Cartilage Therapy*? The books claimed that consuming shark cartilage was the cure for cancer. I was still in medical school when the first book came out, and it certainly got me wondering. Could eating shark cartilage really cure cancer? Now, decades later, I wish it were that simple. But of course, don't we all?

I bet many other people still remember the shark cartilage saga—and still believe it. And that is why this enduring pseudoscientific myth is so dangerous. According to an article published in the journal *Cancer Research*, faith in the falsehood that shark cartilage can cure cancer can divert patients from genuinely effective treatments.[1] The real danger then is that some potentially curable cancer patients, who might otherwise benefit from proven therapy, will unnecessarily die. So, in essence, shark cartilage could be deadly.

This saga is sadly reminiscent of the snake-oil salesmen of yesteryear. People with cancer can understandably become desperate, frantically grasping at straws without support, in a final hope that *something* will work. The danger is that cancer patients might prematurely grasp at these "miracle cures" as an *initial* natural remedy instead of seeking possibly curative conventional treatment for early stage cancer. We are all aware of this; but no one knows this better than the medical mountebanks with their countless concoctions. Shark cartilage just happened to be a craftily camouflaged version of snake oil, specifically aimed at the most vulnerable victims of all—those diagnosed with cancer.[2] Sadly, there are far too many despicable charlatans out there who are eager to prey on the desperate and sell false hope in a bottle (often at significant financial gain) via untested, unproven pills and gadgets.

Time to set the story straight. We've been misinformed. Sharks *do* get cancer. And eating their cartilage does not cure cancer.

So, how did this myth get started? Perhaps the first bit of legitimate scientific data that became perverted into this fallacy came from some high-quality research done by Henry Brem and Judah Folkman.[3] Just as background, Moses Judah Folkman (February 24, 1933–January 14, 2008) is considered by many to be the founding father of the field of tumor angiogenesis—the study of new blood vessel formation in cancer. (The growth of new blood vessels is called "angiogenesis," and the newly formed vessels are the "neovasculature.") Folkman was a world class scientist and served as chief of surgery at Boston Children's Hospital. Arguably, he should have been awarded the Nobel Prize for some of his work. In 1971, he reported in the *New England Journal of Medicine* that malignant tumors are "angiogenesis-dependent." Simply stated, with their plethora of rapidly proliferating cells and inordinate demands for nutrients, tumors need a reliable blood supply. If tumor-feeding blood vessels are not there to begin with, the tumor will seek new blood. It will generate its own new blood vessels, that is, it will induce the development of neovasculature.

Conversely, given the nagging needs of tumors, couldn't one conceivably starve a rapidly proliferating tumor by pinching off its blood supply? Wouldn't that cause the cancer to wither and die? Rather than actively block off an existing blood supply, what would happen if one simply prohibited a cancer from creating a new and needed one? Shouldn't this limit the tumor's growth and prevent it from becoming a real threat?

With these fresh hypotheses in mind, explorations began. Henry Brem, as a brilliant young graduate student in biochemistry at Harvard University, worked under the guidance of Dr. Folkman. Brem later became the

Harvey Cushing Professor of Neurosurgery at the Johns Hopkins University, director of the Department of Neurosurgery, and neurosurgeon-in-chief at the Johns Hopkins Hospital. At the time they worked together, Brem and Folkman made a dynamic duo indeed.

Based on Folkman's hypothesis that tumors, in their hunger for nutrients, enticed the growth of new blood vessels to feed themselves, Brem and Folkman began searching for substances that would inhibit this angiogenesis. Their early work involved rabbit cartilage.[4] They suspected that since cartilage is devoid of blood vessels, there might be something in cartilage tissue that is actively abolishing angiogenesis. Was there an anti-angiogenesis substance to be discovered in cartilage?

If so, they could be onto something big. Angiogenesis, it turns out, is one of the *hallmarks of cancer*, which we will discuss in a later chapter. Tumors, as mentioned, have exorbitant demands for nutrients. Thus, it should come as no surprise that they emit molecular "want ads" that entice nearby blood vessels to sprout new branches that will feed the mother mass and carry away waste products. In addition to facilitating growth of the primary tumor mass, in principle, these new blood vessels might facilitate the spread of metastases to distant tissues. Naturally then, targeting angiogenesis by blocking these tumor-secreted molecular signals would appear to be a very reasonable cancer-therapy strategy. In fact it is. Drugs such as the angiogenesis-inhibitor bevacizumab (Avastin) specifically hinder new blood vessel formation and thereby potentially starve cancers of their needed nutrients. Bevacizumab is designed to bind to and inhibit a protein called *vascular endothelial growth factor* (VEGF, often pronounced "veg-F"), one of the principal molecular signals sent out by cancers in their cry for new blood. By binding up VEGF, bevacizumab hampers the creation of cancer-feeding capillaries.

Incidentally, beyond cancer therapy, anti-angiogenesis agents are finding applications in another, seemingly completely unrelated area—management of certain eye diseases. A leading cause of vision loss in the elderly, with estimates ranging up to ten million afflicted people in the United States, is macular degeneration. There are two broad categories of macular degeneration: the "wet" and "dry" types. The wet subtype of macular degeneration is caused by a proliferation of abnormal new blood vessels in the retina. As in cancer, this neovasculature is spurred on by VEGF. These unnecessary and unwanted new blood vessels are fragile, frequently bleeding and leaking proteins, which leads to retinal scarring.

With time, these abnormal new blood vessels lead to irreversible damage to the retinal photoreceptors, culminating in progressive vision loss. As bevacizumab specifically blocks VEGF, the proliferation of these superfluous blood vessels can by thwarted and the wet type of macular degeneration occasionally halted with this treatment.

Returning to our story, Brem and Folkman subsequently made another keen observation—tumor growth in lab animals could be stymied by surgically juxtaposing rabbit cartilage adjacent to tumors. Another puzzle piece was added by Robert Langer and colleagues when they demonstrated similar antitumor effects, this time using shark cartilage.[5] Sharks, along with skates and rays, have entirely cartilaginous skeletons and are members of an ancient class of fishes—the *Chondrichthyans*. The name "chondrichthyan" derives from the Latin words for cartilage and fish. Thus, a *chondrocyte* is a cartilage cell, a *chondroma* is a benign cartilage tumor, and a *chondrosarcoma* is a malignant cancer of cartilage. Quite reasonably, Langer thought that chondrichthyans, such as sharks, could provide abundant amounts of cartilage for their ensuing anti-angiogenesis research.

Meanwhile, a parallel line of work was being conducted by a team including marine biologist Carl Luer, who noticed that sharks, skates, and the like seem to have relatively low rates of many diseases, including cancer.[6] In the laboratory, Luer and colleagues experimentally exposed sharks to high levels of a powerful chemical carcinogen, *aflatoxin B*, to determine the effects. Aflatoxin B, one of the most potent human carcinogens known, is produced by molds (specifically *Aspergillus fumigatus*, for those who are interested) that often grow on improperly stored grains. Incidentally, besides possible long-term consequences such as cancer, acute poisoning with aflatoxin B can be deadly: in 2003, 120 people died after eating contaminated corn in Kenya. It works as a liver poison, or *hepatotoxin*, first causing severe acute liver damage (hepatic necrosis) and later on cirrhosis and/or in some cases, primary liver cancer (also known as *hepatoma*, or *hepatocellular carcinoma*). At the molecular level, aflatoxin B prefers to target specific sites in the DNA of a key cancer-preventing molecule called p53. Whenever the *TP53* gene is altered by mutation, it spells bad news—cancer often follows. Curiously, aflatoxin B tends to knock out the third base of codon 249 in the human *TP53* gene. Just why aflatoxin B likes this particular site on the *TP53* gene is still not fully understood, but the outcome—liver cancer—is a common consequence.

This particular experiment, using this particular carcinogen (aflatoxin B), and using sharks as the test animals, yielded few tumors. But before jumping to any concrete conclusions based on these data, one must consider that different carcinogens can have different cancer-causing potential in different animals. The saccharine scare is a classic example of how one might easily be fooled by this fact. In rats, the artificial sweetener clearly is carcinogenic and consistently causes bladder tumors. Based on this, some folks jumped to conclusions regarding risks in humans, and for a while, the FDA banned saccharine.[7] It wasn't until years later that researchers appreciated that subtle metabolic differences between rats and humans were to blame. Metabolism of saccharine in rats yielded different biochemical end products in the rat bladder than in the human bladder. Unbeknownst at the time of the studies, people jumped to conclusions that were not valid. Saccharine is now back on the market, and no studies have ever confirmed a real link to bladder or any other cancers in humans.

The point is that overinterpreting early data and rushing headlong to unsupported, erroneous, or illogical inferences is a hazard of hastiness. And that is exactly what author I. William Lane did. Acquainted with the research by Brem and Folkman, as well as the work by Langer and Luer, Lane perhaps disingenuously jumped to conclusions. His book *Sharks Don't Get Cancer* became a blockbuster bestseller, and the concept that shark cartilage could cure cancer drew inordinate and inappropriate attention. But despite outcries from the scientific and medical communities, a sequel—*Sharks Still Don't Get Cancer*—came out only four years later! The consequences have been catastrophic. First, it is likely that no one was directly cured of cancer by shark cartilage. Second, no one can say how many potentially curable patients chose the shark cartilage route and deferred conventional treatments until their cancer had progressed to an incurable stage. If this happened to just one person, it is a tragedy. Third, in addition to this clinical disaster, the ecological effects on the shark population were devastating.

A lucrative world market for shark cartilage exceeding $30 million annually arose in the mid-1990s, prompting even greater harvesting of sharks. By some estimates, North American populations of sharks decreased by up to 80 percent as "cartilage companies" cropped up and hunted nearly two hundred thousand sharks *per month* in US coastal waters alone. A single operation in Costa Rica harvested an estimated 2.8 million sharks per year during its peak periods. Worldwide, an estimated one hundred million sharks per year are *still* killed by humans. Since sharks and other chondrichthyans are generally slow-growing species, they cannot replenish their drained populations fast enough to survive such sustained, intense hunting. Alarmingly, one in six species of sharks, rays, and skates are listed as threatened by the International Union for the Conservation of Nature.[8]

During my own literature search on cancer in sharks and other chondrichthyans, the oldest report I came across described a tumor on a skate written in 1853. Thus, scientists have known for more than 150 years that chondrichthyians can indeed get cancer.[9] The first tumor actually found on a shark was reported in 1908. Scientists have since observed benign and malignant tumors in eighteen different species of sharks. Rather than a genuine scarcity of tumors in sharks, a scarcity of shark oncology studies is what has perhaps permitted the myth to be perpetuated for so many years. Of the myriad fish tumors in the ichthyology collections at the Smithsonian Institution, only about a dozen are from sharks and their relatives. But in recent years, more and more reports have been amassed that contradict the myth. For instance, a presentation by John Harshbarger and Gary Ostrander provided details on forty benign and cancerous tumors found in sharks.[10] In one interesting report, a couple of sharks were found to harbor *multiple* tumors. This leads one to speculate that these two animals might have had a genetic predisposition toward cancer, similar to the breast cancer and ovarian cancer predisposition found in carriers of *BRCA1* and *BRCA2* mutations. Alternatively, these two sharks could have been exposed to extremely high levels of environmental carcinogens. Either way, they clearly had cancer. A well-documented report revealed that the liver and gonads of an accidentally caught, adult male blue shark, *Prionace glauca*, were riddled with cancer. In another highly publicized case, a malignant tumor was removed from the mouth of a captive sand tiger shark, *Carcharias taurus*. Most ironically, there are even cases of sharks with chondrosarcoma—sharks can get cancer of the cartilage! The bottom line is that these and other anecdotes definitively dispel the myth that sharks are immune to malignancy.

Sharks are indeed prehistoric; they have been swimming the seas since the "age of fishes"—the Devonian period—as far back as four hundred million years ago. Nevertheless, surprising and fresh findings could yet be reaped from studying sharks, some of which might shed new light on cancer research. Sharks and their relatives not only have the primitive *innate immune system* of cells that attack and swallow invading bacteria, they are the earliest evolutionary lineage to have an *adaptive immune system*—that aspect of immunity that can adapt to new challenges not previously encountered through antibodies generated by B cells and through aggressive T cells that can be trained to recognize, attack, and kill intruders. For instance, our acquisition of long-term immunity to hepatitis B or the mumps thanks to vaccinations are examples of adaptive immunity.

Like us, sharks possess antibodies (immunoglobulins), T cells, T cell receptors, major histocompatibility complex proteins (MHCs), and other components of the immune system—and curiously they do it without bone marrow. Given their completely cartilaginous skeletons, they are without true bone—and therefore are bereft of bone marrow, the supplier of almost all of human immune system cells. Sharks instead possess two still poorly understood but unique immunological organs—the Leydig's organ and the epigonal organs—which spawn the cells of their immune systems. One could argue that studying the shark immune system is an important avenue toward understanding immunity in general and the adaptive immune system in particular. Perhaps studying shark immunology could elucidate unknown mechanisms, which might be valuable in the human quest for a cure to cancer.

But what about the idea that consuming shark cartilage can conquer cancer in humans? Unfortunately, but not unexpectedly, this is not the case. And shark cartilage did get a fair trial. Several in fact.[11] A 1998 American study involving sixty patients showed that powdered shark cartilage supplements had no real clinical benefit, did not improve quality of life, and cured no one of cancer. A purified shark cartilage extract called Neovastat was clinically tested in the United States between 2007 and 2010.[12] It was given along with chemotherapy and radiation therapy for advanced lung cancer patients. It didn't extend survival. The National Center for Complementary and Alternative Medicine reported results of a trial of BeneFin, a brand of shark cartilage, in 2005. Eighty-eight people with advanced breast or colon cancer were given standard cancer care either with or without BeneFin. The shark cartilage preparation did not improve quality of life or extend overall survival. The final results of three prospective, randomized, clinical trials published in 1998, 2005, and 2010 conclusively demonstrated no substantial benefit associated with consumption of shark cartilage when it comes to curing cancer.

Just as disappointing and maybe more infuriating, the US Federal Trade Commission discovered that many over-the-counter products don't actually contain much shark cartilage in them. What you see is *not* what you get when it comes to these commercial shark cartilage preparations.

And what ever happened to author I. William Lane? The Federal Trade Commission fined him $1 million and barred him from asserting that his supplements could prevent, treat, or cure cancer.[13] In an interview for *Discovery News*, shark researcher David Shiffman provided the best quote yet regarding all of this: "Sharks get cancer, but even if they didn't get cancer, eating shark products won't cure cancer any more than me eating Michael Jordan would make me better at basketball."[14] Enough said.

Chapter 15

WHO TRULY DOESN'T GET CANCER?—MEET THE MOLE RATS

So if sharks do indeed get cancer and are not our salvation, is there anywhere else we can turn to in the animal kingdom for cancer-curing insights? Are there any animals that do indeed seem to be resistant to cancer? Well, in fact, yes, there are. And it turns out that there might be people who are genetically resistant as well.

Thanks to their short lifespans and easily manipulated biology in the lab, rats and mice have long held the spotlight for cancer research and all sorts of other animal studies. But perhaps it is time for them to step aside and make room for *Heterocephalus glaber*, the naked mole rat.

While there is serious competition in this space, the naked mole rat is a solid contender for the title of "World's Ugliest Animal." It has variously been described as a miniature pink walrus, a saber-toothed sausage, or a bratwurst with teeth—all of which I feel are generously flattering descriptions! Nevertheless, it most definitely has its place in the Zoological Hall of Fame, for no member of the species has ever been found to have cancer.

Naked mole rats are so named because (like humans) they have a paucity of body hair, which allows their pale, wrinkled, pinkish/yellowish skin to shine through. These subterranean natives of the Horn of Africa are not true rats, nor are they moles. They are more closely related to porcupines and guinea pigs and are most interesting critters. For one, they are *eusocial* meaning that, like many insects, they have a caste system in which individuals are assigned a certain role in the society. In certain ant colonies for instance, there are sterile female workers and soldiers, male drones, and a queen. Naked mole rat colonies similarly have a sole reproductive female and most of the dozens of individuals in the colony cooperatively care for the brood of this solitary queen.

Another highly unusual feature boasted by this species is its rather low metabolic rate and respiratory rate, two characteristics that allow naked mole rats to endure their low-oxygen, extraordinarily harsh underground environments. Compared to a similarly sized mouse, a naked mole rat might have a 30 percent lower oxygen consumption and, most unusually for a mammal, can slow its metabolism to cope with rough times such as famines and droughts. While their other oddities (such as a variable body temperature and skin insensitivity to pain) are remarkable, for our purposes, it is their "immunity" to cancer that is most curious.

How is it that naked mole rats don't get cancer? One explanation might be their production of copious quantities of a molecule called *high-molecular-mass hyaluronan*, or HMM-HA. While we humans also have hyaluronan in our cartilage and connective tissues, the naked mole rat version of this molecule is about fivefold larger than ours. This discovery, made by a team from the University of Rochester, was deemed significant enough to capture the cover of the journal *Nature*.[1] What seems to be significant is that naked mole rats don't degrade or break down this huge version of the molecule as fast as other animals who own ordinary sized hyaluronan molecules. The net result is that the "mega-hyaluronan" molecules tend to build up in the spaces between cells. It is theorized that this intercellular accumulation of hyaluronan might keep the cells from clumping together and forming tumors. This serves as a physical barrier to excessive cell contact but naked mole rat cells also exhibit "behavioral" barriers to tumor-favoring cell contact.

In a laboratory dish, normal human cells are disinclined to pile up on one another; they prefer to keep a respectful distance between each other. When normal cells multiply to the point where they start crowding each other and their edges are in contact, they cease reproducing—a phenomenon called *contact inhibition*. On the contrary, contact inhibition is *not* observed in cancer cells. With an "every man for himself!" attitude, cancer cells disregard each other's territory, eagerly crowd one another, and ambitiously compete for any available space. The idea of contact inhibition and controlling their rate of reproduction to avoid conflict simply doesn't occur to them.

In normal human cells, the process of contact inhibition is governed by a protein called p27. But while human cells display contact inhibition via expression of p27, naked mole rat cells go one step beyond and use a double-whammy defense against excessive cellular proliferation. In addition to the standard contact-inhibition protein p27, they produce another peculiar anticrowding mechanism that also might help them avoid cancer—their special *p16* protein. Expression of p16 could be the naked mole rat's secret weapon.

It appears that the p16 protein kicks into action at a much lower cell density than p27. The net effect is that naked mole rat cells are *truly* sensitive to contact inhibition. In contrast to self-centered cancer cells that rudely elbow their neighbors to get more room for themselves, naked mole rat cells are polite and respectful to one another and provide each other with ample personal space. Naked mole rat cells are at the opposite end of the spectrum from discourteous cancer cells in this regard. Whereas malignant cells don't seem to mind congested conditions and invariably wind up in a big, old, ugly pile in a lab dish thanks to their disregard for the contact inhibition rule, it is almost as if naked mole rat cells actively shun contact and go out of their way to obey the contact inhibition policy. Thanks to this exaggerated contact inhibition in the lab, naked mole rat cells seem especially primed to steer clear of one another in real life. In this manner, they minimize their probability of clumping together in a tumor.

In addition to their massive hyaluronan molecules that physically space their cells apart plus their peculiar p16-mediated aversion to cellular crowding, naked mole rats have yet one more cancer-fighting trick up their sleeves. It appears that the biochemical protein synthesis mechanisms in naked mole rats are unusually error-free. Mistakes in protein synthesis result in mutant proteins, which in turn might lead to cellular errors. Over time, sloppy controls over protein synthesis can culminate in cancer. Errors in protein synthesis are occasionally observed in normal human cells (as well as in just about all other animal cells). Such errors occur because the message from the DNA gene is not faithfully *transcribed* into RNA and that RNA is not properly *translated* into the desired protein product. Somewhere along the chain (described in chapter 13) from DNA → RNA → polypeptide → protein, something goes wrong and the final protein product just doesn't get the job done. Such molecular mistakes are nearly nonexistent in naked mole rats however. The high-fidelity protein synthesis machinery in naked mole rats could be yet another explanation for their cancer resistance—as well as for their extraordinary life spans. While ordinary mice might last less than a year in the wild, and maybe two to three years in captivity, it is not uncommon for a naked mole rat to make it to age thirty!

Together, these features catapulted the lowly naked mole rat to *Science* magazine's coveted "Vertebrate of the Year" title in 2013.[2]

While that is all great for the naked mole rats, the pressing question for us is whether using heavy hyaluronan in the clinic could possibly prevent human cancers and/or extend our lives (preferably without having to live underground in a queen-serving colony). Researchers are actively investigating.[3]

But while the *naked* mole rats may be the sexy stars getting all the media attention these days, they are not the only mammals with a low incidence of cancer. *Blind* mole rats (*Spalax golani* and *Spalax judaei*) also appear to be "immune" to cancer, but by completely different mechanisms. And unlike their homely distant cousins, these guys are actually quite cute. They can aptly be called "little furballs" thanks to their vestigial eyes being completely grown over and covered with fur. True to their name, these natives of southeastern Europe, Mediterranean North Africa, and the Middle East are unable to see. And with skin and fur overgrowing their ears as well, they also appear to be close to deaf.

Confusingly, the thirty or so different species of mole rats are neither moles nor rats—and despite their names, blind mole rats are not closely related to naked mole rats. Blind mole rats share closer kinship to mice, rats, and particularly Chinese hamsters than they do to naked mole rats. In fact, a recent examination of the entire blind mole rat genome indicates that the genetic lineage between naked mole rats and blind mole rats split somewhere around seventy-one million years back during the Cretaceous period, just about six million years before the great dinosaur extinction. The one thing they do share in common, however, is a love for tunneling. So it seems that their parallel evolution of long lives and cancer resistance are attributable more to their corresponding burrowing lifestyle choices rather than kinship.[4]

Examination of the blind mole rat genome reveals some remarkable findings. For instance, it contains 259 defunct genes, including twenty-two involved in vision and the development of the eye. Apparently, abandoning the need for vision and hearing has led to the degeneration of the corresponding genes (alternatively, perhaps loss of these genes left their ancestors with no choice but to go underground, and once underground, a domino effect of further losses ensued). In any case, these sensory-deprived rodents appear to have more than compensated when it comes to the gene that encodes *interferon-beta1*, a cancer-fighting *cytokine.* (Very simply, cytokines are chemical compounds produced by immune-system cells that stimulate or inhibit other cells of the immune system. Thus, interferons are examples of cytokines. Recall that in chapter 1 I briefly mentioned another interferon [interferon-alpha] as an early form of cancer immunotherapy.) As we shall see, it turns out that the abundance of interferon-beta in blind mole rats offers a most unique (and convoluted) means of warding off cancers.

To explain, let's start out by mentioning that fibroblast cells are commonly used in biology experiments for a variety of reasons. Fibroblasts are the most common connective-tissue cells and are involved in the production of collagen and extracellular matrix. They are instrumental in wound healing and provide structural support in various connective tissues. In some ways, wound healing and cancer are very similar processes; the difference is that one knows when to stop and the other doesn't. In any case, studies on blind mole rat fibroblasts have found that instead of the normal process of cell death called *apoptosis*, a different mechanism called *necrosis* seems to dominate. Apoptosis, also called programmed cell death, is a means of eliminating cells that are present in excess, irreparably stressed, or that are diseased. Apoptosis is quite normal in embryonic development and many other biological processes. Apoptosis helps sculpt organs into their normal final shapes and sizes. Importantly, in apoptotic cell death, the cell is an active participant in its own demise, and the process is highly orchestrated at the molecular level. Necrosis, on the other hand, is relatively sloppy and appears haphazard at the molecular level. In contrast to the clearly controlled apoptotic mode of cell death, in necrotic cell death, the cell appears to be caught off guard. The cell's main membrane (the plasma membrane that encases the entire cell) breaks down, releasing cellular contents (including the highly inflammatory lysosomal enzymes). In a sense, apoptosis is a neat process of dissolution whereas necrosis is a cell's way of unwillingly spilling its guts.

This curious difference in apoptotic cell death versus necrotic cell death in blind mole rat cells is believed to be an adaptation to the low oxygen concentrations encountered in their subterranean tunnels. Mammalian cells such as ours, when stressed to the max by exposure to very low oxygen conditions, will ordinarily tend to undergo apoptosis. The under-oxygenated cells begin to deteriorate and die in a very regimented and predictable fashion with very specific steps that are faithfully followed. But since blind mole rats are constantly exposed to low oxygen levels and must thrive in such hypoxic environments, constant apoptosis would present a serious survival disadvantage. Having one's cells dying all the time would be a counterproductive biological strategy indeed! It appears that blind mole rats have evolved an effective and unique way of coping with the lower oxygen levels in their subterranean abodes through a mutation in their *TP53* gene.[5] (We previously encountered the *TP53* gene in chapter 12 when we discussed cancer-resistance in elephants through amplification of this particular gene.)

The multifunctional p53 protein, which is produced by the *TP53* gene, is one of the most crucial of all cellular regulatory proteins. It is charged with a variety of key functions including the repair of DNA damage and the initiation of apoptosis if the damage is just too extensive. In this way, if the cell's DNA is damaged beyond repair, rather than allow that cell to live and procreate, possibly leading to mutants and malignant descendants, the p53 protein initiates a "self-destruct sequence" and eliminates that cell. In this fashion, under normal circumstances, p53 protects us against cancer by causing heavily damaged cells to self-destruct through apoptosis rather than risk propagating themselves as a cancer. When p53 itself is mutated however, it can no longer carry out the function of preventing damaged cells from proliferating. Not surprisingly then, mutant *TP53* genes are associated with a wide array of cancers.

At first blush, one might reasonably predict that blind mole rats should be *more* susceptible to cancer because of their mutated *TP53* genes. After all, since p53 protein is so instrumental in the repair of DNA damage, and is also critical in initiation of the self-destruct sequence in cells "too far gone," it would seem that blind mole rat cells should be more prone to formation and propagation of malignant mutations. But this is where the hero cytokine interferon-beta comes into action. When blind mole rat cells accumulate because they have failed to undergo apoptosis, they start to produce a surfeit of interferon-beta.[6] Blind mole rats seem to have their own built-in factory for interferon production and tumor eradication. Specifically, an incipient tumor is directed by interferon-beta to undergo *necrosis* and is thereby eliminated. Necrosis, while not as nice and neat as the programmed cell-death process of apoptosis, still gets the job done and handily eliminates undesirables.

If this is indeed the true mechanism through which blind mole rats avoid cancer, in contrast to their naked mole rat cancer-free compatriots, blind mole rats can truly be considered "immune" to cancer since it is in fact an *immunological* mechanism that seems to be fighting off cancer in these animals.

Is the list of cancer-free critters limited only to blind, deaf, or homely rodents who must live out their years dwelling in dark, dank, oxygen-deprived caverns? Well, no. The cancer-free club is certainly not limited to naked mole rats and blind mole rats. Interestingly and relevantly, there are some people who seem to be highly resistant to cancer as well. But before meeting them, let's visit an ordinary mouse with some extraordinary powers.

Chapter 16

PAR FOR THE COURSE

Admission to the cancer-free critter club is not restricted to naked mole rats and blind mole rats. In recent years, several others have joined the menagerie. Researchers have managed to breed certain lab animals that are practically invulnerable to cancers—even some extremely aggressive cancers. One particular strain of mice that is impressively impervious to neoplasia, was genetically engineered to overexpress a cancer-fighting gene called *PAR4*.

Originally discovered by a team of investigators led by Vivek Rangnekar of the Department of Radiation Medicine at the University of Kentucky in 1993, *PAR4* stands for prostate apoptosis response-4.[1] Functionally, *PAR4* (also sometimes known as *PAWR* for PRKC apoptosis WT1 regulator) serves as a *tumor-suppressor gene*, particularly in the prostate gland. Increased activity (often referred to as *up-regulation* in the parlance of molecular genetics) of the *PAR4* gene yields higher levels of its protein product, Par-4 protein. (Incidentally, nomenclature rules in genetics dictate that human genes be italicized and given in *ALL CAPS*, while protein gene products be given just a capitalized first letter and not be in italics. Most of the medical and scientific literature doesn't conform to this rule, however, and there is much variability.)

In cancer cells, the Par-4 protein induces apoptosis or programmed cell death (discussed in the previous chapter as it pertained to blind mole rats). Thus, Par-4 protein can be described as a "pro-apoptotic" protein.

Further study showed that the function of *PAR4* is not limited to the prostate gland but rather is expressed in a variety of different cell types; it functions as a cancer-killing gene in a wide array of cancer cells. Conversely, low levels of Par-4 protein are associated with accelerated tumor formation and growth. Inactivity or underactivity of the *PAR4* gene has been linked to a number of human cancers, including those of the breast, kidney, endometrium (uterus), and a pediatric cancer called neuroblastoma. In humans, the *PAR4* gene has been mapped to the long arm of chromosome 12. Perhaps this is why cancers of the stomach and of the pancreas often harbor mutations or show deletions in the *PAR4*-containing region of chromosome 12.

Most unusual and very importantly, the Par-4 protein induces apoptosis *selectively*—it brings about apoptosis in malignant cells but *not* in normal, healthy cells. Therefore, cancer cells expressing the *PAR4* gene are quite remarkably singled out and culled, while normal cells expressing the *PAR4* gene are unaffected. In essence, Par-4 functions as one of the highly sought-after "magic bullets" of oncology—it selectively seeks out and destroys "bad" cells but leaves normal "good" cells unharmed.

Building on these basic observations, in an effort to create a strain of mice resistant to malignancy, the researchers introduced the *PAR4* gene into a mouse egg and implanted that egg into a surrogate murine mother. The resulting offspring indeed expressed high levels of the cancer-fighting gene. In this fashion, scientists were able to establish a colony of cancer-recalcitrant mice. Research later revealed that these super-mice were in fact inhospitable to both spontaneous and experimentally provoked tumors. It seemed nearly impossible to induce malignant tumors in these animals, even upon exposure to some nasty carcinogens.

The bottom line is that mice expressing the *PAR4* gene do not develop tumors!

To further verify the importance of the *PAR4* gene, researchers selectively "removed" the gene from some mice. Stripped of their superpowers, when these mice were subsequently exposed to the same carcinogenic challenges that their *PAR4*-positive siblings were resistant to, they developed an assortment of tumors in various tissues. Sadly, like Sampson after a haircut, they no longer possessed any superstrength or resistance.

Of course, such an advantage conferred by *PAR4* would be negated and useless if it were accompanied by some serious deficiency or deformity. But *PAR4*-positive mice appear quite normal, demonstrating no anatomical defects, developmental abnormalities, adverse behavioral effects, or other obvious deficits. Furthermore, like the cancer-resistant naked mole rats and blind mole rats, the Par-4-producing mice live longer than expected. (Although the magnitude of this life-extension—a few months perhaps—is far less than the order of magnitude longer lives enjoyed by their subterranean cousins). In any case, essentially no side effects were seen in these genetically enhanced mice.

So, could *PAR4* prove to be of benefit to people too? Presently, no one knows. Dr. Rangnekar has speculated that perhaps one therapeutic avenue that could someday be explored would be through bone marrow transplantation—introducing *PAR4* into a patient's bone marrow to fight cancerous cells throughout the body "without the toxic and damaging side effects of chemotherapy and radiation therapy."[2] Fundamentally, this would be the equivalent of taking cancer patients and turning them into *PAR4*-positive people. Given the potency of *PAR4* in mice, might such a strategy be capable of curing cancer? Although one must remain cognizant of the not-insignificant risks associated with bone marrow transplants, this is an exciting concept indeed.

Another, far less risky approach, which might prove useful for tumors that have recurred locally after radiation therapy but are not surgically salvageable, could be to introduce *PAR4* into the tumor via a virus. Animal experiments have shown that tumors can be killed when a custom-made virus carrying the *PAR4* gene (i.e., a *viral vector*) is injected directly into the tumors.[3] Whether such an approach will work for the fraction of men who have a recurrence of prostate cancer despite radiation therapy remains uncertain, but this (or a similar strategy) could certainly be an interesting option.

Secretagogues represent another avenue researchers are exploring to exploit the potential of *PAR4*.[4] Briefly, a secretagogue is any drug or natural substance that stimulates the secretion of another substance. For example, antidiabetic drugs such as the sulfonlyureas (e.g., glyburide, glipezide, and glimepiride) are secretagogues that increase the secretion of insulin from the pancreas. As mentioned earlier, the *PAR4* gene is actually present in all our cells but is not always active. Perhaps someday novel secretagogue drugs will be designed that specifically inspire the secretion of large amounts of the Par-4 protein. Imagine that—taking a daily pill to produce Par-4 and render us impervious to cancer just as though we were *PAR4*-positive mice!

Colloquially, although we tend to say that these cancer-beating supermice are "immune" to cancer, that is a bit of a misnomer. True, the way that *PAR4*-positive mice ward off cancers is both fascinating and potentially promising—but is not genuinely immune-mediated. What I mean by this is that although *PAR4* functions as a unique cancer-selective gene that causes cancer cells to commit suicide, that mechanism does not involve a classic immune response. For mice that are honestly conquering cancer via their immune systems, and therefore are genuinely "immune" to cancer, we must turn to the singular saga of a most unusual superhero among rodents: Mighty Mouse.

Chapter 17

MIGHTY MOUSE TO THE RESCUE!

Although I have written about the Tasmanian devil and now have turned to Mighty Mouse, this isn't a comic book! This chapter is not about the venerated cartoon character. What I am alluding to here is yet another cancer-resistant critter. Here he comes to save the day!

And the mechanism for mighty mouse's form of cancer-resistance might be the most interesting—and relevant—of all.

The discovery of a cancer-rejecting mutant mouse, affectionately known as "mighty mouse" was made somewhat serendipitously. While exploring lipid metabolism, Dr. Zheng Cui and colleagues at Wake Forest University in North Carolina, were conducting research that required injecting large numbers of mice with highly aggressive *S-180* cancer cells.[1] These cells are derived from a Swiss mouse soft tissue sarcoma and are especially potent. One additional feature of this cell line is its ability to grow after transplantation in both inbred and outbred mice. In other words, this tumor is invisible to the mouse immune system. It does not get rejected. This peculiar ability to hide from the immune system is probably due to an absence of MHC (major histocompatibility complex) on its surface.

After receiving an injection of the S-180 cancer cells, ordinarily, none of the mice would survive more than a month or so since these cancer cells were very virulent and consistently lethal to all lab mice. Within two weeks, tumors would appear in the abdomen where they were initially injected, causing intra-abdominal fluid build-up (*ascites*). Shortly thereafter, the cancer would spread to various other vital organs, particularly the liver, kidney, pancreas, lungs, and stomach with death inevitably ensuing.

But one day, a decidedly strange observation was made—one particular mouse had "rejected" the initial injection of malignant cells. Normally, only a few thousand injected S-180 cells is lethal, but the researchers, not wanting to leave anything to chance, injected the animals with around two hundred thousand cells. In one particular mouse, there was no ascites, meaning no cancer. Curious, the researchers re-injected this mouse with another round of cancer cells. And then another and another, Doubling and tripling the dose perplexingly had no effect. He remained impervious. They then increased the dose a full order of magnitude. Nothing! They next escalated the quantity of injected cancer to twenty million cells—one hundred times the standard dose—still no cancer! How about upping it to two hundred million?! It became obvious that this guy was just not going to get cancer this way. In this manner, "mighty mouse"—a mouse naturally resistant to cancer—was discovered.

Mighty mouse has sired many offspring since he was discovered in 1999, leading to multiple generations of cancer-insusceptible descendants. When mighty mouse was bred with normal females, about half his offspring, regardless of whether they were males or females, demonstrated the same cancer resistance, indicating that the genetic trait was passed along in an *autosomal dominant* fashion. Such a pattern usually indicates that a single gene is involved.

Importantly, these mice seem normal in every other way; no deformities or deficits are evident. There does appear to be some slight intermouse variability in the capacity to ward off cancer. Some mice show absolutely no evidence of cancer despite repeated injections of the virulent malignant cells or challenges with other carcinogens, whereas other mice will initially show signs of cancer that subsequently regress. One might note that the latter pattern is somewhat reminiscent of the growth and regression of canine transmissible venereal tumor in dogs.

What seemed to be happening was that some mice initially did develop cancer, as manifested by early ascites, but later underwent spontaneous remissions. Other mice showed no susceptibility to cancer whatsoever; they were completely resistant from day one. Because of this, the proper scientific name for mighty mouse is the SR/CR (for spontaneous remission/complete resistance) mouse strain. I myself prefer the name "mighty mouse."

The mechanism behind the spontaneous remissions (as opposed to the complete resistance) was a slight delay in the immune reaction. It appeared that in some mice the cancer resistance simply kicked in sooner than in others. Initially this subtle difference was thought to be due to some slight genetic variation in the mice, but that proved not to be the explanation. They were in fact all the same genetically. But Cui did identify one variable—older mice tended to react slower than younger mice to the cancerous challenge. Young mice appeared fully capable of rejecting cancers and were thus completely resistant (CR), whereas older mice initially got cancer but subsequently fought it off, thereby demonstrating spontaneous remissions (SR). The mice were all the same; there was just a subtle difference in the speediness of immune reaction that differed as a function of age.

It should be pointed out that, in contrast to the *PAR4* mice of the previous chapter, the immunity to cancer demonstrated by mighty mouse and his descendants proved to be truly *immune*-mediated. By this I mean that the cancer resistance was actually due to his *immune system* recognizing, attacking, and destroying the cancer (as opposed to some form of inherent cancer defiance, such as the heavy hyaluronic acid molecules in naked mole rats or the overactive *PAR4* gene in the lab mice of the previous chapter). This important point suggests that maybe, just maybe, an analogous immune reaction can be mounted in *people* that similarly would be effective against human cancer. This of course would be a far more appealing and practical alternative than trying to genetically engineer cancer resistance into each and every cell in a person's body.

The idea of creating strains of lab animals with hyperactive immune systems that could fight off cancer is certainly not new, but previous attempts were fraught with distressing difficulties. For instance, some mice were specially bred to possess highly energetic, cancer-fighting immune systems, but they exhibited severe autoimmune disorders that proved to be a major shortcoming in this line of research. It appeared to hold little promise since it seemed that this approach would just be trading one deadly disease (cancer) for a host of different deadly autoimmune diseases. So, although an "accidental" discovery, mighty mouse actually turned out to be a vast improvement over prior animal models.

Unlike those specifically bred mice that often developed devastating autoimmune disorders, the mighty-mouse clan exhibits no such predisposition to autoimmune disease. (In people, autoimmune diseases include a wide assortment of conditions caused by one's own immune system attacking one's own tissues and organs. Disorders such as lupus, type I diabetes, ITP [immune-mediated thrombocytopenia], pernicious anemia, Graves disease, rheumatoid arthritis, and many more represent a breakdown of a key aspect of a healthy immune system—the immune system's ability to distinguish "self" from "non-self.") This serious limitation essentially ruled out many previous superimmune lab animals from serving as reasonable models for clinical cancer care in people. Their heightened immunity might have helped them avert cancer—but at too dear a price. Notably, mighty mouse and his kin showed no such shortfalls.

In subsequent experiments, scientists injected normal, cancer-susceptible mice with leukocytes (white blood cells) from the mighty-mouse clan and then challenged those mice with simultaneous injections of cancer cells.[2] Like the naturally endowed mice, recipients of the mighty leukocytes proved to be unreceptive to cancer as well. These experiments proved that it was solely the leukocytes that mediated the cancer-fighting functions. Furthermore (and more clinically significant), the transfer of mighty-mouse leukocytes into mice that *already had* advanced cancers seemed to cure them! This is an absolutely amazing result. It raises the question of whether something similar could work in humans.

Additionally, like the regression seen in dogs with the transmissible canine venereal tumor, after a mouse was injected (just once) with the cancer-killing mighty-mouse cells, it exhibited a long-term remission. Unlike the sad situation in many human cancers where chemotherapy might induce a substantial but brief response, these responses proved to be durable. Again, these findings provide tantalizing hints that the immune system could someday tame cancer for good. And again it leaves us wondering—if it works in a mouse, would it not work in people?

Exactly how do these mighty-mouse white blood cells carry out their cancer-killing functions? And just what type of white blood cells are these? I, and most researchers, would probably have bet on some subset of *T cells*, a class of lymphocytes that is charged with eradicating unwelcomed intruders after being "trained" to recognize them and destroy them. T lymphocytes are thus part of the *adaptive* immune response because they have to first be exposed to these pathogenic trespassers and adapt themselves to the circumstances. They don't have an innate ability to recognize and kill anything.

Most folks would not have guessed that the cancer-killing leukocytes from mighty mouse were not part of the adaptive immune system but rather were *innate* cancer killers. The white blood cells that appear to be carrying out the cancer-fighting functions in mighty mouse are not T cells after all, but instead are *macrophages*, *neutrophils*, and *natural killer cells*—the basic cellular army of the innate (as opposed to adaptive) immune system.

In any case, these observations have broadened our understanding of basic tumor immunobiology and could potentially point toward a useful immunotherapy for human cancers. It now appears that, in sharp contrast to the other animals mentioned (wherein cancer resistance is due to some peculiar cellular, biochemical, or physiological traits characteristic of the animals in question), in the mighty-mouse tribe, the resistance is mediated entirely by the white blood cells—and these cells have proven transferable to ordinary mice. So, unearthing a cancer-intolerant mouse that naturally resists cancer via an immune mechanism is a fabulous discovery.

The immune-mediated mechanism might open many new doors for cancer therapy—and raises numerous new questions. For instance, as Dr. Cui himself has asked, could mighty mouse somehow explain why cancer is generally a disease of aging? Fatal cancers are rather rare in childhood and young adulthood but become increasingly common in later years (although in extremely aged people and very old animals, the incidence again tapers off). The commonly quoted explanation for this increased incidence with age is an accumulation of mutations—the so-called mutation theory of cancer.[3] According to this model, mutations in *oncogenes* and *tumor suppressor genes* (which will be discussed in depth later) pile up over time, first producing a precancerous condition, then a more ominous tumor, and ultimately culminating in a frank malignant cancer. Based on observations in mighty mouse and his descendants, an alternative hypothesis naturally arises—Could cancer evolve not via an accumulation of malicious mutations but rather through a more mundane waning of immune function?

For a quick analogy, immunity to the chicken pox virus typically lasts for decades after exposure, but in many people it eventually wears off, allowing the virus to reemerge as shingles. Could many (most?) people have a natural immunity to cancer during their early years, which similarly wanes away with age? If so, can we somehow reboot our natural immunity?

Even if we don't all possess a natural immunity, could ordinary people someday "train" their own immune systems to recognize and defeat cancer the way mighty mouse and his kin do? Can we somehow grab another gear on our immune systems and thereby acquire seemingly super powers and defeat cancer? Even if we ourselves cannot train our own immune systems to ward off cancer, can others who do have such powers come to the rescue of us mere mortals? Advances are now coming in on a weekly basis and at this point are quite encouraging.[4] As a quick example, one could envision screening individuals from families that have exhibited very few instances of cancer and determining if their white blood cells demonstrate the desired cancer-fighting capabilities. If we do find such a mighty man (or woman), could a blood transfusion from that person help those of us with cancer?

To expound on this concept a bit, could it be that there is a "league of extraordinary gentlemen" out there who, just like mighty mouse, own immune systems that are inherently supercharged and capable of conquering cancer regularly before it becomes a real threat?[5] Most of us either know of, or have heard stories about, ornery old "Uncle Joe" who worked with all kinds of chemicals, smoked like a chimney, and drank like a fish yet never succumbed to cancer. Uncle Joe and his type were rebels. They did what they wished: smoking in the boys' room in school, trying out illicit drugs at an early age, frequently fighting, drinking heavily, handling hazardous materials, and everything else the doctors said was "bad" for them. Yet despite all this, consistent with his defiant personality, his body simply "refused" to get cancer. We have all heard of such people, and I regularly hear from my patients that they don't understand how they could have gotten cancer when they did all the "right" things in life while their Uncle Joe did all the "wrong" things and never got sick and still somehow lived to be ninety. The existence of such individuals has led some people to erroneously grow skeptical of the cancer-causing potential of cigarettes, heavy alcohol consumption, and other confirmed carcinogens. But now I often wonder if the more accurate interpretation is that Uncle Joe, just like mighty mouse, was destined never to get cancer no matter what he did, what he ate, or what he was exposed to. So, are there individuals out there with super immune systems that can vanquish any cancer? And if so, could we someday locate Uncle Joe and other mighty men (and women) who possess cancer-crushing leukocytes and transfer those leukocytes into cancer patients via blood transfusions or a similar method? Would this cure people of their cancers? Cui has asked the poignant question: If cigarette smoking causes a one-hundred-fold increase in lung cancer incidence (from 0.08 percent in the general population to 8 percent in smokers), why do the other 92 percent of smokers not get cancer? Maybe cancer resistance is a biological property among people that we can find in the general population, just as he found among mice in his lab. Cui's concept is to find Uncle Joe, use him and others like him as white blood cell donors, through a cancer treatment strategy he and his team have termed "GIFT" (Granulocyte InFusion Therapy). We all anxiously await their clinical trial results. In any case, the mighty-mouse saga provides a significant ray of hope—in my personal opinion, this strategy might not be a far-fetched idea at all.

It is worth reiterating that lab mice (unlike specially bred lab animals such as the *PAR4* mice and exotic subterranean, oxygen-deprived mole rats) like humans, normally *are* quite susceptible to cancer. For this reason, the serendipitous discovery of a mouse that is *not* susceptible to cancer is most remarkable. It strongly suggests that there could be ordinary-looking people out there who similarly possess this hidden superpower. So, the search for "mighty man" is on. But while we haven't found the Uncle Joe–based leukocyte cure for cancer just yet, we have found some people who do seem insusceptible to cancer. And where do we find these folks? Time for another sojourn—this time to Ecuador. But first let's take a little detour. Not geographically, but through the fourth dimension—about twenty thousand years to be precise.

Chapter 18

FRODO OF FLORES

There is nothing like looking, if you want to find something. You certainly usually find something, if you look, but it is not always quite the something you were after.

—J. R. R. Tolkien, *The Hobbit*

While working with medical students and residents, I may sometimes quiz them and ask things like, "Why do you suppose a radiation oncologist would be reading about hobbits and the latest papers on human evolution?" There are a few clearly *wrong* answers such as: "Because he did not take his medication." In my case at least, the real answer is rather convoluted.

In a discipline historically mired in controversy and renowned for its heated, often acrimonious debates, the latest dispute is particularly fascinating. Boiled down to its bare essence, the question might be phrased, "Were hobbits real?"

In 2004, Australian and Indonesian scientists reported in the journal *Nature,* a new and diminutive hominin species.[1] While numerous human ancestors and relatives have been discovered and reported over the centuries, this one was of particular interest to me. Initially thought to be a child because of its small stature and skull, it was quickly recognized as an adult based on its teeth. Although there was no mention of how furry the toes were, standing at only three and a half feet tall and weighing between thirty-five and seventy-nine pounds, the pint-sized thirty-year-old adult female was nicknamed "the Hobbit" after J. R. R. Tolkien's fabled little folks.

Because the hobbit was discovered in Liang Bua (LB) cave on the Indonesian island of Flores it was given the formal name *LB1*. The LB1 specimen consisted of a nearly complete cranium, a partial pelvis, most of the right arm and clavicle (collarbone), part of the left hand and wrist, and much of both lower limbs including the ankles and feet. Anthropologists assigned a new species to this specimen: *Homo floresiensis*.

An article in the *Journal of Human Evolution* described fossilized skeletal remains of perhaps eight additional similarly small specimens, suggesting that LB1 wasn't an anomalous one-off specimen.[2] The fossilized bones themselves date from about thirty-eight thousand to eighteen thousand years ago, but geological and archaeological analyses of sediment deposits and dating of tools found in the area define the age range of *H. floresiensis* from as early as ninety-five thousand years ago to (possibly) as recent as twelve thousand years ago. It is possible that their disappearance was due to a massive volcanic eruption that occurred on Flores approximately twelve thousand years ago. A layer of volcanic ash dating to this time was found in the Liang Bua cave.

Using 3-D scans of the skull and facial bones and a sophisticated computer-graphics program, researchers at the University of Wollongong, New South Wales, Australia, fashioned fuller facial features for the hobbit compared to previous artistic interpretations. This computer-assisted analysis rendered a less apelike and more humanlike, wider and shorter face. Although she still did not have a humanlike forehead, the nasal details were more human and more modern in appearance than previously believed. Additionally, based on new CT scan data, a 2013 study showed that the LB1 brain was slightly larger than earlier estimates. Although initially calculated to be only four hundred cubic centimeters (cc) in volume, the new study suggested that the hobbit's brain volume was actually about 426 cc. For comparison, the average cranial capacities for adult men are about 1,440 cc and slightly less (about 1,240 cc) for women. It is frequently stated that brain size does not correlate well with intelligence. Those who argue against this would insist that men must be smarter than women given their larger brains. But applying this overly simplistic logic would imply that Neanderthals, with their 1,600 cc cranial capacities, were smarter than all of us. Also note that there is a subtle difference between *cranial capacity*, which simply estimates the volume within the skull, and *brain volume*, which estimates the size of the brain itself.

Initially it was suspected that *Homo floresiensis* was a dwarf descendant of another species of early humans, *Homo erectus,* which was first discovered in 1891 by physician Eugene Dubois in nearby Java. Since the *Homo erectus* specimens from the region ("Java Man") also had relatively small brains (around 860 cc) the new cranial capacity calculation suggested that *Homo erectus* could indeed be the ancestor of *Homo floresiensis*. The leader of the expedition team that discovered LB1—the late Mike Morwood—along with colleagues speculated that a population of *Homo erectus* might have travelled the 310 miles from Java to Flores, perhaps by boat, and upon prolonged isolation grew smaller in both body and brain size. The shrinkage of *H. erectus* could have been a classic case of *insular dwarfism* in which mammals isolated on an island evolve smaller sizes to cope with the limited resources. Although never confirmed in humans, insular dwarfism has been observed in other mammals on other islands, and Flores seems exceptionally ripe for this phenomenon. Flores lies in a geographical region called *Wallacea*, bounded by invisible, imaginary lines like the equator and Tropic of Cancer called *Wallace's line* to the west and *Lydekker's line* to the east. Interestingly, there is relatively little migration between the Asian fauna to the east and the Australian fauna to the west thanks to very strong ocean currents that preclude easy travel. Animals isolated on islands in Wallacea tend to be distinct from fauna either to the east or to the west and have occasionally demonstrated insular dwarfing. On Flores itself, insular dwarfism was evident in *Stegodon*, a pygmy elephant species of less-than-mammoth proportions. (Note that the opposite tends to occur in reptiles, with Komodo dragons and Galapagos tortoises growing considerably larger than their mainland relatives.)

Alternatively but more provocatively, the hobbit could have evolved from *Homo habilis* whose brain was somewhere around 600 cc. If this is correct, it would imply that, in contrast to conventional thought, *Homo erectus* was *not* the first hominin to migrate out of Africa (since *Homo habilis* lived far earlier [2.3 million to 1.5 million years ago] than *Homo erectus* did [1.9 million to 143,000 years ago]). Also, if the hobbit really was directly descended from the rather diminutive *H. habilis* (average height around four feet three inches) instead of the tall *H. erectus* (average height five feet ten inches), it would solve another problem since there would be no need to invoke the insular dwarfing hypothesis.

The "hobbit," as with much in paleoanthropology, has been surrounded by tremendous controversy.[3] Basically, some scientists vehemently reject the idea that the hobbit represents anything more than a diseased example of modern man (or woman in this case). In other words, they assert that LB1 is simply a *pathological* specimen. So, there are two basic camps: one side favoring a separate, new species of hominin (*H. floresiensis*), and the other side espousing the concept that LB1 and others like it are actually nothing more than modern people with developmental disorders. The very recent dates of these specimens (which at possibly only eighteen thousand years ago are far more recent than any other human species, including the Neanderthals who vanished around forty thousand years back) only adds to the controversy—and fascination.

Some scientists contend that the LB1 specimen is nothing more than the remains of a diseased person with microcephaly, a condition that causes short stature, a small brain, and, in about 85 percent of cases, mental handicap.[4] Simply put, microcephaly is a neurological disorder in which the head circumference is smaller than average. Today, microcephaly is rare, occurring in about 1 in 6,200 to 8,500 births. The disorder can be congenital (present at birth) or acquired during the first few months or years of childhood. Acquired microcephaly can appear after birth due to oxygen deprivation, malnourishment, or infection. Most cases of microcephaly, however, are caused by genetic abnormalities or by certain maternal infections, toxins, or use of drugs and alcohol during pregnancy. Infections of the fetus during pregnancy, including toxoplasmosis, cytomegalovirus (CMV), German measles (rubella), and chickenpox (varicella), can result in microcephaly by interfering with the development of the cerebral cortex. Another potential cause of microcephaly is improperly managed phenylketonuria (PKU) during pregnancy. PKU is an inability to break down the amino acid phenylalanine, which if consumed in abundance during pregnancy yields a compound that is highly toxic to the developing fetus's brain. This is why certain food supplements and artificial sweeteners such as aspartame often explicitly state, "Attention phenylketonurics: Contains phenylalanine."

Microcephaly has several other causes. While the above-mentioned causes directly affect growth and development of the brain itself, microcephaly can also be caused by mechanical problems in the growing skull. *Craniosynostosis*, or premature fusing of the sutures between the bony plates in an infant's skull, can possibly lead to microcephaly, but sometimes this can be addressed surgically, allowing normal neurological development and skull growth.

Chromosomal aberrations can also lead to microcephaly. *Down syndrome* is caused by having three copies of chromosome 21 rather than the normal two copies; for this reason it is sometimes called trisomy 21. In addition to Down syndrome, other chromosomal disorders such as *cri du chat syndrome*, trisomy 13, and trisomy 18 can also lead to microcephaly. In fact, as we shall discuss shortly, Down syndrome in particular has been proposed as an explanation for the spate of features associated with LB1.

Not least among the causes of the controversy and confusion over the hobbit is the rather recent age of the LB1 specimen. Recall that similarly sized hominins such as the famous "Lucy" specimen of *Australopithecus afarensis* lived around 3.85 to 2.95 million years ago. Thus it is only natural for skepticism to arise and alternative hypotheses to be proffered.

Yet another suggestion for the LB1 specimen and related fossils was *endemic cretinism*. Cretinism is congenital hypothyroidism. Hypothyroidism in newborns and children can have devastating and permanent effects. In adults, hypothyroidism is often caused by autoimmune disorders such as *Hashimoto thyroiditis* in which antibodies attack the thyroid gland and disable it. Many other cases of hypothyroidism in adults are *iatrogenic* in origin—meaning the hypothyroidism is *caused by treatment* of another disease. That other disease is quite often a different autoimmune disease of the thyroid, namely *Graves disease* which causes *hyper*thyroidism. Treatment for hyperthyroidism might include thyroid surgery or radioactive iodine therapy, resulting in iatrogenic *hypo*thyroidism. In any case, most adults with hypothyroidism can be medically managed with supplemental thyroid hormones and can lead normal lives. Hypothyroidism in infancy on the other hand, if unrecognized and not corrected early, can lead to mental retardation and microcephaly. Uncommonly, cretinism can be inherited in an autosomal recessive fashion. So, hypothetically there could have been higher-than-average rates of cretinism in the isolated population of people on the island of Flores due to an inherited form of the condition. Alternatively, cretinism can result from low levels of iodine in the mother's diet during pregnancy, and if there was a chronic deficiency of iodine in the diet of the Flores inhabitants, cretinism could have been common. Cretinism can also appear when there are developmental abnormalities in the fourth branchial arches of the embryo, resulting in an absent or underfunctioning thyroid. Irrespective of the specific cause, cretinism results in short stature, a small head, and mental retardation—features associated with the purported *Homo floresiensis* specimens. Some scientists have speculated that perhaps there was a whole population of people suffering from endemic cretinism on the island of Flores.[5]

Opponents of this hypothesis assert that the Flores population was *not* just a group of modern people with cretinism or other developmental or chromosomal abnormalities. They have retorted with solid anatomical data that supports the theory that the specimens were indeed a separate new species (i.e., hobbits). One of my favorite counterarguments was a brief abstract written by Bill Jungers and colleagues with the pithy title, "The Hobbits (*Homo floresiensis*) Were Not Cretins(!)"[6] (The exclamation point is mine).

One might hope that detailed computer-assisted cranial analyses would prove definitive and should settle previously acrimonious debates about the hobbit once and for all. Instead, if anything, the arguments have grown more intense. Dean Falk of Florida State University has always been an outspoken proponent of the assertion that LB1 truly represents a genuine new species.[7] Falk and her team generated three-dimensional computed tomographic (CT) endocasts (internal skull contents that reflect brain structure) of LB1 and compared these with CT data from ten normal modern people and nine people with microcephaly. The study aimed to decisively answer the question of whether LB1 was a normal example of a small new species (*H. floresiensis*) or instead was a diseased modern human (*Homo sapiens*) with a pathologically small brain. The group devised various metrics for quantitative and comparative purposes and observed that two particular parameters (cerebellar protrusion and relative frontal breadth) provided excellent discrimination between microcephalic and normal endocasts. Specifically, those with microcephaly had increased caudal cerebellar protrusion and smaller frontal breadths. Falk found that the LB1 endocast ratios were more consistent with a normal, rather than diseased state. She and colleagues thus concluded that LB1 was *not* a microcephalic modern human but rather, a new small-bodied, small-brained separate species—that is, a hobbit.

In response to this, Robert Vannucci and colleagues used the same highly discriminating craniometric metrics on large cohorts with microcephalic and *normocephalic* (that is normal sized and shaped heads) skulls and also used MRI (magnetic resonance imaging) with endocasts.[8] In diametrical contrast, they concluded that the LB1 cranial ratios fell outside the range of normal modern individuals (and *Homo erectus* endocasts) but within the range of people with microcephaly. Their bottom line conclusion was that LB1 *is* a pathological *Homo sapiens* fossil and *not* a new dwarf species of hominin. Naturally, Falk and colleagues did not simply sit back and take this; they retorted with powerful counterarguments—and the battle raged on.

In a 2014 publication in the *Proceedings of the Royal Society B*, Mark Collard and colleagues analyzed a comprehensive data set of 380 skull and dental characteristics of twenty presently recognized species of hominins.[9] Using statistical models, their comparative analysis concluded that *Homo floresiensis* was in fact a distinct species rather than simply a small-bodied or deformed human. But shortly thereafter, a 2015 publication in the prestigious *Proceedings of the National Academy of Sciences* (*PNAS*) claimed that LB1's cranial and limb features were definitively diagnostic of Down syndrome.[10] They pointed out that Down syndrome is one of the most commonly occurring developmental disorders in both humans and other hominoids such as chimpanzees and orangutans, and therefore finding fossilized remains of a modern human with the condition should come as no surprise. However, the polemic grew particularly rancorous at this point. Mark Collard of Simon Fraser University in British Columbia and colleagues rigorously refuted the assertion, arguing that *Homo floresiensis* lacks a defining characteristic of *Homo sapiens*—namely a chin. Therefore the LB1 specimen could not simply be a modern human with Down's syndrome. Unsurprisingly, the other side battled back with a vigorous counterattack and to this day the disagreement has not died down one bit. Falk was quoted saying, "It is interesting that their paper contains no images of skeletons of Down syndrome individuals. If it had you would clearly see that they look nothing like the Flores specimen. The idea is nonsense."[11] But the heated debate only grew more intense. The editors of *PNAS* were rebuked for not insisting on independent objective scientific peer review.[12] One of the authors on the Down syndrome paper was a member of the National Academy of Sciences and allegedly was allowed to select his own referees when submitting the paper. One of the scientists involved in the original discovery of the LB1 specimen, Peter Brown of the University of New England in Australia, said, "This is an outrageous abuse of the peer review process."[13] Professor William Jungers of Stony Brook added, "This is just cronyism."

As if the story weren't weird enough, in yet another bizarre twist, late Indonesian anthropologist Teuku Jacob (who believed the LB1 specimen was an abnormal modern human) confiscated the specimens in early December 2004. Allegedly without explicit permission, Jacob took much of the LB1 material from Jakarta's National Research Center of Archaeology for his own research.[14] On February 23, 2005, Jacob returned the fossils, but some parts were damaged or missing.[15] In a story more like a soap opera than science, reports surfaced that the returned specimens had "long, deep cuts marking the lower edge of the Hobbit's jaw on both sides, said to be caused by a knife used to cut away the rubber mould."[16] Additional accusations were that the chin of a second specimen was damaged and glued back together in a misaligned fashion and at an incorrect angle. (Recall that the chin is a key feature that can be useful when deciding if a specimen is *Homo sapiens* or not.) Furthermore, the LB1 pelvis was allegedly shattered, wiping out crucial anatomical details relevant to gait and evolutionary history. Mike Morwood, the expedition team leader of the original discovery, commented, "It's sickening, Jacob was greedy and acted totally irresponsibly."[17] Not unexpectedly, Jacob denied the allegations and asserted that any damage to the specimen was incurred during transport to Jakarta, before he ever got the fossils. Adding to the suspense and intrigue, the Indonesian government temporarily barred access to the cave, precluding further expeditions. Only in 2007, the year of Jacob's death, were scientists again allowed to return to Liang Bua.[18]

Just where (if anywhere) in the human family tree *Homo floresiensis* fits still remains uncertain, and the whole topic remains highly divisive. Anatomical features of the teeth and jaw bones seem to resemble australopithecines (that is, relatives of "Lucy" and other members of the genus *Australopithecus*), whereas skull bones are more consistent with early *Homo* species. The LB1 foot is quite long relative to the leg, something not seen in other hominins but is a feature found in some modern apes. This finding raises the possibility that the ancestor of *Homo floresiensis* was not *Homo erectus* as initially believed but instead some other, more primitive, hominin. If true, just how they got all the way to Southeast Asia is a major mystery. The *Homo floresiensis* wrist is almost indistinguishable from an ape's or early hominin wrist—and is quite dissimilar from that of modern humans. Even if the specimen were diseased, the wrist would not look as different from a healthy human's since carpal bones (i.e., the wrist) take on their distinctive shapes during fetal development whereas growth disorders and other pathologies affect the skeleton afterward. Thus, the wrist bones are more consistent with another species rather than a diseased, very short person. In 2013, researchers described a second fossilized wrist from Flores that was remarkably similar to that of LB1, further negating assertions that LB1 was simply a sick, small individual.

One very intriguing and perhaps telling item concerns the tools found contemporaneously with LB1 in Liang Bua. Instead of the highly sophisticated stone tools of *Homo sapiens* just entering the *Neolithic* (Late Stone Age) about twelve thousand years back (or the nearly equally impressive *Mousterian industry* of Neanderthals up to forty thousand years ago), the stone tools found at Liang Bua can be described as "*Oldowan*-like" meaning that they were more akin to very earliest *Lower Paleolithic* stone tools found in association with *Homo habilis* in Olduvai Gorge, Tanzania, from 2.6 Ma to 1.7 Ma (mega annum, or million years ago). (*Homo habilis* means the "handy man" since at the time of discovery in 1960, this was the first hominin species associated with stone tools. The very primitive stone tools found at Olduvai Gorge and other regions around two million years back are collectively called the "Oldowan industry.")

Since *Homo habilis* only averaged about four feet three inches tall and *Homo floresiensis* might have been even smaller, one would think the tool sizes on Flores would have to be most modest in order to fit into such Lilliputian hands. If on the other hand, LB1 and friends were just examples of diseased modern humans one might expect to find either some sophisticated tools appropriate for the times or maybe some full-sized tools used by their contemporaries without developmental disorders. In other words, in a tribe of average-sized modern humans, normal-sized and more modern tools would be expected. As far as I know, these tools are not there.

Modern humans arrived in Indonesia between fifty-five thousand and thirty-five thousand years ago and may have possibly interacted with *H. floresiensis*. Although there is presently no evidence of this at Liang Bua, most fascinatingly there are local legends in Flores about the "ebu gogo." In the Nage language of Flores, "ebu" means "grandmother," and "gogo" means "he who eats anything." The ebu gogo were supposedly small, hairy cave dwellers. Such legends surely make one wonder if the "hobbits" survived longer than we currently think and could be the source of these legends.

Returning to the original question posed at the start of the chapter—Why would a radiation oncologist (namely me) be so obsessed with all this? There are two good reasons.

Regarding the recent assertion that LB1 was a modern person afflicted with Down syndrome, professor William Jungers of the State University of New York at Stony Brook once commented, "They say *Homo floresiensis* is similar to a modern person with Down syndrome but no one with that condition has a tiny cranium of only 400cc capacity as floresiensis does, nor do they have the same thick cranial bones as it does. This is shockingly bad science riddled with errors of fact and attribution."[19] Since Bill Jungers (and Susan Larson, who has also played a very large role in the *Homo floresiensis* story) were two of my medical school anatomy professors, this is the first reason I have been following the story along so closely. Perhaps I am biased in favor of my former professors, but regarding the question of whether hobbits were real, I definitely weigh in on the side of the hobbit rather than the abnormal modern human hypothesis.

The second reason for the fascination can best be introduced by a statement in the *Guardian* by Falk: "First they claimed the hobbit is really a modern human with microcephaly—an abnormally small head. We showed that this could not be true. Then they claimed he had *Laron syndrome*, a form of dwarfism. Again my team showed this was not true. Now they are taking a shot with Down syndrome. Again they are wrong."[20]

It was this suggestion of *Laron syndrome* that further fueled my interest. You might ask: Just what is Laron syndrome, and what is so special about it to an oncologist? Good question! People with Laron syndrome are short-statured individuals who tend not to get cancer.

Chapter 19

A CANCER-FREE CLAN?

We are all familiar with the breast cancer-causing BRCA1 and BRCA2 genes thanks to the much-appreciated attention provided by celebrities including Angelina Jolie, Christina Applegate, and others. On the other hand, scientists and healthcare professionals are much less familiar with the inverse: genes that confer *resistance* to cancer. In previous chapters we met members of the animal kingdom with various means of defying cancer. Here we shall take an international tour and for the first time meet a group of *people* with an inherent and inherited protection from cancer. Despite short stature and a tendency toward obesity, it seems that people with Laron syndrome tend to very rarely develop diabetes or cancer—and were it not for an increased frequency of accidents, they might live longer than average.

Laron syndrome, was first reported by Israeli endocrinologist Zvi Laron (along with A. Pertzelan and S. Mannheimer) in 1966 based on observations beginning in 1958 when he first encountered an unusual trio of patients belonging to a Jewish family who recently arrived in Israel from Yemen.[1] Three siblings—two boys and one girl—all had the same general appearance: they were obese, had a protruding forehead, a depressed nasal bridge, relatively sparse hair and showed severely stunted growth (which is medically called "dwarfism"). Since their grandparents were first cousins, Laron wondered if the children might have an inherited defect of the growth hormone gene. Upon laboratory analysis Laron learned something very perplexing. Rather than the anticipated low levels of growth hormone that he thought was the explanation for their appearance, Laron learned that these three individuals had very *high* levels of growth hormone. Laron explored the region for similar cases and identified about twenty more cases by the mid-1960s. But it would be two more decades before the molecular mechanism behind this syndrome, which now bears his name, was finally uncovered.

Before diving in, just what is dwarfism anyway? One definition, provided by Little People of America (LPA) is an adult height of four feet ten inches (147 cm) or under due to a medical or genetic condition.[2] In fact, the actual average height of an adult with dwarfism is just about 4 feet even (122 cm). There are two main categories of dwarfism: proportionate and disproportionate. In proportionate dwarfism, the body parts are shortened but generally all in proportion. Disproportionate dwarfism is the more common form and is usually characterized by a nearly average-sized torso but shorter arms and legs.

There are many causes of dwarfism—over two hundred in fact. Most cases of dwarfism are due to conditions called *skeletal dysplasia*. Skeletal dysplasia is a general term for a group of conditions that cause abnormal bone growth and disproportionate dwarfism. One form of skeletal dysplasia, *achondroplasia*, is the most common form of disproportionate dwarfism; in fact achondroplasia is the most frequent form of dwarfism overall. Individuals with achondroplasia typically have normal length trunks with shortened arms and legs. *Spondyloepiphyseal dysplasia congenita* (SEDC), on the other hand, is a far rarer skeletal dysplasia and is characterized by a shortened trunk with relatively longer limbs. In a way it can be thought of as the mirror-image of achondroplasia. SEDC is inherited in an *autosomal recessive* fashion, meaning that there must be two copies of the mutated gene—one from the father and one from the mother. Achondroplasia, in contrast, is inherited in an *autosomal dominant* manner, implying that only a single copy (from either the mother or the father) of the mutated gene is required. The autosomal dominant inheritance pattern means that a child born to a couple in which both parents have achondroplasia will have a 25 percent chance of normal height. On the other hand however, there is also a 25 percent chance the child will inherit both achondroplasia genes. This double-dominant syndrome is fatal, typically ending in miscarriage. Not infrequently, neither parent of a person with achondroplasia carries the mutated gene, and the achondroplasia mutation appears spontaneously, or "de novo," at the time of conception. This is a rare instance in which a disorder is "genetic" but not inherited. The newly affected individual, however, can pass along the mutation to his or her children.

One cause of *proportionate* dwarfism is Turner syndrome, in which the individual has only one member of the gender-determining XY chromosome pair.[3] Invariably that member is an X chromosome and is therefore sometimes abbreviated XO. Of course, since there is no Y chromosome, all Turner syndrome individuals are female. (Incidentally, there is no corresponding YO condition. The X chromosome contains over one thousand vital genes, and without a copy of this key chromosome, the fetus is nonviable.) As mentioned in the previous chapter, cretinism, or congenital hypothyroidism, is another cause of dwarfism. Unlike Turner syndrome where the females are of normal intelligence, in cretinism there is a high risk of mental retardation. In the past, cretinism was much more common than today and was usually caused by dietary deficiency of iodine. However, thanks to improved maternal nutrition along with newborn screening for hypothyroidism, cretinism is now extremely rare in developed nations.

Disorders that affect the pituitary gland at the base of the brain, whether inherited or not, can cause a form of proportionate short stature through a deficiency of a key pituitary hormone, growth hormone. Some cases of growth hormone deficiency are due to genetic mutations or pituitary injury, but in the majority of cases, no obvious cause is identified. However, even in the presence of normal levels of growth hormone, short stature will result *if the body cannot properly react to the growth hormone*.

Laron syndrome is inherited in an autosomal recessive fashion and most afflicted individuals are generally less than three and a half feet tall. Around the same time as Laron's initial discovery of the disorder in the 1950s, another piece of the puzzle came from research by Dr. William Daughaday of Washington University in St Louis. Daughaday discovered that when growth hormone binds to its receptor on liver cells, it leads to a biochemical cascade that ends in the production of another hormone called *insulin-like growth factor 1*, or IGF-1.[4] In reality it is not growth hormone, but IGF-1 that ultimately causes children to grow. However, growth hormone even in abundance will not produce the normal physiological effect if the molecular receptors for growth hormone on liver cells are not functioning properly. Working with Daughaday, Laron showed that although his patients had normal or even high levels of growth hormone, they were deficient in IGF-1; the growth hormone receptors in the liver cells of his patients were defective. In 1984, Laron published a paper documenting that the short stature and other features of people with Laron syndrome, is the result of an inability to *respond* to growth hormone because of the failure of growth hormone to bind to the growth hormone receptor.

In years to come, Laron identified over sixty patients with this syndrome. Half were in Israel and the other half came from a dozen countries in Asia, Europe, and the Middle East. But it wasn't until the unearthing in Ecuador of the world's largest single cluster of individuals with Laron syndrome that the condition really gained worldwide attention. The story surrounding the discovery of this population residing in remote villages of southern Ecuador is most fascinating.

The first step in this multifaceted discovery story begins with Dr. Arlan Rosenbloom in 1988.[5] As a specialist, he had been working with growth hormone deficiency patients at his clinic at the University of Florida for a while, but this time he elected to see his six Ecuadorian patients at their home, sparing them the expense and inconvenience of traveling to the United States. Making an ultra-long-distance house call, the doctor arrived in Quito, Ecuador, in December 1988. But instead of seeing just six patients, he wound up seeing about a hundred during his stay! In this group of patients, two young sisters stood out. Both girls were less than three feet tall and had prominent foreheads, thin hair, high voices, and depressed nasal bridges. Importantly, their lab results showed that they were not lacking in growth hormone. Instead these sisters actually had high levels of growth hormone, their bodies just weren't responding to it. Rosenbloom concluded that the sisters had the rare Laron syndrome, which was a huge surprise since at that time only about one hundred cases had ever been diagnosed. Little did he know that this was the first step in the discovery of a large hidden population of people with Laron syndrome through a highly productive partnership with an Ecuadorean endocrinologist named Dr. Jaime Guevara-Aguirre.

Dr. Guevara-Aguirre was simply destined to make great scientific contributions in the study of Laron syndrome. Born in the Loja province in southern Ecuador, Guevara-Aguirre occasionally encountered conspicuously short people known as "pigmiettos" (perhaps meaning small pygmies) during his formative years. Guevara-Aguirre's father was the eminently successful, extremely driven owner of a road building company who expected (demanded, in fact) the same success and ambition in his children. Guevara-Aguirre's father directed his two older brothers to become engineers. Later on, he and his younger brother were instructed to become physicians. After Guevara-Aguirre became a reproductive endocrinologist and diabetes specialist, he mentioned to his father that his dream would be to someday start a leading endocrinology research institute in Ecuador. As explained by Gary Taubes in a 2013 *Discover* magazine interview, Guevara-Aguirre said that his father bought him the land, the building, and the needed equipment in order to create the institute with one stipulation: "He asked me for the name of the best medical journal. When I told him the

New England Journal of Medicine he said I had 10 years to publish a paper there or close the place. I was so nervous I couldn't sleep for 15 days."[6]

No small task! But Guevara-Aguirre was more than up for the challenge.

The Institute of Endocrinology, Metabolism and Reproduction first opened its doors in 1986, two years before Rosenbloom showed up in Quito. When Guevara-Aguirre learned that Rosenbloom was there helping patients with growth hormone deficiency and that he had identified two patients with Laron syndrome, the two became friends and began their long-term collaboration. In a manner most uncommon in modern medicine, Rosenbloom returned to Ecuador about six months after his first "house call" to again see and treat his growth hormone deficient patients. By this time, Guevara-Aguirre had diagnosed seven cases of Laron syndrome from the Ecuadorean province of Loja on his own. The pair recognized immediately that, based on these nine recently found cases, there was a very good chance that many more might be scattered about in the general vicinity. In addition, perhaps recalling the short-statured pigmiettos from his childhood days, Guevara-Aguirre suspected that there could be additional cases of Laron syndrome scattered about in southern Ecuador. In his quest to locate them, Dr. Guevara-Aguirre realized that he couldn't rely on the patients simply to come to him—he had to go out and find them.

Sometimes traveling on horseback where there were no roads and other times travelling with Julio Lozada, a driver, skilled mechanic, and employee of his father's company, Dr. Guevara-Aguirre began voyaging to remote villages in Loja and nearby El Oro province. During the many medical expeditions, he met with many families, asked many questions, and took many blood samples and would regularly update Rosenbloom in Florida on what he had discovered. Within a year they had found a total of twenty cases, nineteen of whom oddly were women. But this was enough to fulfill his father's stiff stipulation—with several years to spare! In 1990, he and Rosenbloom published "Little Women of Loja" in the prestigious *New England Journal of Medicine*.[7] Over the next half decade, Guevara-Aguirre identified an additional forty-five individuals with Laron syndrome in Ecuador; interestingly, although more males were discovered in Loja province, the male to female ratio remained 1:4, while in neighboring El Oro province, there was an even sex ratio. In this epic manner, the world's largest population of people with Laron syndrome (over one hundred people) was discovered.

But the story was far from finished. Genetic analysis revealed a specific mutation in the growth hormone receptor these individuals from southern Ecuador all shared in common. This common mutation strongly suggested that they all were descended from a single person, a phenomenon known in genetics as the *founder effect*. Another curious observation was that one of Dr. Laron's patients from Morocco had the very same specific mutation. Too much to be a coincidence, the team of Laron, Rosenbloom, and Guevara-Aguirre speculated that the person with the founder mutation arrived in Ecuador from the Iberian Peninsula five centuries earlier and was probably a member of the *conversos*.

The conversos were a group of Sephardic Jews in Spain and Portugal who were forced to convert to Christianity in the late fifteenth century. Many conversos later fled during the Spanish Inquisition. Some went to Morocco, while others went to different Mediterranean countries. Still others made it to the New World where they settled in the Andes; their descendants form part of the population of southern Ecuador to this day. Perhaps one of these conversos happened to carry the Laron syndrome mutation and served as the founder of the current Ecuadorean cohort. Over the generations, as the result of intermarriage within the isolated population isolates, carriers of the recessive mutation eventually produced affected offspring. Supporting this idea is the fact that in the villages where many of the Laron syndrome patients live, most people are of native descent, but those with Laron syndrome often have Spanish surnames typical of conversos.

While this was already an amazing enough story, the real journey was just about to begin.

It was in 1994 that Guevara-Aguirre first suspected that his Laron syndrome patients were not developing cancer at the same rate as others in the community.[8] He noticed that although cancer was not uncommon in the general Ecuadorian population, among those with Laron syndrome it was essentially unheard of. He obtained detailed histories from his patients and their families and failed to find *any* cancer in those with the condition (although their normal-height relatives had high rates of cancer possibly related to the environmental effects of gold-mining). Over the years, only one patient has developed cancer (a case of fatal ovarian cancer).[9] Furthermore, although many of them were obese, diabetes (a condition linked to obesity) also appeared to be conspicuously absent.

None of this really made sense until 2005 when Guevara-Aguirre hooked up with Walter Longo, a researcher on aging at the University of Southern California (USC). Longo was studying some unusual yeast strains and had discovered a "dwarf" strain of yeast with oddly small cells. Stranger still, these dwarf yeast lived three times longer than ordinary yeast. Additionally, these miniature yeast cells seem impervious to most causes of DNA damage and cellular aging. He then discovered that these yeast cells harbored a mutation affecting a biochemical pathway very similar to the one that was affected in Laron syndrome. Based on his yeast studies, Longo speculated that the low levels of IGF-1 were the secret that kept Laron syndrome patients protected from cancer and diabetes.[10]

Meanwhile, equally interesting research was being conducted by biologist John Kopchick at Ohio University. Kopchick was carrying out experiments on mice that had a mutation in the growth hormone receptor (making them exactly analogous to the situation in Laron syndrome). These mice might rightly be referred to as "Laron mice."[11] In fact, Kopchick and Laron coauthored a 1999 paper in *Molecular Genetics and Metabolism* titled, "Is the Laron Mouse an Accurate Model of Laron Syndrome?"[12] Notably, these mice do seem to have a reduced susceptibility to malignant disease. Kopchick's Laron mice were living on average 40 percent longer than typical mice—the equivalent of people living to an average of 110 years old. Another scientist, Andrez Bartke won the 2003 Methuselah Foundation Award for his contributions to life extension research.[13] Bartke had one particular Laron mouse that lived almost 5 years—about twice as long as usual—the equivalent of around 150 human years!

Mice and yeast were not the only ones having their lifespans lengthened in the lab. In the roundworm *Caenorhabditis elegans*, a mutation was found that blocked a biochemical growth pathway similar to the one affected in Longo's dwarf yeast and also led to an increased lifespan. In nematodes (roundworms), the receptor for IGF-1 is called DAF-2. Research going on at the University of Colorado by geneticist Thomas Johnson showed that when the DAF-2 gene is knocked out in roundworms, the lifespan is approximately doubled.[14] Thus, IGF-1 appears to be part of a very ancient aging-related biochemical-signaling pathway that exists in some form or another in organisms as distantly related as people, mice, yeasts, and roundworms.

Longo reasoned that the genes that regulate aging in dwarf yeast, as well as long-lived mice and roundworms, could be the same that affect aging in man. Patients with Laron syndrome would provide a "natural experiment." These individuals have an equivalent interference in IGF-1 signaling and, according to theory, should have some protection against age-related diseases or live longer than average. Longo learned about Guevara-Aguirre's work and invited him to give a talk at USC. In an April 2006 presentation, Guevara-Aguirre mentioned that people with Laron syndrome do indeed live to a ripe old age. He also said that if they weren't so susceptible to accidents, alcohol-related problems, and convulsive disorders, they might live even longer than average, since up to that point he hadn't ever encountered a single case of cancer in any person with Laron syndrome. Longo realized right away that Guevara-Aguirre was on to something. Longo obtained serum from Laron syndrome patients through Guevara-Aguirre and bathed human cells in this serum in his lab. The cells were subsequently exposed to certain cancer-causing, DNA-damaging chemicals. The Laron syndrome serum brought about two remarkable effects. First, the serum seemed to protect the cells from chemically induced genetic damage. Second, in any cells that *were* damaged, the serum seemed to encourage self-destruction through apoptosis. As mentioned earlier, apoptosis is a mechanism that prevents damaged cells from perpetuating themselves and turning into a full-blown cancer. It would thus appear that any putative epidemiological studies suggesting a reduction in cancer incidence in Laron syndrome could indeed be real and have a sound scientific basis. In 2011, Longo and Guevara-Aguirre published a report indicating that of just about a hundred people with Laron syndrome there was only a single case of cancer recorded.[15] In contrast, in relatives who were of normal height, one out of every five developed cancer at one point in their lives. This was consistent with the general background rate of cancer in the region. It is worth keeping in mind that today in the United States (despite the decades-old "war on cancer") about one in every two will be diagnosed with some form of invasive cancer at some point during their lives, and one in four people will eventually die of cancer. Most interesting, studies suggest that in women, the likelihood of developing cancer increases about 16 percent for every extra four inches (10 cm) in height above average, and there is a similar trend in men. This has led some to speculate that the increases in average height of people during the twentieth century could partly explain the observed increase in cancer incidence over time.[16]

A few years back, I came across a paper written in 2007 by Laron and Orit Shevah.[17] They published their survey of 169 individuals with Laron syndrome and 250 of their relatives from the Middle East and Europe. They found no cancers at all in the Laron syndrome subjects, however they did find an unsurprising 24 percent incidence of cancer in their relatives. The researchers also examined thirty-five individuals with a closely related condition called IGHD (*isolated growth-hormone deficiency*) and eighteen more with another related condition (*growth-hormone-releasing hormone receptor defects*) and found that none of these individuals had any cases of cancer either. I followed up with Dr. Laron via e-mail a few years later to enquire if he had any additional data to corroborate this astonishing claim. He kindly provided me with a paper hot off the presses in the *European Journal of Endocrinology*.[18] In this larger study, Rachel Steuerman, Orit Shevah, and Laron observed that among 230 patients with Laron syndrome, there was again not a single incident of malignancy. And among 116 people with IGHD, only one person had a diagnosis of skin cancer—but that patient also had a very rare cancer-causing condition called xeroderma pigmentosum. The rates were also very low for patients with the related growth-hormone-releasing hormone receptor defect condition (three cancers among 79 individuals) and congenital multiple pituitary hormone deficiency (three cancers among 113 individuals).

So, it does indeed seem that Laron syndrome confers resistance to cancer. But that resistance to cancer does come at a price. First of all, there are ample challenges imposed by obesity and a stature of under four feet. Furthermore, at this time there is still debate about whether Laron syndrome truly confers resistance to diabetes. It may be that not all Laron syndrome mutations are created equal in this regard; the jury is still out. Additionally, despite all the supportive laboratory data on life extension, it remains unclear whether or not people with Laron syndrome truly have longer lives. Some studies suggest they might actually suffer premature aging, but it is difficult to sort out the differences between what is due to Laron syndrome and what is due to heavy alcohol consumption. What is known is that people with Laron syndrome have an unusually high rate of death from accidents, alcohol-related disease, and convulsive disorders.

Also while the low levels of IGF-1 might explain the reduction of cancer in Laron syndrome, whether *high* levels of IGF-1 correlate with increased cancers in humans is uncertain. While some studies suggest that high blood levels of IGF-1 may increase the incidence of breast or prostate cancer, the results are far from clear at present. Conversely, very low levels of IGF-1 in otherwise ordinary people might increase the risks of cardiovascular disease, dementia, and muscle-wasting. Ironically, low levels of IGF-1 are common in cancer-associated muscle wasting (*cachexia*, discussed in an earlier chapter in the context of brown fat activity), which is often encountered during the final phases of cancer. Nevertheless, there is ongoing research to synthesize medications that might block the growth hormone receptor, thereby pharmaceutically doing what the genetic defect in Laron syndrome naturally does.[19]

While some small amount of IGF-1 seems necessary to ward off heart disease, for a number of reasons the average American might have higher than appropriate levels of this hormone. So, after one has attained full stature, what would happen if we then blocked IGF-1 signaling? All this has certainly gotten some people thinking. One might speculate that drugs that reduce or interfere with IGF-1 could prolong life span and/or reduce the risk of cancer. There are already a couple of drugs out there that might do the trick. One such medication is called *pegvisomant* and is used to treat a life-shortening and diabetes-inducing condition called *acromegaly*. Acromegaly is caused by an excess of growth hormone in adulthood (after the bones have fully grown, sealed off, and can no longer lengthen) and is usually due to a pituitary tumor. Nonetheless, if all this research comes to fruition and something truly useful emerges on cancer (or diabetes or life extension) from the Laron syndrome saga, it would be thanks to a miraculous constellation of good fortune. Rosenbloom himself never imagined that traveling to Ecuador to see patients there, rather than have them come to Florida, would have led to meeting Guevara-Aguirre and ultimately culminating in all the astonishing findings made in the past two and a half decades. As intimated in Gary Taubes's *Discover* magazine article, perhaps the stars have aligned just right.

Finally returning to the LB1 fossil from the previous chapter, scientists still debate whether this was a human with Laron syndrome or a diminutive new species of hominin (a hobbit). As mentioned in the previous chapter, I personally favor the view that this and the other specimens do in fact represent a separate species. But assuming for now that it was a different hominin, my still unanswered question is whether they were resistant to cancer? This may forever remain a subject of speculation and debate for us all (unless of course someone finds a fossilized skeleton with a bone tumor!).

Chapter 20

CANCER: A DISEASE OF IMMUNE FAILURE?

As I have said repeatedly, in many ways one can conceivably think of a malignant tumor as a unique new life-form—a brand new distinctive species. In this context, this new living entity cannot be faulted for striving to do what all life-forms aim to do: survive. Thus, instead of being content to reside within the confines of the person, plant, or animal it arose in and dying with that corpus, a cancer might yearn to break free of its shackles and journey onward in its quest for immortality. As we know, such renegades do not attain their goal of living forever. In fact, in foolhardy fashion they kill their hosts prematurely, thereby ending their own quest for eternal life the hard way. Of course there are exceptions.

In early 1951, a young African American lady by the name of Henrietta Lacks went to the Johns Hopkins Hospital complaining of a "knot" in her pelvis. It was soon learned that the knot was a symptom of cervical cancer. Today, the five-year survival rate for stage I cervical cancer is over 90 percent, and with modern Pap smear testing and HPV testing, over 90 percent of cervical cancers can be detected at an early stage.[1] (Disturbingly, despite these encouraging facts, approximately one-third of eligible women in the United States do not undergo Pap and HPV testing, which accounts for the higher than expected death rate from this highly curable disease.) Sadly for Ms. Lacks, who was born Henrietta Pleasant in Roanoke, Virginia, in 1920, back in her days, detection and treatment of cervical cancer was not what it is today. She died on October 4, 1951, at the young age of thirty-one, with widespread metastatic cancer. But while she died, her tumor lived on—and still lives to this day.

During her cancer therapy, two samples of Henrietta's cervix were removed—a healthy part and a cancerous part—without her explicit permission or knowledge. The cells were given to Dr. George Otto Gey who attempted to culture them in the lab. The normal cervical cells died after several rounds of cell divisions. But to Gey's surprise and delight, the malignant cells lived on. Those cells were destined to become the famous—and immortal—"HeLa cells," named in honor of Henrietta Lacks. Actually, the contribution of Henrietta Lacks to medical history was forgotten for a while, and when I was working with these cells myself in graduate school, I was told they were derived from a mythical patient named Helen Lane. I could not find out much about her history or the history of those curious cells back at that time—and I was not alone. In the 1970s, Henrietta's family began getting strange requests from researchers for blood samples so they could better understand the genetics behind the celebrated HeLa cells. It was only upon probing into the historic journey of those storied cells that her family learned about their removal from Henrietta's body years before.[2]

Since then, there have been a number of tributes to Henrietta Lacks. In the late 1990s, a BBC documentary on Henrietta Lacks and the HeLa cells line won the Best Science and Nature Documentary at the San Francisco International Film Festival. In her 2010 book, *The Immortal Life of Henrietta Lacks*, Rebecca Skloot[3] resplendently recounts the full fascinating story of Henrietta, her family, and the HeLa cells that changed cellular biology. Henrietta Lacks was inducted into the Maryland Women's Hall of Fame in 2014.[4]

With their immortality, HeLa cells marked a revolution in human cellular biology. Around the world and over the decades, scientists have grown around twenty tons(!) of HeLa cells since they were taken from Henrietta's cervix. Normal human cells live only a certain length of time *in vitro* before succumbing to "senility" and ultimately dying. They get to a certain point in time and then cannot divide any further. They have reached the end of the line. They are mortal. This concept of cellular mortality—the *Hayflick limit*—was advanced by biologist Leonard Hayflick in 1961.[5] Hayflick demonstrated that a culture of normal fetal human cells in the lab will divide somewhere between forty and sixty times before growing senile and dying.

One might be surprised to hear that the Hayflick phenomenon and the HeLa cells cultured by Gey were not immediately heralded as monumental discoveries. This lack of appreciation was attributable to an erroneous belief held at that time that all normal cells had an unlimited potential to replicate. That misconception was perpetuated by surgeon and Nobel Laureate Dr. Alexis Carrel, who allegedly kept a culture of cells derived from chicken hearts growing in his lab for thirty-four years.[6] Today, Carrel's results are suspected to be due to the daily addition of "nutrients," which were contaminated with embryonic chicken cells. These supplemental new cells might have continuously replenished the culture, apparently keeping it going all those years. Alternatively, Carrel's cell culture could have contained some *stem cells*, which were capable of avoiding the Hayflick inevitability if given proper nutrients. We will never know as those cells were unceremoniously tossed in the trash in 1946, two years after Carrel's death. In any case, we now know that normal cells do *not* have unlimited replicative capacity in the lab, but cancer cells do.

The Hayflick limit appears to be due to the progressive shortening of the ends of chromosomes upon each cell division. Each time a cell reproduces or undergoes mitosis, the tips of its chromosomes—the *telomeres*—get clipped off a little bit. Thus, daughter cells have shorter telomeres than their parents and granddaughters have shorter telomeres still. This inexorable shaving of telomeres works like a cellular clock. Eventually there comes a point of no return: the telomeres are too short, the clock has run out, and the cells enter a phase called *senescence* and soon can no longer divide. The cells—and the entire cell colony in culture—then die. Cancer cells, including those taken from Henrietta Lack's cervical tumor, are different. Thanks to an enzyme called *telomerase*, which adds length back onto the telomeres following each cellular division, the telomeres in cancer cells do not inexorably shorten over time.[7] Like ordinary cells, their telomeres get a haircut after each cellular division, but unlike normal cells, their telomeres fully regrow in short order thanks to the telomerase enzyme. The bottom line is that their cellular hourglass, in essence, never runs out.

Returning to the concept of a cancer as a new organism yearning to break free of its corporeal yokes and continue its life anywhere it can, we have just encountered one successful example: HeLa cells have been cultivated in labs across the world since 1951. And along the way, we have encountered other successful examples: the Tasmanian devil transmissible cancer that has been fiendishly jumping from devil to devil for a couple of decades, the recently discovered contagious clam cancer that no one knows the real age of, and the ultimate survivor—the canine transmissible venereal tumor cells that have been around for over eleven thousand years. These cancers have somehow learned to live outside their original hosts and have continued their existence in novel ways such as bounding from body to body. (Along this same line of reasoning, one might argue that HeLa cancer cells have also figured out a great way of thriving outside their original host—they have been manipulating laboratory scientists into keep them alive, nourished, warm and well since 1951.)

Soon we shall describe a few highly disturbing "near misses" where human cancer cells almost figured out how to perpetually propagate from person to person. But first, revisiting CTVT for a while, let us reexamine a few apparent parallels between the normal pattern of regression observed in CTVT and the abscopal phenomenon. In CTVT, the cancer may initially grow rapidly without restraint; it may even begin to spread. This is quite like the situation in untreated human cancers. But then, in most cases CTVT begins an abrupt spontaneous regression and completely vanishes, never to return. In the abscopal effect, the human cancer, even if it has already spread, similarly regresses (although a key difference between the abscopal phenomenon and most cases of CTVT is that abscopal effects may or may not be permanent whereas most cases of CTVT regression are). The fact that CTVT tends to never comes back is a crucial point. In fact, subsequent exposures to CTVT via sexual contact with a new dog or dogs will *not* lead to reinfection. The dog has become permanently immune; the cancer has been banished forever. This pattern has all the hallmarks of a vaccination.

This makes perfect sense. After all, this is what would be expected with any transplanted tissue from another animal. Consider the situation in human patients in need of heart, lung, liver, or kidney transplants. In order for organ transplantation to work, the donor and recipient must be good matches—they must be genetically compatible. As mentioned earlier, compatibility is dictated by a set of genes generally called the MHC, or major histocompatibility complex (*histo* comes from the Greek word for *tissue*; thus *histology* is the examination of tissue samples in the lab). In humans, the clinically relevant MHC genes are called the HLA (for human leukocyte antigen) genes. For successful organ transplantation, there needs to be a good degree of HLA compatibility between donor and recipient. The more HLA antigens shared by donor and recipient, the greater the odds of success. Also, the specific organ in question makes some difference. For example, a near-perfect HLA match is required for bone marrow transplantation whereas a less perfect match may be sufficient for a kidney transplant. Not surprisingly, a higher probability of a good match can be found in first-degree family members (parents, siblings, and children) who on average share about 50 percent of HLA genes.

Outside the rare situation where the HLA match is perfect (as is the case with identical, or *monozygotic*, twins), following a transplant, the recipient typically must be on lifelong immune-suppressing medication to prevent rejection of the new organ. This of course can have dire consequences. As one might imagine, frequent infections may occur, often with unusual and exotic pathogens that people with normal, healthy immune systems don't commonly get.

Crucial to our story is the fact that with transplants there is also a significant risk of developing *cancer*. Over the years, ample data has confirmed an average two- to three-fold increased risk of cancer after an organ transplant. Based on a study published in *JAMA* in 2011,[8] organ transplant recipients are at increased risk for at least thirty-two different types of cancer. This particular study, led by researchers from the National Cancer Institute, was the largest of its kind and included a sufficiently large sample to draw firm conclusions. Data from 175,000 transplant recipients (representing approximately 40 percent of all US organ recipients), substantiated previous smaller studies that showed an increased cancer risk. The study also showed that the chances of certain *specific* cancers are highly amplified. Among the types with exceptionally elevated odds, several are associated with infectious agents such as Kaposi sarcoma (which is linked to a virus aptly, if not imaginatively, known as *Kaposi sarcoma associated herpes virus*, or KSHV), anal cancer (which is linked to strains of *human papilloma virus*, or HPV), and liver cancer (which is known to be linked to *hepatitis B virus* and *hepatitis C virus*). On the other hand (and key to our story), there is also an increased risk for many cancers *not* known to have an infectious cause, such as melanoma and thyroid cancer. This raises the question of whether immune suppression alone is a risk factor for cancers in general (as opposed to those related to viruses).

The cancer with the *most* magnified risk (by some estimates as high as sixty-five-fold) is one often overlooked: skin cancer (specifically, nonmelanoma skin cancer). Nonmelanoma skin cancers must first be distinguished from melanoma, a far deadlier disease. At nearly 3.5 million cases annually in the United States, there are far more cases of nonmelanoma skin cancer than all other types of cancers combined.[9] Additionally, these cancers are normally easily cured by simple surgery. Distressingly however, nonmelanoma skin cancers in the post-transplant setting do not behave like ordinary skin nonmelanoma cancers.

There are two subtypes of nonmelanoma skin cancer: *squamous cell carcinoma* and *basal cell carcinoma*. Basal cell cancer predominates in the general population, but after an organ transplant, squamous cell carcinoma is more common. The chances of getting squamous cell carcinoma skin cancer are proportional to the intensity and duration of exposure to the immune-suppressing drugs (and mysteriously, also to the organ transplanted: heart and lung might predispose one more than kidney and liver).[10] There appears to be a geographical variation, too. In Australia, the risk of developing squamous cell skin cancer rises to nearly 80 percent twenty years after an organ transplant.

Squamous cell skin cancers are occasionally associated with viruses, specifically certain strains of human papilloma virus (HPV)—the same viruses that cause cervical, anal, and penile cancers. HPV comes in well over a hundred different strains, most of which do nothing more than cause common warts (which, after all, is what the virus is named for: *papilloma* is just a fancy medical word for wart). Incidentally, at this time there is a noticeable increase in cases of *head and neck cancers* (e.g., tonsils, pharynx, larynx, and base of tongue) that are linked to HPV.[11] Exactly what is responsible for this "epidemic" is not clear. But in addition to the traditional heavy drinker/heavy smoker, older male patients, nowadays we are seeing a higher number of much younger folks with no history of heavy drinking or smoking. Fortunately, these virus-associated cases seem to have a better prognosis than the classic cases. A good deal of research is ongoing to elucidate just what the cause of the increase is, and how HPV-positive head and neck cancers should be best managed.[12]

Returning to the post-transplant situation, the skin cancers in these individuals behave quite differently from normal squamous cell carcinomas—they are rather aggressive. The odds of a nonmelanoma skin cancer spreading to the lymph nodes, bones, and beyond in nonimmunosuppressed people is very low: the cure rate is close to 100 percent. Sadly, this is not true in the post-transplant setting where this type of skin cancer does occasionally recur, despite our best surgical and radiotherapeutic efforts, and eventually metastasizes just like other dangerous cancers. The fact that squamous cell skin cancer behaves so drastically different in the immunosuppressed population might not be all that surprising since there is precedent. A rare genetic condition called *epidermodysplasia verruciformis* exists that is associated with frequent and aggressive infections from the HPV virus. Rather than simply causing warts as in the rest of us, the HPV infections in epidermodysplasia verruciformis patients more often worsen into full blown cancers.

Although details are not fully understood yet, the disease does appear to be linked to defective cell-mediated immunity. [Epidermodysplasia verruciformis appears to be a genetic condition with various modes of inheritance (primarily autosomal recessive). There are a couple of genes linked to the disease (*EVER1* and *EVER2*) but in about a quarter of patients there is no family history of the condition and the *EVER1* and *EVER2* gene mutations are not present.] Curiously, many patients with what appears to look just like epidermodysplasia verruciformis are immunosuppressed thanks to organ transplantation or HIV infection.

Aside from nonmelanoma skin cancers, other cancers occasionally encountered after an organ transplant are non-Hodgkin lymphoma, lung, liver, and kidney cancers. The incidence of non-Hodgkin lymphoma (NHL) is elevated more than sevenfold in transplant recipients.[13] For reasons unclear, the incidence of post-transplant NHL is highest in lung recipients and lowest in kidney recipients. NHL is known to be related to both immune suppression and infection with Epstein-Barr virus. Analogous to the situation wherein squamous cell skin cancer is virally induced and high in the post-transplant setting and in people with the unusual disorder epidermodysplasia verruciformis, NHL is also virally induced and high in the post-transplant setting and in people with another unusual disorder: Duncan's disease.[14] The disease is named for the Duncan family where the condition was first recognized as the cause of death in six of eighteen young males, but it also goes by the name *X-linked lymphoproliferative disorder*. Being X-linked means that only males develop the disease, which manifests as an odd inability to mount an adequate immune response to Epstein-Barr virus. In people with normal immune systems, Epstein-Barr virus is often "silent," meaning that it produces no obvious symptoms. In other cases, the virus might cause mononucleosis, particularly during adolescence (explaining the common name "kissing disease"), but with no long-term consequences. In boys with Duncan's disease on the other hand, Epstein-Barr virus is not adequately controlled and is followed by a general decrease in immunity manifested by diminished antibody production (hypogammaglobulinemia), resulting in various catastrophic, opportunistic infections. Alternatively, the Epstein-Barr virus itself might directly contribute to death due to NHL or liver failure caused by severe and uncontrollable mononucleosis.

As mentioned, several studies have confirmed that transplant recipients are more likely to develop cancer than the general population. The risk of developing any cancer following an organ transplant is inversely related to the age at which the transplant was done with a fifteen-fold to thirty-fold increase in children who had a transplant versus a twofold risk in organ recipients beyond age sixty-five. And those who do get cancer face a higher risk of dying of that cancer (than others in the general population with the same cancer). As Dr. Joseph Buell (who is a past director of the Israel Penn International Transplant Tumor Registry at the University of Cincinnati and presently with Tulane University and Louisiana State University Health Sciences Center in New Orleans) pointed out in the *Journal of the National Cancer Institute,* in kidney and heart transplant recipients who make it beyond the first three years, cancer becomes the leading cause of death over the next two decades.[15]

For this reason new antirejection drugs are desperately needed. One promising class of drugs is *sirolimus* and related agents. Sirolimus was originally discovered as a biochemical byproduct of bacteria growing on Easter Island. It was initially approved by the FDA for organ transplantation immunosuppression in 1999. Although sirolimus was originally developed as an antifungal medicine, its more promising role in preventing organ rejection quickly overshadowed that original aim. More recently and quite ironically, however, sirolimus and its derivatives such as temsirolimus have found newer and even more promising roles as none other than cancer medicines! These new agents hold promise as cancer therapeutics because they inhibit one of the many cancer-related signaling pathway molecules, namely mTOR. (mTOR stands for mammalian target of *rapamycin*. Confusingly, rapamycin is another name for sirolimus. mTOR is a cancer and growth-related protein kinase complex that is part of a signaling pathway for cell division.)

Thus sirolimus, tacrolimus, temsirolimus, and related drugs which inhibit mTOR and have anticancer properties, intuitively seem to be a wise choice of immunosuppressing drugs for organ recipients who are at risk of cancer. Clinical trials are now underway. As with all medications, however, the devil is in the details—sirolimus can cause serious lung side effects (interstitial pneumonitis), especially in patients with underlying lung disease. Whether or not these new antirejection drugs will significantly reduce future cancers remains to be seen.

In addition to post-transplant patients on chronic immunosuppression, people with HIV (human immunodeficiency virus) infection also have a higher likelihood of developing infection-related cancers (and infection-unrelated cancers for that matter). According to Eric Engels, who led the post-transplant investigation discussed above, the cancer incidence among organ recipients is similar to that seen in people with HIV. Both populations have elevated risks due to their immunosuppression. As in the post-transplant situation, HIV patients have higher incidences and/or increased severity of several virus-associated cancers including Kaposi sarcoma, NHL, and invasive cervical cancer. In fact, these three cancers are "AIDS-defining" conditions, meaning that a person with HIV infection would now be defined as having full-blown AIDS upon coming down with one of these three cancers (among about two dozen other AIDS-defining conditions). Just as with post-transplant patients, people with HIV have higher odds of several other virus-induced and non-virus-induced malignancies including Hodgkin lymphoma, anal cancer, lung cancer, liver cancer, head and neck cancers, testicular cancers, colon cancer, angiosarcoma (a type of cancer that begins in the lining of the blood vessels), melanomas, and nonmelanoma skin cancers. Lymphoma might be the most problematic malignancy for HIV/AIDS patients since up to 10 percent of people with HIV/AIDS eventually develop NHL, most often of a fast-growing, high-grade B cell subtype. Unfortunately, *primary central nervous system lymphoma* (lymphoma involving the brain or spinal cord) accounts for nearly 20 percent of all NHL cases in this population. Sadly, this form of lymphoma often defies treatment and has a very poor prognosis.

In both the post-transplant and HIV/AIDS populations, the underlying theme appears to be *chronic immunosuppression*. But some have raised the question: Could the increased cancer incidence be due to the medications themselves? In other words, are the drugs themselves carcinogenic? Historically, immunosuppression drugs for organ transplants have included steroid medications such as prednisone, "purine antagonists" such as azathioprine and mycophenolate, and "calcineurin inhibitors" such as cyclosporine; more recently, sirolimus and tacrolimus have been added to the list. It would be difficult to conclusively prove whether the drugs directly caused a cancer in a given person or if they worked via suppression of the immune system to produce that cancer. Given the peculiar array of cancers that particularly plague people on chronic immunosuppression (such as the increased incidence of virally associated cancers), it seems more likely that the immune suppression is the real reason and that the medicines themselves are not carcinogens. But when one thinks about this question, it becomes evident that the term "carcinogen" deserves deeper thought.

A carcinogen, by definition, is a drug or physical agent that can produce cancer. Thus, since immune-suppressing drugs appear to increase the odds of cancer, they could rightly be considered carcinogens—even if they are not directly DNA-damaging, as is customary for classic carcinogens. In the context of virally induced cancers in the immunosuppressed population, one might consider the virus as "necessary but not sufficient" for producing a cancer whereas the immunosuppression, whether caused by HIV/AIDS or by drugs, enhances the effect of the virus. The combination leads to the formation of a clinical tumor. In classic cancer biology, an "initiator" starts the carcinogenic process by causing a mutation but frequently cannot finish the job alone. Another agent, the "promoter," is often needed to magnify the effects of the initiator. So, a promoter *cannot* cause cancer directly but rather augments the effect of an initiator thereby facilitating the formation of a clinical tumor. Together the initiator/promoter pair *can* lead to a tumor. In the present context, the virus could be considered an initiator and the immunosuppression (whether caused by HIV/AIDS or by drugs) augments the effect of the virus and can be considered the promoter. The virus (for instance, Kaposi-sarcoma-associated herpes virus), which doesn't produce cancer in people with normal immune capacity, is necessary but not sufficient. It needs a promoter, namely, a weakened immune system, to complete the job. Although this is not the conventional way of thinking about initiators and promoters, it is logical.

How then do we explain the increased incidence of *non*-virally associated cancers in the immunosuppressed population? Clearly our already modified concept of initiators and promoters will need some further revision. One could logically conclude that the observed melanomas, sarcomas, and the thyroid, lung, colon, and testicular cancers that are *not* known to be virally associated (yet are seen with increased regularity in the immunosuppressed population) are the result of the deficient immune system "promoting" problems caused by other initiators such as smoking, pollution, genetics, or diet. Although this is now clearly outside the realm of conventional thought, this still seems logical and perfectly plausible.

Now, for the sake of argument, let's extend this logic to all of us (not just those with recognized immunosuppression from an organ transplant or HIV). Obviously, there are plenty of cases of virus-associated cancers out there such as cervical cancer, liver cancer, and NHL. And one might wonder if in these cases, a necessary-but-not-sufficient viral infection was augmented by a subtle drop in immunity, perhaps due to age or disease. In these cases, the virus was the initiator and the drop in immunity due to age or other causes was the promoter. We are still working within a perfectly reasonable and logical framework.

But then analogously, might the various *non*-virally caused cancers (that is, the overwhelming majority of regular old cancers) also be due to some subtle and heretofore unrecognized alterations in immunity? We know that most cases of cancer appear later in life. So, could it be that the countless cases of cancer in the United States are caused not just by smoking, drinking, or environmental/occupational carcinogens alone but instead are caused by a one-two punch of initiator (smoking, drinking, carcinogens) *plus* subtle immunosuppression that appears later in life? And if so, what are the causes of that late-onset immunosuppression that enables cancer to manifest in our later years? Might stress be one such cause? And regardless of the cause of such postulated cancer-allowing immunosuppression, how can we reverse it?

The above line of reasoning is certainly a minority (if not solitary) view, and to date there is no solid scientific proof of the hypothesis. But based on much of what has been discussed in this chapter and throughout the book, this remains, at least in my opinion, a very credible concept. We will revisit the idea again in a later chapter.

In any case, it is clear that immunosuppression, whether caused by disease such as HIV/AIDS or by drugs, can increase the chances of developing cancer. (And maybe not just virus-related cancer but cancer in general.) Therefore, it is very reasonable to conclude that the immune system plays an active role in fighting off cancer under ordinary circumstances. We shall later explore the key role the immune system plays in "immunosurveillance" of newly developed cancers. According to this concept, cancers crop up all the time (daily?), but the immune system is constantly on guard and readily quashes these unwelcomed newcomers. Failure, for whatever reason, of this routine erasure of incipient tumors is what leads to a clinical cancer.

In my own opinion, the most valuable lesson learned from the transplant clinic is what happens when a donated organ harbors some undiagnosed "undesirables" that tag along for the ride. The amazing story of what happens with accidentally transplanted cancers might hold the key to a more comprehensive understanding of the role of the immune system in cancer.

Along the way, we will encounter what I believe is the strongest proof of the power of the immune system in holding cancer at bay.

Chapter 21

MALIGNANT CARGO

The plot was rather simple:

Skillful surgeons transplant a kidney into a desperately ill young woman thereby saving her life.

The woman appears to be doing well for a while but develops cancer months later.

Based on a hunch, her clever doctors insist on getting chromosome counts in the tumor cells.

The tumor cells all contain an X and Y pair of sex-determining chromosomes proving that they came from the male donor's kidney, since all her own cells, being female, are XX.

Her quick-witted doctors discontinue her immune-suppressing antirejection medications thereby leading to a prompt rejection of her donated kidney—but also the rejection of her transplanted cancer.

Her transplanted kidney is removed, and she is put on dialysis for a short while, and miraculously, her cancer is cured! Her own immune system has defeated the foreign cancer that came along for the ride.

Her heroic doctors quickly find a new, cancer-free kidney and transplant it into the now cancer-free young woman.

She lives happily ever after.

Oh and there is one more thing—the author of the script for this television episode is paid a million dollars and he, too, lives happily ever after.

Sadly for that author (me), the 2009 medical drama television series *Three Rivers* was cancelled after only eight episodes. The above proposed episode was never accepted nor developed. In fact, I suspect they never even read it . . . and now here I am recounting the tale in this present narrative. In a way, however, although I was sad to see the program terminated, I am glad my particular plot was never accepted and aired; things are always far more complicated than the television fairy-tale versions.

I have learned much since then. In actuality the lady's outcome could very well have ended on a more somber note. Just why this is so is most enlightening.

Over the past twenty-five years, well over half a million people received organ transplants, and thanks to this, an estimated two million years of life were saved. Nevertheless, the need for healthy organs greatly outpaces donations. At any given time, over one hundred thousand people are on waiting lists for organ transplants in the United States, the majority awaiting kidneys. Regrettably, fewer than thirty thousand transplants are performed annually. An estimated eighteen Americans die every *day* while waiting for an organ transplant.[1]

Thus, one must keep things in perspective—it is clear that organ donations save lives, and there are far too few available organs for the number who need them. Given this enormous disparity, patients and physicians must carefully weigh the risks and benefits of donations from less-than-ideal donors. In the overwhelming majority of cases, the benefits outweigh the risks, but inevitably a few suboptimal organ donations will sneak by.

In an effort to weed out marginal donations, all organs are routinely screened for contamination with common infections and diseases. Still, data from the US Centers for Disease Control and Prevention shows that 1 percent to 2 percent, or up to 560 organ recipients annually in the United States, contract a concealed infection or disease from their transplanted organs. In rare instances, even cancer has been transmitted via transplanted organs.

As previously discussed, cancer is not an uncommon condition in transplant recipients. There are four general ways in which cancer can develop in someone who has had an organ transplant. First, the organ recipient might have had a previous cancer and that cancer could be revived through the immunosuppression needed to prevent rejection of the newly transplanted organ; in other words turning off the immune system can resuscitate existing quiescent cancer cells. Second, as described in the previous chapter, the obligatory immunosuppression puts organ recipients at risk for the development of various new cancers. Third, the transplanted organ, like any organ, could develop a malignant tumor. For instance, if a lung recipient were to smoke heavily or get exposed to carcinogenic vapors, his or her new lung could develop a cancer just as readily as any other lung.

The final possibility is the particularly unfortunate situation that arises when an organ is transplanted from a donor who unknowingly has cancer.

This last example very rarely occurs now but was more frequent in the early days of organ transplants. Today, potential organ donors are screened and those with *active* cancer are generally excluded. Of course, given the dearth of available organs, there must be a careful balancing of the risks and benefits in a given situation. The waiting list is so long that under certain circumstances the risk might be deemed worth taking. Certain cancers, such as primary brain tumors are considered to be at very low likelihood of transmission. Thus the patient and transplant team might be willing to proceed with an organ donation from such a patient when the risk-to-benefit ratio appears acceptable. On the other hand, certain other cancers have a considerable propensity of being transmitted along with a donated organ. According to the Israel Penn International Transplant Tumor Registry (IPITTR), melanoma was transferred nearly 77 percent of the time. Other cancers not infrequently passed along included lung cancer (41%), prostate cancer (29%), breast cancer (26%), and colon cancer (19%).[2]

Upon reviewing three decades of transplants, the Israel Penn International Transplant Tumor Registry found that when a donor definitely had cancer, that cancer was transmitted to the recipient an astonishing 43 percent of the time. Although several other studies have shown far lower rates (and one study put the risk of cancer communication as low as two in ten thousand transplants, or 0.02%), there are countless case reports in the medical literature, raising a slew of questions.[3]

We learned the hard way that cancer passed along in this fashion behaves especially virulently. In the way of statistics, one study of over a hundred cases of cancers carried in by transplanted kidneys found that nearly three-quarters of melanoma recipients had died within three years. Two-thirds of patients in whom lung cancer was transferred died within twenty months. Why the dismal prognosis? For one, the immune-crippling antirejection drugs hamstring any hopes that the host's immune system might mount an effective counter attack. It must be kept in mind that people who are on immune-suppressing drugs to prevent organ rejection are very vulnerable to a wide array of diseases that would promptly be overcome under ordinary circumstances where the immune system was intact.[4] In one large study highlighting this general susceptibility, when the donor had a transmissible virus, fungus, or parasite, over half of the time the infection was transmitted—and over a quarter of those recipients died from the transmitted disease. Many of these infections would readily be repelled in individuals with healthy immune systems.

Treating patients with transplanted cancerous organs is particularly vexing if the organ is absolutely essential such as the heart, liver, or lungs. In all transplanted cancer cases, the ideal first line of treatment would be to stop the immunosuppression, remove the contaminated organ and desperately try to find another organ to transplant promptly. Given the scarcity of available organs in the first place, the odds of immediately finding a replacement organ in such short order is highly unlikely. Therefore heart, lung, and liver (but not kidney) recipients often must retain their dubious transplanted organ and simply hope for the best. Regrettably, the prognosis in patients who must keep their tainted organ is quite grim if a transplanted cancer awakens. In the case of kidney transplants, however, there is a second chance. For such individuals, first the immunosuppression can be discontinued allowing the host to reject the bad kidney—along with the stowaway cancer. Then, after rejection, that kidney and its cancerous cargo can be removed, and the patient can be sustained on dialysis until a new kidney can be found. This strategy is sometimes successful. On the other hand, sometimes the outcome is dreadfully disturbing.

The fact that transplanted cancers can survive at all in a new, unrelated, immunologically hostile host is unexpected. The fact that it happens as frequently as it does, and that the cancers not only survive but thrive, is both astonishing and disquieting. What can be gleaned from all this?

First, let's ask again why the failure rate is so dismal. One is initially tempted to attribute the grim prognosis solely to the compromised immune systems in these organ recipients. However, this *cannot* be the entire explanation since, as mentioned, in kidney transplants the immune-suppressing, antirejection drugs can be discontinued and the patient can be placed on dialysis. Under such circumstances, in the absence of active immunosuppression, the resuscitated immune system s*hould* reject the foreign kidney—and eliminate the foreign cancer that has tagged along for the ride. Yet, perplexingly this sometimes does not work. There have been numerous documented cases where the revived immune system rejects the donated kidney—but *mysteriously grants the transplanted cancer a free pass*. Remarkably and unexpectedly, the revitalized immune system—which has confirmed its competency by rejecting the transplanted kidney—sometimes still seems unable to "see" the transplanted cancer.[5]

Admittedly, when I first learned of this, it was one of the most confusing things I had ever heard. The transplanted cancer is an "allograft"—an organ or tissue donated from another individual. As such, it should be subjected to all the rules imposed on any donated organ or tissues. Like any allograft, the donated tumor should be rudely rejected. Yet, this is often not the case. Why? How does the cancer effectively emulate *The Invisible Man* and remain hidden from the immune system?

One possible explanation is that the host's recently revived immune system, while now strong enough to reject foreign *but normal* kidney cells, is not yet strong enough to expel a foreign *tumor*. An alternative possibility, and one I am more inclined to believe, is that the cancer itself is somehow *actively concealing itself.* Even though it is literally alien, coming from someone else's body, the malignant tumor wields an effective "cloaking device," right out of *Star Trek*, which renders it invisible. This shield, in essence, grants it immunity from immunity.

Certainly not all cancer biologists accept this wild idea, but I personally find this to be a perfectly logical conclusion. And if cancer has obtained a Klingon cloaking device it would explain much. For one, this could aid a tumor immensely in terms of its ability to survive. And if a tumor is so elusive that it can hide from the immune system of *another person*, it bodes badly for any hope that the original host's immune system could ever recognize and reject it. It would appear then that this clever and ominous "invisibility act" is an effective trick indeed—and this would be especially true if this is a general characteristic of cancers. Yet, as will be reviewed in the next chapter, there is room for cautious optimism here. For in some scenarios, our immune systems can *definitely* "see" the cancer—and either eliminate it outright or hold it at bay indefinitely.

Chapter 22

THE POWER OF THE IMMUNE SYSTEM

In order for a cancer to be transferred from one individual to another by an organ transplant and cause trouble in the new host, several things must fall into place.

First, the donor must unknowingly have cancer. As mentioned, many transplant teams will screen out and disqualify potential donors with active cancer, so a donor must not have obviously died from widespread cancer. Nevertheless, there must of course be cancer somewhere in the donor for it to eventually be transferred into the recipient. Second, the cancer in the donor must have spread to the organ being transferred. Even if the donor has pervasive cancer, if it doesn't involve the particular kidney, lung, liver, or heart that is donated, there is no risk of transmission. The reality is that given the limits of current medical science it is impossible to be 100 percent certain that a donor doesn't have cancer and that the organ in question isn't involved with that cancer. Next, the cancer, once in its new host, must survive any initial assaults from the new immune system, and it must be capable of thriving in the face of a hostile immune environment. While the recipient's immune system will be compromised because of the transplant medications, there could still be some reaction of the host's immune system toward this foreign tissue. Finally, the malignancy must remain invisible to the new host's immune surveillance even after the immune-suppressing medications have been discontinued.

With all these barriers, it should be almost impossible for cancers to be transmitted through this mode, yet there are ample cases found in the medical literature. Let us explore just why this is so. Without constant immunosuppression, poor organ donor/recipient matches are quickly and uncompromisingly rejected no matter what, and even good matches may be summarily rejected as well. This is because, in addition to the *major* histocompatibility (MHC) antigens, there are dozens of *minor* histocompatibility antigens that, while not quite as crucial for transplant considerations, are still of some import. Yet, contrary to expectations, *cancers* that get inadvertently transferred along with the donated organ occasionally go unmolested irrespective of the degree of immunosuppression and to an extent, irrespective of the degree of donor/recipient match.

Cancers in organ recipients introduced by contaminated lung, liver, and heart transplants are almost invariably fatal. But as mentioned in the previous chapter, thanks to hemodialysis kidneys are not completely indispensable, and in principle, the immunosuppression could be halted. The patient could be put on dialysis while awaiting another kidney, and the expected outcome is that the transplanted kidney, along with its cancerous cargo, should be rejected by the awakening immune system. This frequently works, but on occasion it does not. Every so often, after the immunosuppression is ceased, the *normal* tissues of the kidney are indeed aggressively attacked by the immune system, but the *malignant* passenger is unaccosted and unharmed. It appears that the immune system is blind to such cancers.

So, how is this happening?

The rare instances in which immunosuppression was discontinued and the host rejected the transplanted kidney yet failed to reject the transplanted cancer, tells us that malignant tumors have immune-evading capabilities that normal tissues do not have. In other words, cancer can use that invisibility gimmick we talked about in the previous chapter. Obviously cancer has quite a few tricks up its sleeve. Based on such observations, it might seem as if cancer is just so nefarious and resourceful that the immune system could never hope to vanquish it. What hope do we have?

Perhaps a great deal.

Our story has a silver lining. Despite the tragic cases, there are a few inspiring scenarios that show us that in some cases the immune system *can* indeed recognize cancer as an enemy. In such people the cancer was belligerently confronted by the immune system and, if not flat-out killed, "imprisoned" for a life sentence. So where can we find proof of such police power? Back in the transplant clinic.

As mentioned, organ donors are carefully screened to avoid accidental transfer of malignancy to a recipient. In one highly instructive case study, an organ donor had cancer in the distant past but was declared cancer-free since so many years had passed.[1] Nevertheless, the kidney from this supposedly cancer-free donor somehow slipped a cargo of malignant melanoma cells into the recipient. The amazing thing about this case, however, is that the donor was considered cured of melanoma *over three decades ago*!

The apparently disease-free organ donor was diagnosed with melanoma thirty-two years prior to the kidney donation and showed no signs of cancer for decades. He was considered cured of cancer. For all intents and purposes, he was. The odds of any dangerous cancer cells concealed in his kidney after all those years were considered nil.

Nevertheless, although this individual had no trace of any cancer for over three decades, he still sheltered malignant cells in his kidney. Although it is impossible to prove today, it is quite likely that the cancer had metastasized to multiple organs; if the kidney was involved, odds are that the donor's liver, lungs, and other organs similarly harbored microscopic malignant melanoma cells, which were also being kept at bay for decades. This indicates that despite metastasis, melanoma cells can be restrained within an immunocompetent host for decades—and possibly forever. Upon transfer into an immunocompromised recipient however, these previously imprisoned cancer cells go on the rampage. In my opinion, this is one of the most remarkable observations in all of cancer biology and medicine.

Although there are various possible explanations, such as angiogenesis inhibition or inherent tumor cell dormancy, my interpretation is that the original donor's immune system was effectively keeping these cancer cells in check all those years.

And if I am right—and if the immune system of an immunocompetent individual can keep metastatic cancer at bay for over thirty years—it makes me wonder if it could have done so indefinitely. Was that individual in essence "cured" of his cancer? I would assert that by any practical definition, he was.

This case, although extreme at *over three decades* between cancer diagnosis and transmission to another person, is far from unique. There have been other documented cases in which cancer was unintentionally transferred from persons who were understandably declared cancer-free since they were over ten or twenty years out from their cancer diagnosis and treatment.[2] Such individuals showed no signs or symptoms of cancer and could rightfully be considered cured. Nevertheless, despite being clinically free of cancer, obviously there was still some trace of the malignancy in their bodies, and that malignancy was capably checked by their immune systems.

And when that strict and competent guardian immune system in the donor was replaced by a weakened, naïve, and inept immune system in the medicated organ recipient, the cancers retaliated with a vengeance. Importantly, melanoma is not the only cancer transmitted unwittingly in this fashion. Even lung cancers have been transferred from donors who died with no outward signs of cancer. Only after their death from other causes was their cancer ever discovered.[3] This makes one wonder if such donors could have unknowingly had cancer for many years but their malignant disease was kept under control somehow by their highly capable immune systems.

Perhaps when considering cancer the term "cured" has to be redefined. In a sense, what happened in the donor was the equivalent of life imprisonment of his cancer. In sentencing a dangerous mass murderer, the killer can no longer harm society, whether imprisoned for life or executed. Analogously, the malignant cells cannot harm a person if imprisoned by the immune system indefinitely or eliminated. In a sense then, a patient can be considered "cured" of cancer either if the neoplastic cells are completely erased or if they are held in check by the immune system forever.

This analogy may be quite apt since that crazed assassin, if set free, could once again go on a killing spree. This is what apparently happened when the malignant payload was unwittingly transferred into a defenseless organ recipient. Unlike the powerful immune system of the donor, which was adroitly controlling the cancer for over thirty years, the new kidney recipient's weakened immune system was unable to keep the killer confined with the same dexterity.

This scenario makes one wonder what would have happened to the original donor if he ever became immunocompromised. Could that cancer have then overpowered the debilitated prison guards and escaped after all those years? And this concept makes one wonder if *we all* might have cancer cells somewhere in our bodies that are being kept sequestered by our immune systems and out of harm's way. Do we all have "cancer could be" cells in our bodies, successfully being kept in solitary confinement by our adept immune systems?

Another important lesson to be learned from this case is that even though one individual's immune system might be capable of recognizing and "imprisoning" the malignancy indefinitely, that same cancer, when released into unguarded territory, quickly returns to its old shenanigans. In the new organ recipient's body, the cancer again dons its veil of invisibility and spreads like wildfire, unconfronted by the recipient's naïve and debilitated immune system. To me, this tenable explanation diminishes the plausibility of the oft-quoted alternative assertion that the malignant cells were simply lying "dormant" in the donor all those years. I think the fact that upon release these supposedly dormant cells ran amok indicates that they were anything but docile and dormant. They were violent criminals kept restrained by a powerful police force—the human immune system.

Interestingly and again instructively, the scenario has been seen in reverse—presumably cured cancer patients who needed organ transplants years after their own cancer was apparently beaten, have had their old cancers recur with a vengeance upon the immunosuppression required to keep their new organs. In one case, there was a very rapid recurrence of a relatively rare kind of cancer (endometrial stromal sarcoma) after a kidney transplant.[4] The recipient had undergone cancer treatment around seven years earlier and appeared fully cured of her cancer. But soon after her kidney transplant and the commencement of immune-suppressing antirejection drugs, her long-gone cancer proved to be anything but gone. Obviously, there were still malignant cells lurking somewhere in her body, kept under guard for the last seven years. But upon letting her guard down, through immunosuppression, these baleful cells escaped following her kidney transplant. Returning to the prisoner analogy, in this case the criminal cancer was not imprisoned in the donor's body and let loose in the recipient's body but instead was kept under control in the recipient's body alone all those years. Only when the guards were put to sleep via immune-suppressing medications was the cancer capable of breaking loose.

The experience from the transplant clinic provides strong evidence in support of the concept of "cancer immuno-editing"—that the immune system is regularly confronting and doing battle with incipient (and even more mature) malignancies.[5] Using the prisoner analogy again, the convicts are always contriving a means of escape, and the prison guards are always sharpening their skill at keeping these inmates imprisoned. In this way, however, the skills of both sides constantly escalate. In the current version of the immuno-editing hypothesis, this constant honing of escape skills by the prisoners might lead to their someday outwitting the guards and making a successful getaway.

Frankly, I think that our current comprehension of immuno-editing is being refined regularly, and eventually we will have a better understanding than we have today. But I feel that our present level of understanding remains only rudimentary.

The seeds of the immuno-editing idea were originally planted by Paul

Ehrlich just around the dawn of the twentieth century. Ehrlich, born in 1854 in East Prussia (now Poland) discovered the first effective treatment for syphilis and coined the name "chemotherapy" for drugs that specifically aimed to disarm certain diseases.[6] He refined this concept and later proposed the idea of "magic bullets" that would selectively seek and destroy pathogens. If a compound could be created that selectively sought out the disease-causing entity, perhaps that compound could carry a toxin or other agent that would eliminate that disease causing germ. In cancer treatment, *monoclonal antibodies* (antibodies that are all identical and come from the same antibody-producing cells) are the closest thing to Ehrlich's magic bullets. Perhaps these cancer-specific, engineered antibodies could have a toxin or radioisotope attached to them and be set loose in a cancer patient's bloodstream. In principle, they would seek and destroy the targeted cancer cells—and only the cancer cells. Adding the toxin or radioisotope sounds like an attractive double whammy, but we have learned that the double whammy isn't always necessary. In some instances, the antibodies alone can inflict enough damage to be of clinical benefit. Research on monoclonal antibodies has been going on for a while, and in modern cancer care, monoclonal antibodies are now frequently used for treating certain breast, colon, and head and neck cancers, as well as lymphomas.

In any case, Ehrlich was perhaps the first to venture the *cancer immunosurveillance hypothesis*, which was later redeveloped by Lewis Thomas and Sir MacFarlane Burnet in the 1950s.[7] A key concept of the hypothesis holds that the immune system constantly is on the lookout for new cancers, and when these new cancers are found and recognized, they are promptly eliminated. In fact, MacFarlane explicitly said in 1957, "It is by no means inconceivable that small accumulations of tumour cells may develop and because of their possession of new antigenic potentialities provoke an effective immunological reaction with regression of the tumour and no clinical hint of its existence."[8] In other words, according to this constant immunological patrol, small tumors may come and go regularly (daily?) leaving no trace of their previous existence.

In its modern iteration, the cancer immuno-editing hypothesis holds that the immune system is constantly seeking and destroying incipient cancers. This continuous patrolling by the immune system is the first of the "3 *E*'s" of immuno-editing: elimination. There are occasions in which, rather than elimination, the cancer and the immune system are at *equilibrium*—the cancer is not fully eradicated but the immune system is still capable of controlling it. This second *E* of immuno-editing might be what happened in those cases where donors who were without *evidence* of cancer for years, or even decades, passed on payloads of malignant cells upon organ donation. Finally, the third *E* of immuno-editing is *escape*, wherein the cancer either eventually outwits the immune system or the immune system's guards snooze and lose.

As mentioned, although there are alternative interpretations, I would suggest that the extremely protracted latent cancers in the organ donor scenarios described above offer proof that the "equilibrium" concept of the immuno-editing hypothesis is valid. And if the immuno-editing hypothesis is correct, how can we cope with cancers that have evolved beyond equilibrium and gotten all the way to stage three, escape? This might depend on just how it is that the cancers managed to escape. If they have escaped through constant reshaping and refinement through a never-ending battle with the immune system, this could bode badly for us. The cancers in this case have finally evolved to the point where they have outmaneuvered immunity and won the war at last. If, on the other hand, the cancers have gained the upper hand because the immune system is simply sleeping on the job, there is hope indeed. In this case, a rebooting of the immune system might be all it takes to reestablish equilibrium, which, as we have seen, can keep even the most unruly cancers at bay for decades.

Chapter 23

MAN DIES OF OVARIAN CANCER

No, this is not some bogus news headline from the *Onion*. Nor is it a case involving a transgender individual. It is the sad but true story of a thirty-two-year-old man who received a kidney from a fifty-eight-year-old woman who died following cardiac arrest.[1]

The previously healthy-seeming donor had no known history of cancer nor did she show any signs or symptoms of cancer upon her death. She had no evidence of cancer at the time her organs were harvested for donation.

The young man received her right kidney and was placed on some of the newer transplant medicines mentioned in an earlier chapter, namely, tacrolimus, mycophenolate mofetil, and prednisone, to minimize rejection. A few months later, he appeared to be doing well aside from needing some dose adjustments. But around six months after his transplant, he developed fevers, his appetite diminished, he felt nauseated, and his belly became enlarged (in other words he developed *ascites*, an ominous sign that was discussed in chapter 17. CT and MRI scans showed swollen lymph nodes plus a large mass near his transplanted kidney. Alarmingly, this mass was positive on PET scan, raising the suspicion of cancer. Post-transplant non-Hodgkin lymphoma, a fairly common malignancy in the setting of organ transplants was suspected. The immune-suppressing medications were discontinued and his transplanted kidney was removed shortly thereafter.

To everyone's disbelief the pathology returned a diagnosis of ovarian cancer. This was confirmed by karyotyping of the chromosomes, which demonstrated the malignant cells were XX, indicating they were female in origin, whereas all his own cells were XY. Blood tests showed elevation of certain "tumor markers" called carcinoembryonic antigen (CEA) and CA125. While the first marker is elevated in a variety of cancers, the latter is fairly specific for ovarian cancer. A chest CT scan showed that the cancer had already spread to his lungs. Hoping for natural rejection of the cancer, he initially declined specific cancer therapy, but despite the discontinuation of most of his immunosuppressants the cancer progressed relentlessly, eventually spreading to his liver. He had surgery and began chemotherapy with carboplatin and Taxol, a regimen commonly used for ovarian cancer. The malignancy marched onward, and he ultimately died. The autopsy confirmed widespread ovarian adenocarcinoma.

Adding to this tragedy, the other kidney had been donated to another woman, and despite discontinuation of immunosuppression, removal of the contaminated kidney, and initiation of chemotherapy, she, too, developed widespread metastatic disease and succumbed to the ovarian cancer—within just weeks of the death of the male recipient. At the time in 2009, it was unclear why, despite the withdrawal of immunosuppression and removal of their transplanted kidneys, the patients did not reject their transplanted cancers. It was fully expected that both patients should have developed a robust immune response mediated by their *cytoxic T cells*. However, if any such T cell activation occurred in either patient, it was obviously far from robust and was completely inadequate in halting tumor progression and spread. The complete absence of any meaningful anticancer responses despite withdrawal of immunosuppression and removal of the transplanted kidneys portended a dire prognosis for both organ recipients.

As in the previous chapter, one has to wonder how such shocking and appalling occurrences are biologically possible. Returning to concepts from chapter 22, it could be that the original host's immune system had to deal with and "imprison" the original cancer and that a "guards versus prisoners" war had been raging on for years in the donor. A long and drawn-out *immuno-editing* process could have been taking place in the original host—her immune system and the cancer might have "danced" together for years. And each time the cancer learned a clever new move, the host's immune system smoothly kept pace, never letting the cancer dictate the pace or take control. In contrast, the new hosts (irrespective of the immunosuppression) hadn't danced the dance to the same point of maturity. In fact, this dance was brand new and unfamiliar to both recipients' immune systems. So, in these terms, the cancer cells had the clear upper hand because they were "leading" in a dance completely unknown to the new host's immune system.

Ovarian cancer is not the only gynecological cancer transmitted to a man via organ transplantation. In a highly publicized case that evolved into a legal suit, a man contracted endometrial (uterus) cancer following a kidney transplant from a woman with no known history of cancer.[2] As in other cases, the kidney came from a woman who unknowingly had cancer (in this case of the uterus), and the donated kidney appeared healthy and normal. The fifty-year-old donor died of a stroke, and no one knew or suspected she had cancer according to news reports. But autopsy results later showed that she had uterine cancer and that the cancer had spread to her lungs. After the transplant team concluded that the risk of actually transmitting the gynecological cancer was low, the thirty-seven-year-old male recipient elected to keep the donated kidney. Regrettably, he died with cancer about seven months after the transplant.

While men dying of metastatic gynecological cancers is about as weird as it gets, if one scans the medical literature, it becomes painfully apparent that there are copious numbers of cases of cancers transmitted through transplants—including some malignancies that supposedly shouldn't even be possible. For instance, at one time I and many others were under the mistaken impression that malignant brain tumors did not metastasize. Brain tumors include *astrocytomas* (so named because of the star-shaped presumed precursor cells called astrocytes). Astrocytes are one of several types of *glial cells*, which work alongside the more well-known *neurons* in the brain. There are several types of tumors that appear to be derived from glial cells, and these are collectively called *gliomas* (including *astrocytomas, oligodendrogliomas*, and *ependymomas*). The most aggressive primary brain tumor—and perhaps the most malignant of all human tumors anywhere since the prognosis is so abysmal—is a glioma called *glioblastoma multiforme* (GBM). It was long assumed that GBM did not metastasize. It was (and sadly still is for the most part) locally uncontrollable, meaning that despite extensive surgery, intensive chemotherapy, and highly focused radiation therapy, it inevitably recurs right where it started, ultimately taking that person's life. Although there has been some improvement in median survival over the last decade or so, the prognosis remains dismal, and the majority of patients pass away despite all efforts. Many of us in oncology were under the impression that the cancer was fatal entirely because it was affecting the most vital organ in the body—the brain—but was not spreading elsewhere. Once again, the transplant experience has demonstrated the fallacy of this long-held assumption.

Because people who died of brain tumors were considered (and still are considered) at very low risk of transmitting their brain tumors onto an organ recipient, the risk-to-benefit ratio has often been considered acceptable. It was believed that GBM killed patients in such a rapid and decisive fashion that the malignancy didn't even have time to metastasize. Thus, patients with even this most malicious of malignancies were considered acceptable organ donors.

Nevertheless, upon review of the literature I found undeniable proof that GBM and other brain tumors do occasionally metastasize.[3] During an autopsy performed on a young man who died of a malignant astrocytoma, a normal-looking piece of rib was *randomly* sampled and examined. Despite appearing entirely normal, this rib specimen was unequivocally involved with metastatic astrocytoma. The author commented that "what is possible for a rib is possible for a kidney"[4] and advised against accepting donors with brain tumors. This was published in 1981. Despite this stern warning, in a few instances malignant brain tumors have been transferred to new hosts via organ transplants. I came across a case from 1993 in which a donated kidney transmitted a GBM, another report from 1996 describing a GBM transferred via a liver transplant, and then in 2004 a case from a lung transplant. Clearly, GBM and other brain tumors can and do spread beyond the brain—and in rare instances they have even spread to other people.

But clinical examples of primary brain tumors spreading beyond their intracranial confines and metastasizing to extracranial organs are relatively rare. What is not rare is the reverse: tumors of the lungs, breasts, colon, and elsewhere spreading to the brain. Melanoma is one such cancer that has a strong penchant for metastasizing to the brain, and, as mentioned, melanoma has been transmitted to other people via organ donations. Another cancer with a proclivity of spreading to the brain and a high risk of transmission to other people via organ donation is choriocarcinoma. And in many ways the title of "most unusual cancer" has to go to choriocarcinoma.

Before getting to that however, it might be worth pointing out that organ transplantation is not the only way that people have gotten cancer from others. There are even odder ways.

In a recent letter published in the *British Medical Journal*, researchers have warned that the condition they call "XM," or *extracorporeal metastasis*, which has been largely overlooked or dismissed as implausible, might actually be of real concern.[5] Beyond the Tasmanian devils, dogs, clams, and transplant cases lies XM. XM has been observed under often heartbreaking circumstances. For example, in early experiments that today sound downright barbaric, cancer was intentionally inoculated into other people in an effort to induce immunity. It was hoped that such inoculated individuals would generate anticancer antibodies that could then be given to cancer patients and cure them. Most such inoculations failed to do anything. But, to everyone's distress and dismay, in some cases the cancer "took" and ignited in the new host.

One particularly disheartening yet astounding story involved a fifty-year-old lady who was diagnosed with malignant melanoma in 1958.[6] The cancer returned in 1961, and despite a battery of treatments including chemotherapy and a blood transfusion from a patient previously treated and cured of melanoma, her cancer progressed. As a last resort, a tiny piece of her cancer was taken from under her skin and transplanted onto the abdominal muscle of her eighty-year-old, but otherwise healthy, mother. At that time it was believed that the cancer posed little risk of spreading in the mother since it was a transplant from another person (that is, an allotransplant) and should be rejected. It was hoped that the rejection process would generate antibodies that could then be extracted from the mother's blood samples and used to cure the daughter. However, the daughter unexpectedly died the very next day due to a perforated bowel. The mother, having no need for the transplanted tumor, had it surgically removed twenty-four days after implantation. However, adding to this heartrending tale, she developed metastatic melanoma eighty-six days after the procedure. Widespread metastatic cancer ultimately killed her 451 days after the transplant. One curious observation was that for a brief interval, the mother experienced a transient remission where she seemed to be gaining strength and weight and looked to be on the road to recovery. During this short window of remission her gamma-globulin levels (blood proteins that are part of antibodies) were markedly elevated, and one wonders if for a short period of time she actually was generating anticancer antibodies or demonstrating some sort of effective immune reaction. As mentioned, however, the remission was short-lived. Medical science learned the hard way that cancer rarely follows the rules.

Speaking of not following the rules . . . our next chapter will recount some of the famous (and infamous) studies conducted by one of cancer immunotherapy's most controversial pioneers, Dr. Chester Southam.

Chapter 24

MAN'S LIFE SAVED BY MOSQUITO BITE

With a title like "Man's Life Saved by Mosquito Bite," I might have some explaining to do.

After all, throughout history, mosquitoes have caused countless millions of deaths worldwide through familiar mosquito-transmitted diseases such as malaria, yellow fever, dengue, West Nile virus, La Crosse encephalitis, and St. Louis encephalitis, as well as some lesser-known diseases including chikungunya. Even today, malaria alone is responsible for around five hundred thousand deaths annually among a staggering two hundred million cases per year. All told, around seven hundred million people a year contract mosquito-borne illnesses and over one million of those die. If mosquitoes are such a medical menace, how then can a mosquito bite ever *save* someone's life?

Our story begins in the 1950s, when there was still a great deal of mystery surrounding immortal cancer cell lines. As first discussed in chapter 20, unlike normal cell cultures, which live only briefly in the lab, cancer cell lines have no expiration date; cancer immortal cell lines used back then are still in use today. The now-famous HeLa cell line, which was born in 1951, was one such new cell line, and understandably, many people were wondering about the hazards of working with HeLa and similar cancer cell cultures.

Could a lab worker wind up with cancer after a needle stick?

To answer this and many other questions regarding the immunology of cancer, Dr. Chester Southam and collaborators embarked on a series of probing experiments.[1] One early experiment involved injecting terminally ill cancer patients with HeLa cells and other living cancer cells. Although the published paper suggested otherwise, in years to come, the big ethical conundrum was whether these patients *fully* understood what the experiment was really about and what it entailed. In other words, did they truly provide *informed consent* for the clinical trials they participated in?

In the experiments, some *normal* human cells (fibroblasts) were injected into some of the volunteer cancer patients; not surprisingly the cells did nothing. These normal but foreign cells did not "take"—they were promptly rejected by the subjects' immune systems. In contrast, the injected *cancer* cells *did* take and multiply to some degree in most subjects. In some of these subjects, small tumors initially grew but soon vanished on their own by four to six weeks thanks to immune rejection. In other volunteers, however, the tumors grew large enough to require surgical removal. Unfortunately and unexpectedly, in four of the subjects, the tumors grew back despite the surgery. Most tragically, two volunteers ultimately died due to metastases from the injected cancer cells only forty-two and fifty-seven days after the implantation.

These unforeseen results were explained away by the idea that these volunteers had highly advanced cancer—and therefore had severely compromised immunity. According to this concept, those with active, advanced cancer have increased susceptibility to transferred cancers, whereas healthy people do not. Southam and others next tested this hypothesis on a group of volunteer inmates at Ohio State Penitentiary and published the results in the prestigious journal *Science* in 1957.[2] The prison-inmate subjects did know they were having live cancer cells injected into them. Interestingly, neither of the two HeLa cell implants "took" in any of the healthy inmates whereas two out of three HeLa cell implants did establish tumor nodules in the previous experiment involving advanced-cancer subjects. Upon injection of some other cancer cell lines, tumors did begin to grow in the prisoner volunteers just as in the cancer patient subjects. In sharp contrast, however, in these healthy inmate subjects without cancer, the tumor implants invariably started regressing spontaneously by three weeks and completely vanished by four weeks. No uncontrollable, catastrophic cancers materialized in this group.

The results were consistent with Southam's hypothesis that the normal and strong immune systems present in healthy prisoners could readily fend off the injected cancers, whereas the enfeebled immune systems of cancer patients could not. Although there were alternative possible explanations, the observations were compatible with the "immune theory of cancer," which was first introduced in chapter 20 and shall be described in greater depth later.

Although I was aware of Southam's cancer cell injection experiments through mandatory medical ethics education, up until recently I was unaware of his involvement in another experiment that might be of interest to the average American today—especially in the summertime.

In the summer of 1999, a mysterious new disease suddenly struck New York City. It was not until late September of that year, after the death of three people, that the cause was identified as the mosquito-transmitted *West Nile virus*.[3] Infectious-disease specialists and epidemiologists promptly and decisively declared that this exotic disease had never before been seen in the United States. But this was not entirely true. America's first cases of West Nile actually occurred in the 1950s. Specifically, the "outbreak" took place on the Upper East Side of Manhattan in what is now known as the Memorial Sloan Kettering Cancer Center. In this case, the vector was not a mosquito but none other than Dr. Chester Southam.

Southam was intrigued by some studies suggesting that infection with the Russian spring-summer encephalitis virus could cure experimental mouse cancers. Wanting to test this idea in humans but realizing that this particular virus was far too risky for use in people, he began seeking safer alternatives. Eventually, he settled upon the then recently discovered West Nile virus, which at that time he believed was a rather innocuous virus. The experiments were done in special rooms separated from the rest of the hospital by screens to reduce the chances that an errant mosquito might sneak in, become infected, and later escape, thereby setting off an epidemic in New York. In the clinical study, Southam injected West Nile virus into over a hundred terminal cancer patients with the hope that an immunological reaction to the West Nile virus would also result in remission of cancer in these very ill individuals. The real impact on cancer was a bit overblown and sensationalized by the popular press with a *New York Times* headline on April 15, 1952, announcing, "Deep Cancers Temporarily Shrunk by Rare Nerve Virus from Africa." On the other hand, the impact on the cancer patient volunteers was understated, for they were most impressive—in a negative way.

Southam selected the West Nile virus since cases in Africa were known to only cause a brief, mild fever. But that was the situation in healthy African individuals; in this population of immunologically weakened, advanced-cancer patients, things panned out quite differently. A few folks became seriously ill with full-blown West Nile *encephalitis*, or brain inflammation. "Rare Nerve Virus" indeed! Paradoxically, lymphoma patients appeared to show slightly higher temporary remission rates than other cancer patients—but on the same token it was the lymphoma patients who also tended to develop the most serious cases of West Nile itself. The operative word here is "temporary," since none of the cancer patients exhibited any durable remissions, and overall the experiment was deemed unenlightening.

Southam died in 2002 at the age of eighty-two. Although in his West Nile medical papers he stated that "all patients were volunteers and were informed of the experimental nature and the infectious nature of the virus inoculation"[4] much of his work today bears a black mark because of the questions regarding truly informed human subject consent.

Allegedly, in at least one of his experiments in which volunteers were injected with living cancer cells, the volunteers purportedly did not know they were receiving live cancer cell injections, and supposedly Southam did not want to frighten them. In 1963, while he was conducting research at the Jewish Chronic Disease Hospital, colleagues allegedly refused to continue the collaboration without the patients having full understanding of the research and giving their consent. His colleagues complained to the Regents of the University of the State of New York and an investigation ensued. Southam's defense was that he believed the patients faced no real risk of harm and that his tests would provide invaluable medical knowledge. The court disagreed, and in 1964, Southam was found guilty of "fraud or deceit" and unprofessional conduct. This case, and his name, remain forever in the national debate about proper protection of human volunteers.

As described in a 2009 *New York Times* article by Dr. Kent Sepkowitz of Memorial Sloan Kettering Cancer Center, Southam occupies an unenviable position in the history of medicine.[5] Today, Southam is not known so much for his groundbreaking scientific contributions to cancer immunology but rather for drawing public attention to the ethical challenges of clinical research and the safety of human research subjects.

In any case, it was perhaps most fortunate that Southam chose a syringe and needle rather than a mosquito vector in his experiments involving cancer patients. It is now well known that mosquito bites sometimes provoke a severe and extreme reaction in a minority of people who have certain malignancies. Thanks to this phenomenon, however, when we see and treat patients who have odd reactions to arthropod bites and stings, we may uncover the first sign of a serious underlying disease. For example, a large overexaggerated welt in response to an ordinary mosquito bite in somebody who previously did not exhibit any such severe reactions is almost *pathognomonic* (i.e., diagnostic) for chronic lymphocytic leukemia, or CLL. Given how relatively rare the phenomenon is, it was most ironic that I saw this reaction very recently in a patient I was treating for prostate cancer while I was writing this chapter. He was doing fine with regards to his prostate cancer radiation therapy, but one day he came in with a huge welt on his shin. He told me that this has been going on for the past couple of years—ever since he was diagnosed with CLL. CLL is a relatively indolent leukemia, meaning that it often has a slow, smoldering course that might cause few symptoms over many years. People diagnosed early with the slow, smoldering form can generally expect to live over ten years. But CLL is not the only malignancy in which patients exhibit hyperreactive responses to ordinary mosquito bites; amplified reactions have also been reported in mantle cell lymphoma, acute lymphocytic leukemia, and large cell lymphoma among other hematological malignancies, including some that can progress quickly and be rapidly fatal if unaddressed.

In an amazing case from Israel, a twenty-one-year-old previously healthy male soldier got what seemed to be a few mosquito bites on his feet. His doctor first prescribed an anti-itch cream and some antihistamine tablets for his intense itching and swelling. As there was no improvement, antibiotics were started. Despite the antibiotics, the painful itching and swelling spread from his feet into his legs, and he eventually wound up in the hospital emergency room. A thorough physical examination revealed an enlarged spleen, which is always worrisome. Blood tests revealed a very high white blood cell count suspicious for leukemia. Further tests were performed, including a bone marrow biopsy that showed the characteristic "Philadelphia chromosome," or, more precisely, a t(9;22) chromosomal translocation, which will be described in greater depth in chapter 29. The Philadelphia chromosome is often associated with a specific type of leukemia, *chronic myelogenous leukemia*, or CML. As we will discuss in an upcoming chapter, CML was thrust into the limelight when NBA basketball legend Kareem Abdul-Jabbar publicly disclosed that he was diagnosed with this form of leukemia in 2008.[6] CML, previously a fatal leukemia, now has a high cure rate. This high cure rate is thanks to new *molecularly targeted therapies* that work on hyperactive *tyrosine kinases* triggered by the mutant Philadelphia chromosome. In this fashion, a grossly exaggerated reaction to a plain old mosquito bite led to the early diagnosis of a once fatal, but now highly treatable malignancy.

In any case, Chester Southam's original questions and concerns about the potential hazards of working with immortal cancer cell lines ultimately proved to hold merit. Many years ago, at the National Institutes of Health (NIH), a healthy young woman accidentally pricked her hand with a needle that was previously used to draw up living cells from a human colon cancer cell line.[7] The puncture wound seemed superficial, but a couple of weeks later a small nodule formed at the site. Whether her immune system would have eventually completely rejected this transplant of foreign cells is unknown—but given that they were from a malignant cell line, who in their right mind would take chances?

The lump was removed and analysis of the nodule confirmed that it was indeed from the colonic adenocarcinoma cell line she had been working with. Upon examination under the microscope, this tumor nodule mysteriously showed no evidence of inflammation. One would have expected to see *intense* inflammation associated with this infusion of foreign cells. The absence of inflammation suggests that there was in fact no immune rejection of this foreign transplant. She probably did the right thing by having it surgically removed.

This case again highlights the extraordinary capabilities of cancer cells to conceal themselves from immune detection and destruction. I am not aware of any instances in which HeLa cells in particular were accidentally inoculated into a lab worker but this case (and the experiments by Southam on the cancer patients) makes us wonder about the possibility. Frankly, I was completely ignorant of any potential dangers when I was working with HeLa and similar cell lines way back when.

The lab is not the only place where such extracorporeal metastasis or accidental transmission of cancer can happen. The operating room has proven to be another opportunity for cancer to jump from one person to another. Because the prestigious *New England Journal of Medicine* explicitly states that it does not publish case reports, I knew that this one exception had to be most unusual. In this now well-known case, a surgeon nicked his left hand during an operation on a patient with a malignant fibrous histiocytoma (a type of sarcoma).[8] Five months after the incident, a hard, 3.0 cm (1.2 inches), tumorlike swelling was apparent on the surgeon's left palm. The lump was surgically removed, and histological examination (i.e., examination under the microscope) revealed that it, too, was a malignant fibrous histiocytoma. The pathologist who examined the patient's tumor also happened to examine the surgeon's tumor and astutely noticed that their microscopic appearances were uncannily similar and therefore initiated an investigation. As surprising as it was, the detailed analysis confirmed that the surgeon's sarcoma was indeed transferred to him from the patient.

Notably, the surgeon had neither immunodeficiency nor any genetic relationship to the patient. Additionally, the MHC alleles (described in our earlier chapters on organ transplantation) between the patient and the surgeon did not appear well matched. This mode of transmission thus would be entirely unanticipated.

Further adding to the complexity and mystery, the surgeon's transplanted tumor proved to differ slightly from the patient's original tumor and also differed from the surgeon's normal cells. The new tumor was an odd "chimeric" constellation of chromosomes, meaning that some of its chromosomes were contributed by the original tumor and other chromosomes derived from the surgeon himself. This weird chimerism implied that the final cancer cells possessed features of *both* the surgeon's normal cells and the original patient's sarcoma cells.

One has to wonder if this chimerism—this hybridization—could have served as a means through which the cancer cells evaded immune detection and elimination.

After all, through this hybridization, the tumor cells were not *entirely* foreign—they were now also partly his own normal cells. Perhaps his immune system failed to attack and destroy them thanks to this "part-self" aspect. I sometimes wonder if regular cancers (i.e., not those transplanted from one person to another) might somehow use chimerism as a means of escaping detection from our immune systems. For instance, genetic chimerism, wherein a tumor cell appears to have fused with another cell such as a macrophage, might be of great significance in the process known as EMT, or *epithelial-mesenchymal transition*. Some scientists, such as Dr. John Pawelek of Yale, believe fusion between a cancer cell and a macrophage could have an important role in the transformation from a homebound benign tumor into a malignant metastasizing cancer suffering dangerous wanderlust.[9]

Returning again to Chester Southam, his interesting yet controversial studies showed that very advanced, incurable cancer patients did not seem to be capable of rejecting transplanted tumor cells whereas prisoners without cancer readily rejected transferred cancer cells. His overall results were consistent with the concept of some degree of cancer resistance in healthy, cancer-free individuals but a loss of cancer immunity in those with cancer. This leads us to wonder—Could it be that cancer patients are cancer patients *because* they lost their immunity against cancer? An interesting hypothesis.

It now seems that Southam may have been many decades ahead of his time. What he might have discovered was not just a generalized weakening of immunity due to progressive cancer, but perhaps something far more specific. Today in modern immunology we sometimes talk about *Th1*to *Th2* transitions in T cell polarity.[10] We shall describe this in further detail later, but briefly a Th1 to Th2 transition reflects a change from an alert and aggressive state of immune function to a passive state of comparative immunological apathy. In simple terms, the Th2 condition is one of permissiveness, and tumors that *should* be rejected are paradoxically accepted when the Th2 polarity prevails. Although we may never know the cellular and molecular details of what actually allowed Southam's subjects to accept the transplanted cancers, this seems like a plausible mechanism.

Similar findings were observed in rats whose thymus was surgically removed at birth. The thymus plays a major role in the immune system during fetal development and in early life. (In fact the "*T* cells" we have been alluding to throughout the book are named for their embryological site of development, the *T*hymus). Lab rats whose thymuses were removed allowed transplanted tumors to take and grow for a while. In contrast, transplanted tumors would not take and grow in rats with normal intact thymuses. Apparently removal of the thymus of a rat in early life somehow altered its immunity to cancer. While this work was done almost sixty years ago, it remains of great interest to me personally. For well over a decade, I have been trying to figure out the link between the thymus and cancer susceptibility.

About fifteen years ago, I, with my good friend and colleague Dr. Steve Howard along with others at the Johns Hopkins Hospital, published a short paper in *JAMA* (the *Journal of the American Medical Association*) showing that patients with thymoma (a tumor of the thymus) appeared to have a significantly increased risk of various cancers.[11] What was missing from our report, however, was a plausible explanation for this increased risk. We did not have a detailed immunological mechanism to explain our results. Therefore, our findings have always been viewed with some skepticism. Could the surgical removal of the thymus as part of the treatment for thymoma have left these individuals at elevated risk for developing subsequent cancers? Although this initially seemed plausible, the timing was off. In some cases, the additional malignancies appeared very soon after or even before the thymus was surgically removed, making that mechanism unlikely to be the explanation.

However, in the summer of 2015 researchers reported in the *Journal of Immunology* that some thymoma patients have a specific defect in their T cell receptors.[12] The researchers believe that this molecular defect renders them immunologically deficient in certain ways. For instance, patients with thymoma have a higher susceptibility to selected infections, including the parasitic disease leishmaniasis.

Steve and I suspect that while it is probably true that the recently discovered molecular defects in T cell receptors in thymoma patients explains an increased susceptibility to severe cases of unusual parasitic infections such as leishmaniasis, this T cell receptor defect may also be the explanation for our findings of increased *cancer* in thymoma patients. We are presently optimistic that the recently discovered T cell receptor defect found in certain thymoma patients finally provides the sought-after molecular mechanism behind our observations. Additionally, if confirmed, this could prove clinically useful for thymoma patients who possess the particular "*gamma-delta*" T cell receptor defect, since they could undergo more frequent cancer screenings (such as colonoscopies) and thereby detect and cure any cancers they develop before they become problematic.[13]

It is true that transplanted cancer in healthy humans is exceedingly rare and documented by only a small handful of cases. I once wrote that based on the rarity of such cases, it seems that friends and family members of cancer patients, as well as medical caregivers of cancer patients, "need not be unduly concerned with the remote possibility of catching cancer."[14]

Nevertheless, today I would certainly not advocate a cavalier attitude. Alluding to the surgeon and the NIH lab worker who in the routine course of their occupational activities, acquired transmitted tumors (or as they call it, extracorporeal metastases, or XM) Drs. Yuri Lazebnik and George Parris of Yale and Montgomery College respectively, mentioned in the *British Journal of Cancer*, "such accidents—pricking yourself with a needle or scalpel previously exposed to cancer cells—are not common, but by no means extraordinarily rare in the operating room or the laboratory, implying that these two reported cases of transmission may be exceptional only in that XM was noticed, documented and communicated to warn the broader biomedical community."[15] Of course, recommending caution in the operating room and in the lab might seem like a common sense suggestion warranting no need for repetition. We all know that lab workers and surgeons encounter far more contagious things they could contract than transmissible cancers. However, I recall one example of cancer communication that was *most* unexpected. And it was far more distressing, since no matter how careful we might be, encounters with this next mode of transmission are all but inevitable.

So, in my opinion, the *most* bizarre and disconcerting mode of cancer spread of all was through mosquito bites! There is one example in which an outbreak of a leukemia (or "reticulum cell sarcoma," as it was described) mysteriously began spreading through a colony of lab rodents at the NIH about fifty years back.[16] Upon closer inspection, it appeared that the epidemic was yet another example of transmissible cancer. Apparently clams, dogs, and Tasmanian devils are not the only victims of contagious cancers. But this cancer was not spread by sexual contact or fighting. Instead, it appeared that several members of this highly inbred colony of Syrian hamsters contracted the leukemia through mosquito bites. If this is really what happened, one could imagine that in a highly immunosuppressed individual (e.g., an AIDS patient or an organ transplant recipient on high-dose immune suppressing medications) the same bizarre mode of transmission might happen in humans as well. Although this particular strain of lab rodents was highly inbred and the whole colony was very closely related genetically, it nevertheless raises some scary questions. For instance, if you have an identical twin who happens to have leukemia and who gets bitten by a mosquito, what might happen if that same mosquito then bites you? Can you get that same leukemia? What if the mosquito bit your non-identical twin sibling who has cancer? What if it were just a regular old (non-twin) brother or sister? What about just a random person with malignant cells in his or her blood? Could a bug bite transfer those cancer cells to us somehow, and through some immunological fluke, could those cells survive and give us cancer? While this has never been reported, the Syrian hamster colony case gives one pause. Maybe it *has* happened. We just never connected the dots and solved the cold case.

Presumably, the malignant cells were transferred from rodent to rodent as passive passengers on the mouth parts of the mosquitoes (rather than internalizing and going through a complicated replication cycle within the mosquitoes as is the case for malaria parasites). Regardless, the prospect of cancer being spread from animal to animal or person to person via mosquito bites has to take the cake as the most perplexing and disturbing mode yet encountered!

While transmission of cancer via bug bite has to be the most frightening concept in this chapter on transmissible cancers, I said earlier that the title of "most unusual cancer" goes to *choriocarcinoma.* In its "ordinary" form, choriocarcinoma can affect men as a testicular cancer and women as an ovarian cancer. In rare instances, it can arise in other places including the brain (above the pituitary gland or near the pineal gland) where it is a type of intracranial *non-germinoma germ cell tumor*. Even less frequently, choriocarcinoma might originate in the chest, lungs, or elsewhere. Regardless of it specific origin, choriocarcinoma ranks high among the most rapidly growing cancers, perhaps taking top honors. Choriocarcinoma cells produce large amounts of a chemical called *human chorionic gonadotropin*, or HCG, which serves as a "tumor marker" that doctors can measure in the blood. Such tumor markers can be monitored following cancer therapy to determine if the treatment has worked. For example, physicians might follow blood levels of HCG for a number of years after curative chemotherapy for an HCG-producing choriocarcinoma. Some readers might be familiar with HCG as the compound that is also detected in pregnancy tests. But there is one variant of choriocarcinoma which, in my opinion, is the weirdest of all cancers. This form of cancer develops from what was supposed to be a fetus.

Chapter 25

MOLES, MOLES, AND MORE MOLES

It seems that we have been talking an awful lot about moles in this book!

For starters, Daniel of chapter 1 had a **mole** on his skin that had gone bad and became a malignant melanoma. Next, the abscopal effect, which was also discussed in chapter 1, was first described by Dr. **Mole** in 1953. Additionally, in previous chapters we went into some depth about cancer-resistant rodents called **mole** rats. Although we haven't discussed it much yet, readers will recall from chemistry that a **mole** is Avogadro's number (6.02×10^{23}) of atoms or molecules. In this chapter, we will be discussing yet another form of "mole"—and this mole might be odder than all others.

One day during my medical school clerkship in obstetrics and gynecology, I was reviewing my morning sign out and learned that I would be meeting a most interesting patient that day. She was a thirty-five-year-old woman from Ethiopia diagnosed with a *hydatidiform mole*. Having little knowledge of the condition, I figured I had better read up on this material quickly before meeting her. What I was about to learn was among the most fascinating diagnoses I had ever encountered in my medical training to that point.

Pregnant women occasionally can develop familiar cancers such as breast cancer, cervical cancer, melanoma, or lymphoma along with a variety of others. Management of these common cancers during pregnancy can be particularly challenging since the standard treatment regimens (which might include high dose radiation therapy, for example) may not be appropriate due to the need to protect the developing fetus. In addition to this information about cancer in pregnancy, I learned that pregnancy can be associated with a very different and most uncommon medical condition—something called *gestational trophoblastic disease* (GTD). It happens when a fertilized egg (a zygote) doesn't become a fetus. Instead it becomes a neoplasm.

Learning that a fertilized egg can turn into a tumor rather than an embryo or fetus was mind-boggling enough. But additionally I learned that although most cases of GTD were benign, sometimes the neoplasms were aggressive and malignant. The most common type of GTD is called a "molar pregnancy," or *hydatidiform mole*. This is what I was told my patient had. Upon meeting her and interviewing her I learned there was far more to the story. First, I found out that she was quite familiar with this condition since two of her sisters had molar pregnancies. In fact, at the time, she knew a whole lot more about hydatidiform moles than I did! She told me that she had had a hysterectomy in Ethiopia a short while ago for a "complete mole," but her HCG levels remained disturbingly high[1] and they recommended methotrexate chemotherapy. Apparently, her case of gestational trophoblastic disease was a bit more advanced than initially suspected. Since she had relatives in the United States and was planning to move here, this is where she elected to have her treatment.

Just what is gestational trophoblastic disease? Well, first, let's review what trophoblasts are. The development of an embryo begins when a sperm fertilizes an egg to create a zygote. This fertilized egg, or zygote, then undergoes several cell divisions to develop into a solid ball of eight cells called a *morula* at about three or four days post-fertilization. Cell division continues and by around five days post-fertilization, the next stage is reached. This is called a *blastula*, which, in addition to being larger, differs from a morula in being hollow. In addition to a hollow cavity called the *blastocoele*, the blastula has two main identifiable components: an *inner cell mass* (or ICM) and an outer layer of cells collectively called the *trophoblast* that surrounds the inner cell mass and the blastocoel. The inner cell mass is destined to become the embryo and the trophoblast gives rise to the placenta.

While this is going on, hormonal changes in the mother prepare the endometrium (the inner lining of the uterus) to receive the blastula. Incidentally, fertilization itself normally does not happen in the uterine cavity but instead usually occurs way up near the very beginning region of the fallopian tube not far from the ovary. From this site of fertilization, the zygote or morula are propelled by tiny, wiggling "whips" (cilia) on the cells lining the fallopian tubes. The constant flailing of these cilia produces a current directed toward the uterine cavity, which ushers the zygote, morula, or blastula along. Typically, by about five to six days following fertilization (which is the time it usually takes to reach the uterus), the fertilized egg has developed beyond the zygote and morula stages and has become a blastula, which begins to embed itself into the endometrial lining of the uterus proper. This is its new home and this is where all subsequent development occurs.

This process of the blastula embedding itself into the maternal tissues is called *implantation*, and as we will discuss in the next chapter, implantation has some striking similarities to malignant processes. For one, a key aspect of implantation is *invasion* into the maternal tissues—and invasion into neighboring tissues is a key characteristic of cancer. But while in cancer, invasion is the root of all evil, invasion during early pregnancy is necessary for success. The invasion of the trophoblast into maternal uterine tissue is a critical phase in early pregnancy. In fact, certain serious disorders of pregnancy trace back to this early step. Failure of the trophoblast to adequately invade deep enough seems linked to some cases of *preeclampsia* (which will be described in greater depth in a later chapter), whereas too firm an attachment may lead to *placenta accreta* (or *placenta increta* and *placenta percreta*, if the depth and degree of invasion is even greater); these latter conditions can be associated with severe hemorrhaging during delivery.

As mentioned, the trophoblast (derived from Greek *tropho*: to feed, and *blastos*: germinator) is the outer layer of the blastula and, as the name indicates, provides nutrients to the developing embryo; it later develops into a large part of the placenta. The trophoblast cells are the first cells to differentiate from the fertilized egg. Gestational trophoblastic disease develops when something goes seriously wrong with this process. Instead of forming a placenta, the trophoblast becomes an abnormal growth.

GTD is actually a spectrum of related disorders. The most common type, the hydatidiform mole is rare in the United States at about 1 in 1,500 pregnancies but can be tenfold more frequent in certain other regions such as Southeast Asia.[2] Hydatidiform moles tend to be more common in teens and in older women of childbearing age. Also a diet low in carotene might be associated with an increased risk of molar pregnancy. In a hydatidiform mole, instead of becoming an embryo and placenta, the abnormal trophoblast cells grow into a mass of tissue resembling a cluster of grapes that fills the uterus. Most molar pregnancies are benign, and the woman can be cured by evacuation. Some moles, however, show clearly invasive and aggressive behavior. Unlike normal, well-behaved trophoblasts that develop into a healthy placenta, in *malignant trophoblastic disease* (also known as *gestational trophoblastic neoplasia*), the invading cells just don't know when to stop. Such conditions are called *invasive moles* if they just invade locally but do not spread. On the far end of the gestational trophoblastic disease spectrum is the type of molar pregnancy that evolves into a clearly invasive, aggressive, and metastatic cancer—this is choriocarcinoma. Untreated, this cancer will spread very quickly, often involving the brain and leading to death.

What seems so truly strange about this whole situation is that many molar pregnancies contain *no maternal DNA*. In other words, they are *completely paternal* in origin. In fact, when this is the case and the mole is completely paternal, it is termed a *complete mole*. This can occur when two separate sperms fertilize an "empty" ovum (an empty ovum is a maternal egg which has no viable maternal DNA) or when a single sperm fertilizes an empty ovum and duplicates. An *incomplete* or *partial mole* on the other hand, does have maternal DNA but is fertilized by two sperm, leading to *triploidy* (three copies instead of the normal *diploid* condition of two copies of each chromosome).

Most cases in which remnants of the mole keeps on growing after evacuation or hysterectomy are derived from complete moles. It turns out that about 15 percent to 20 percent of complete moles will evolve into malignant gestational trophoblastic disease. Put another way, some cases of choriocarcinoma are *completely paternal* in origin. This means that although they are entirely foreign, they are quite capable of invading, overtaking, and metastasizing widely throughout another person's body (namely the mother). Apparently, if foreign male tissue is offered to a woman's body in the form of a fertilized egg (even if that egg doesn't have any of her own DNA in it), that male tissue is granted unrestricted license to do what it wishes—even if what it wishes to do is become a dangerous cancer.

Just how cells derived from another person's body can go immunologically unchecked and kill a woman is still a mystery. It is understood that under the ordinary circumstances of pregnancy, it is imperative that the mother turns off her immune reaction and allow the fetus to develop. Such an immunological holiday is necessary for the survival of the embryo and fetus. But outside the context of a pregnancy, foreign male cells should be vehemently rejected. Nevertheless, when stealthily snuck into a woman via conception (abnormal as it may be), these male-derived cells pretending to be a normal pregnancy can turn malignant and effectively evade the woman's immune system, potentially overrunning her body and killing her.

For benign molar pregnancies (that is, hydatidiform moles) close follow-up is essential. This is because after suction evacuation or even hysterectomy, there is the possibility of residual disease—as was the case with my patient. Technically, she had *persistent trophoblastic disease*. Her mole was invasive, and it was suspected that she might even have choriocarcinoma given her persistently elevated HCG levels. This was many years back, and PET scans were not yet available. But her chest x-rays and CT scans did not show any extracranial metastases, and a brain MRI did not reveal brain metastases. We gave her methotrexate chemotherapy, which has a high rate of success for all forms of malignant trophoblastic disease. Her HCG levels began falling quickly, and she seemed well. Although I will never know for sure since I had to move onto my next medical school clerkship, it appeared that she was cured of her disease.

Let's return to why choriocarcinoma gets my vote as the world's weirdest cancer. As I mentioned, some cases of choriocarcinoma are completely paternal in origin. This means the cancer started out from a man's sperm that fertilized an "empty egg" that had none of the woman's DNA in it. Again, it is absolutely amazing that such a cancer, being completely foreign and derived exclusively from another individual's body, can be tolerated at all, let alone spread uncontrollably in an unrelated woman's body. In this bizarre situation, all the MHC barrier rules are tossed out the window! A husband cannot simply donate a kidney to his wife unless they are a close MHC match, so why can his sperm-derived choriocarcinoma be so readily passed along? As we shall see in an upcoming chapter, some very strange things happen during pregnancy, and there are several uncanny parallels to what goes on in cancer. In the context of gestational trophoblastic disease, it seems that as long as it was initially disguised in the form of a *conceptus* (a fertilized egg), alien and malignant male tissue remains completely invisible to the woman's immune system. But some choriocarcinomas have taken this invisibility advantage to the extreme and have gone even further in their journeys.

There have been a few cases of choriocarcinoma (which recall, arose from a man's sperm cell) that grew unchecked and uncontrolled in a woman's body by masquerading as an embryo and ultimately killed that woman through metastasis to the brain, and then managed to get into other people's bodies via organ transplants! In one report, a twenty-six-year-old woman died at seven months into her pregnancy from a massive cerebral hemorrhage (a type of stroke).[3] The fetus had multiple malformations and did not survive. Inspection of the donor's heart, liver, pancreas, and kidneys did not reveal any abnormalities, and these five organs were transplanted into four separate recipients. All four recipients soon developed choriocarcinoma, which was traced to the female donor. Quite probably, she had brain metastases from her choriocarcinoma that hemorrhaged and led to her demise. The liver recipient developed intestinal metastasis and died from hemorrhage despite an initial response to chemotherapy. The heart recipient also had an initial remission but later progressed to widespread metastatic disease. The recipient of one kidney quickly had that kidney removed and was treated with actinomycin chemotherapy. She was still in remission two years later when the authors put the report together. The recipient of a combined pancreas-kidney transplant was also still in complete remission two years after chemotherapy without removal of the grafted organs.

This was not the only instance of a choriocarcinoma making a getaway in this fashion. However, one important common denominator was that in every case where choriocarcinoma was transmitted via organ transplant, the donor died of cerebral hemorrhage. Therefore, before organs are harvested and transplanted from a woman of childbearing age who died of hemorrhagic stroke, choriocarcinoma must be ruled out.

What seems to have happened in these tragic but mesmerizing cases is that rather than a sperm uniting with an egg to form an embryo, placenta, and fetus, the sperm-only derived conceptus instead developed into a fatal malignant cancer that the woman's immune system just could not see. In addition to masquerading as an embryo and fooling the mother's immune system, the cancer also disguised itself as a stroke and managed to fool doctors who enabled it to sneak itself into other people's bodies. In this manner, a human cancer, choriocarcinoma, has (*almost*) done what devil facial tumor disease, canine transmissible venereal tumor, and the hemic neoplasia of clams have succeeded in doing—jumping from one body to another in its quest to evade death and attain eternal life.

In an upcoming chapter we will explore just how it is that cancer can so effectively conceal itself from our immune systems. What is its secret for so cleverly evading immune detection? Well, as we discussed in this chapter, there is one other biological entity that can hide from our immune systems—the fetus. We shall soon delve into that a bit deeper. But before going there, I wish to digress just a bit to discuss choriocarcinoma's nearest rival for the title of world's weirdest tumor. Although I have voted for choriocarcinoma to take the prize, another truly crazy case of cancer recently gave it a serious run for the money. As in the choriocarcinoma case, this cancer managed to wrangle its way in and grow unbridled in another person's body. This time however, rather than starting in another *person's* body, the cancer started in another *animal's* body.

Chapter 26

TUMORS THROUGH THE WORMHOLE

In chapter 6, we discussed "perfect parasites" and noted peculiar parallels between tumors and parasites. Cancer, most obviously when contagious, behaves remarkably like a parasite. Immortal communicable cancers that jump from host to host, transiently using one body after another before moving on to the next victim, invite legitimate comparisons to genuine parasites. But what if we turned the scenario around? What if a parasite got cancer?

In chapter 13, we saw that clams, of all creatures, can get cancer. Well, if a clam can, can a worm? And what if that worm was a parasite and got cancer while inside a person's body? I wonder about such scenarios often. If you do too, you needn't wonder much longer.

Although tapeworms are fascinating in their own right (to some people), it turns out that tapeworms tie in directly to the main theme of our book—cancer and the immune system.

As has been discussed repeatedly, people with enfeebled immunity are more susceptible to certain cancers. In addition, they are of course highly susceptible to various infectious processes, including unusual parasitic infections. One such example is echinococcosis, infection with the dog tapeworm *Eccinococcus granulosus*. As the name indicates, the definitive host normally is a dog or some other canid; sheep are common intermediate hosts. By definition, the worms mature into reproducing adults within the gut of their definitive host, the dog, which acquires the infection by consuming intermediate hosts that harbor larvae in their organs. Within the intestines of definitive hosts like dogs, the tapeworms cause few problems. In contrast, in intermediate hosts life-threatening disease may develop. It turns out that man can serve as an intermediate host for this type of tapeworm.[1]

Unlike the enormous fish tapeworm of chapter 6, with thousands of proglottid segments, the adult dog tapeworm is a relative runt at around a half centimeter in length with only three puny proglottids. In humans, as intermediate hosts, the worms never attain even this adult stage. The worms remain as larvae, typically ensconced in the liver and lungs, just waiting until the day that a definitive host eats this temporary transportation vehicle. Fortunately, dogs don't often eat people. Unfortunately, however, this allows the cysts to slowly enlarge within a person's liver or lungs to the point where they can become symptomatic. While the growing cysts can provoke pain and discomfort in the upper abdomen or chest (along with nausea, vomiting, or coughing), the real risk is associated with rupture of a large cyst, which can lead to extreme allergic reactions or even death.

One curious commonality between cancers and dog tapeworm infestations is that in order to thrive, both the *E. granulosus* parasites and malignant cells subvert our immune systems to maintain a "favorable" (to them, that is) immunosuppressive environment in their vicinity. What is even more fascinating still is that the molecular mechanisms of the immunosupression used to yield this tolerance are quite similar between the worms and tumor cells. In essence, liver cancer and cystic echinococcosis in the liver both seem to flourish in the liver upon reduction of *Th1-mediated immunity* (which will be described in greater detail in a later chapter). Experiments have demonstrated that the simultaneous coexistence of *E. granulosus* infection and breast cancer in the same animal transforms the liver into a preferred site for metastasis of breast cancer.[2] It appears that the parasites produce copious quantities of certain inflammation-related chemical signals that allow breast cancer cells to flourish within the liver. An alternative explanation is that breast cancer cells could be encouraged to travel to the liver more readily upon exposure to such parasite-induced chemical mediators. Either way, in these animal experiments it did seems likely that the increased frequency of breast cancer metastases to the liver were attributable to the inflammation associated with dog tapeworm cysts in the liver.

On the other hand, and in stark contrast, other researchers have recently argued that infection with the dog tapeworm might in fact elicit a *protective* effect against cancer, especially during the early phases of tumor development.[3] This could be the result of an old trick backfiring. Parasites, and other infectious organisms, often attempt to trick our immune system by mimicking normal human host cells. In this fashion, the immune system gives them a pass and doesn't harass them. It appears that certain cancers share surface features, or antigens, that are quite similar to dog tapeworm antigens. In this manner, the *cancer* might get a pass by mimicking the tapeworm! Whether it is the tumor mimicking the worm or the worm mimicking the cancer is uncertain but the end result is the same—the immune system doesn't aggressively attack either. The same group of researchers had earlier shown that the coexistence in a given person of both cancer and dog tapeworm liver cysts was very rare.[4] To explain this, the researchers proposed that antigens found on both the worm and the cancer cells incited an immune reaction that might not have otherwise materialized. While the cancer cells and the worms might each ordinarily fly under the radar, the combination of the two of them might alert the immune system to something fishy going on. And in this manner, a cancer that otherwise would have gone unnoticed gets eliminated. An interesting idea indeed.

This could possibly prove to be of importance if we could figure out just how tapeworms and tumors subvert our immune systems and quell immune reactions against them. And if we learn how to reverse parasite-induced immunosuppression, perhaps a clinically valuable anticancer therapeutic avenue will open up.

Speaking of tapeworms and cancer, there was a recently reported medical curiosity that might top them all. Basically, a patient had a tapeworm infection and the *worm* got cancer![5] The forty-one-year-old patient in Medellín, Colombia, had poorly controlled HIV infection with a low CD4 count and a high viral load. Clinically, he had what surely seemed to be cancer. He had weight loss, fever, and fatigue for several months and had multiple swollen lymph nodes in his neck. A CT scan showed numerous masses in his lungs (some measuring up to 4.4 cm in diameter), along with nodules in his liver and adrenal glands. The scan also revealed suspiciously enlarged lymph nodes in his neck, chest, and abdomen consistent with widespread metastatic cancer. Biopsies of a bloated neck node and a lung mass showed something startling. Although they had many of the classic characteristics of cancer, the "tumor cells" ultimately proved to be nonhuman in origin! Detailed examination revealed that they were bizarre tapeworm stem cells with multiple mutations—a worm cancer of sorts. And that tapeworm cancer behaved as all cancers are wont to; it spread throughout the patient.

The patient was infected with *Hymenolepis nana*, the dwarf tapeworm. The dwarf tapeworm is the most common human tapeworm with up to seventy-five million carriers, and in some regions, the prevalence among children ranges to as high as 25 percent.[6] Fortunately, most infections are asymptomatic, but the dwarf tapeworm is a rather unique tapeworm species that can complete its life cycle in the small intestines of humans without the need for an intermediate host. This "autoinfection" process (where the worms eggs hatch in the gut, larvae develop and mature in the gut, and mature adults reproduce new eggs in the gut) can go on for years and lead to a very high parasite burden. This is particularly true in immunocompromised hosts. The patient in question was one such individual, as he had a weakened immune system due to advanced HIV infection. Somehow, one of the tapeworms got "cancer," and its transformed cells went haywire in this immunosuppressed host. One would expect that invading cells from another species (especially if as foreign as a playhelminth invertebrate) would be promptly rejected by the human immune system. Of course, all bets are off when the cells in question possess malignant features that aid immune evasion, especially if the host's immune system is severely compromised by HIV/AIDS. As the authors mentioned, "Human disease caused by parasite-derived cancer cells is a novel finding. The host–parasite interaction that we report should stimulate deeper exploration of the relationships between infection and cancer."[7] I most certainly agree. I would also contend that another area that should be explored in greater depth is the relationship between pregnancy and cancer. So that is where we shall go next.

Chapter 27

THE IMPOSTER

Mr. Lee: "A lot of people have cast this endeavor as something terribly monstrous—a startling example of how science and medicine have simply gone too far. From my perspective, however . . . there is nothing more natural and beautiful on this earth than that. This is something that I've always wanted to do."

Janice: "But surely you understand why some people find the idea of a pregnant man disturbing?"

—from *The First Male Pregnancy*, online at http://www.malepregnancy.com/mingwei/

Readers might not initially appreciate the connections between cancer and a pregnant man.

Similarly, the lessons we can learn about cancer from a cow serving as a surrogate mother for a buffalo might not be obvious at first blush. However, I believe there are very important and clear clues in the obstetric clinic that could ultimately prove revolutionary in cancer medicine. In a nutshell, there may very well be multiple underappreciated and unclarified protective mechanisms whereby the fetus is protected from the immune system at all costs, and cancer has stolen the playbook.

Many call it "the miracle of life." Whether tenderly alluded to as "bun in the oven" or less charmingly, a "gravid uterus," pregnancy represents continuity. Survival of the species. But others express a less adoring view—and I won't repeat these less than flattering descriptions of pregnancy easily gathered during a five-minute Internet search! One common disparaging theme I found most unsympathetically alludes to the fetus as an invasive "parasite" that greedily takes advantage of the mother's generosity.

Is it fair, or even reasonable, to mention a fetus in the same sentence as a parasite? One represents the future of humanity, while the other is the bane of our existence. What, if anything, can possibly be gained from making such uncomplimentary comparisons? Well, since pregnancy is the key to survival of any species, protection of the embryo has always been a biological prime directive. Therefore, it should come as no surprise that successful parasites have learned to emulate the immune-evading strategies present in pregnancy. For an infectious agent to survive in a new host, it has two broad options. It can choose to directly fight and overpower the immune system or it can try to hide from it. Given the strength of the immune system, it might seem that directly facing such a foe is an unwise approach for all but the most robust pathogens. And as we have seen, even Ebola first hides from our immune systems and later, when the immune system finally targets its enemy and charges, Ebola uses a judo throw to use the immune system's momentum against us. On the other hand, given that the immune system has two deep-rooted vulnerabilities—that it won't attack a fetus and that it must distinguish "self" from "non-self"—pathogens might be expected to exploit these chinks in the armor. Indeed, many pathogens do evolve along these lines. If one thinks about this, what better strategy is there in order to conceal itself from immune attack than to pose as an innocent fetus?

Now if malignant tumors employ these same traitorous tactics, we would have a serious fight on our hands indeed.

But is this just nonsense? Do tumors really conceal themselves from the immune system? And if they do, are they successfully concealing themselves by impersonating an embryo? A quick glance at the names of some common blood tests might give us a candid and chilling clue. The most common pregnancy test measures levels of a hormone called HCG (*human chorionic gonadotropin*) in the blood or urine. Another blood test routinely done in pregnancy is AFP, or *alpha feto-protein*, which, as the name indicates, is something that might be elaborated by a developing fetus. High levels of AFP correlate with certain birth defects including spina bifida and other abnormalities of the spine, brain, and belly. Conversely, lower than normal levels are associated with Down syndrome.[1]

I routinely order HCG and AFP blood tests. But as I am an oncologist, not an obstetrician, one might wonder why. The answer is that these two blood tests have dual citizenship. They are big players in both the obstetrics clinic and the cancer clinic. HCG, for instance, is a "tumor marker" elaborated in abundance by certain cancers. I might measure HCG levels to determine if any cancer was left behind following surgery for testicular cancer. Additionally, HCG levels can be followed year after year to verify control or confirm recurrence following curative chemotherapy or radiation therapy for certain germ cell tumors. Similarly, blood levels of AFP can be monitored in other cancer patients (e.g., liver cancers and certain testicular or ovarian cancers). And there are many others besides just these two tumor marker examples. For instance, *CEA* levels are routinely monitored in patients with colon cancers; it doesn't require much imagination to envision a cancer-embryo connection with a tumor marker called CEA, or *carcino-embryonic antigen*.

Based on this simple circumstantial evidence I think one can make a reasonable argument that there must be some profound connection between embryogenesis and carcinogenesis. But one will need far more than just crude circumstantial evidence to venture a truly convincing argument. So, let's consider another line of reasoning.

It is intuitively obvious to everyone that the immune system is indispensable. We are constantly barraged by bacteria, parasites, fungi, and viruses that might otherwise be deadly were it not for our body's defenses. As we are painfully aware thanks to the relatively recent worldwide pandemic of HIV/AIDS, a weak or inadequate immune system can render us susceptible to many common infections that never bother us under ordinary circumstances. Diseases such as cryptosporidiosis, uncontrolled toxoplasmosis or candidiasis, and cryptococcal meningitis are conditions that HIV patients are all too aware of. However, these are not caused by bizarre, unfamiliar foreign germs. The germs that cause these conditions are all around us every day.

What is different in those with HIV/AIDS is that the defense we have in place against these ordinary agents—the immunity we take for granted—has become incapacitated. Most of us encounter the fungus, *Candida albicans* ("yeast") every day. But ordinarily our immune systems eagerly and easily fight it off. Babies and those taking high doses of steroid medicines, however, cannot repel the daily onslaught and are susceptible to *thrush*, or oral yeast infections. As babies grow older and stronger or when patients discontinue steroids, their immune systems gain or regain normal function. And a normal immune system readily vanquishes the daily challenge from this ubiquitous fungus. Regrettably, because of the HIV/AIDS pandemic, we are all now keenly cognizant of just how critically important an intact immune system is. We need that army of immune cells to keep us safe and healthy.

Then again, an overzealous army of potent but mindless warriors can be dicey. Obviously, strict rules and regulations must be imposed on that army of immune cells. Without restraint, the immune system, always ready for battle, might go about "looking for a fight." And if it can't find one, it might start one. In this way, a hyperactive but undisciplined immune system could lead to autoimmune disease. Autoimmunity occurs when impetuous immune defenders, who are always poised and ready to fight, turn against their own kind. Common examples of autoimmune diseases include ITP (*idiopathic thrombocytopenia*, a cause of platelet deficiency that leads to easy bruising and bleeding), *Graves disease* (a common cause of hyperthyroidism), *Hashimoto disease* (a cause of hypothyroidism), multiple sclerosis (a neurological disorder), and "lupus" (also known as *systemic lupus erythematosis*, or SLE). In these, and dozens of other autoimmune disorders, our indispensable friend—the immune system—has turned against us and become a bitter enemy. At this time, understanding autoimmune disorders and quelling fervid immune systems though "biologics," or modern immune-modifying drugs, is one of the most exciting and rapidly evolving branches of medicine.[2]

All this leads to a notion we don't really think about until things go wrong. Another essential function of the immune system is preservation of the species.

Aside from warding off unwelcomed invaders, the immune system must also have a means of keeping the always-alert soldiers from attacking the one thing that *must* be spared: the embryo. For the survival of the species, if anything must be safeguarded from an overzealous, temperamental, and indiscriminate immune system it is the developing embryo. The developing embryo must bear a stern warning sign blatantly shouting "*Noli me tangere*!"—touch me not! Based on this line of reasoning, the mindless, attacking, *army* branch of the immune system, as essential as it is for defending us against pathogenic intruders, must take a back seat to an even more powerful *regulatory* branch of immunity. Could there be an elite squad of stalwart "royal guardians" in the placenta whose sole purpose there is to keep the peace between the aggressive soldiers and the foreign fetus? If so, for survival of the species, these guardians/regulators *must* be mightier than the army.

Given that a fetus is only "half-self" (meaning that it is half-mother but also half-father) one might predict that the mother's immune system would mount an aggressive and unrelenting attack. After all, this semiforeign body is invading its home turf. Thankfully it does not. Precisely why not, and exactly how the embryo is granted sanctuary, is of great interest.

For years it was believed that the womb or uterus itself was what granted the embryo protection from the immune system. The womb was thought of as an immunologically "privileged" site. The immune system was not allowed to trespass here. Some still think that this is the key reason why the embryo and fetus remains free from immune system attack. But if the womb were the sole explanation for how the embryo/fetus escapes immunological destruction, one would be hard pressed to explain *ectopic pregnancies*.

Ectopic pregnancies are pregnancies in which the embryo is not located in its designated home, the uterus. Rather than implanting itself in the welcoming uterine lining, or endometrium, the newly fertilized egg embeds itself someplace else. There are many places where ectopics can implant but 95 percent are found in the fallopian tubes. Ectopic pregnancies can also grow in the ovaries, the cervix, or even the abdominal wall lining (the peritoneum). Although these erroneous sites are foreboding territories in contrast to the hospitable uterine lining, somehow the developing embryo commandeers the entire environment and still makes a home for itself. Seemingly summoned, new blood vessels work their way there to feed the demanding new embryo. All might seem normal save the unusual location. But make no mistake about it—ectopic pregnancies not normal. They are true surgical or medical emergencies. Unfortunately many end very badly. The growing embryo demands, and receives, nourishment and oxygen from the mother's body by inducing new blood vessels. These abundant new blood vessels would be readily accommodated in the normal womb environment, but anywhere else they pose a significant risk of hemorrhage. The uterus was made to expand and accommodate the ever-enlarging fetus. The fallopian tube, in stark contrast, cannot. It is a stiff little cylinder that does not stretch to accommodate anything. Instead, with an expanding embryo inside, it catastrophically ruptures. With all the new embryo-feeding arteries and veins in place, uncontrolled bleeding is inevitable. If unrecognized and unaddressed, ectopic pregnancies can result in the death of the mother. Similarly, the cervix, ovaries, and other sites of ectopic implantation are not built to stretch and accommodate a growing fetus; without emergency surgery they, too, can tear and hemorrhage. Although rare and (and definitely not medically recommended) some ectopic pregnancies have survived to full term and viable fetuses have been extracted from the abnormal implantation site by the equivalent of cesarean section.

The path of the egg or ovum out of the ovary and into the uterus is surprisingly circuitous. Near the end of each fallopian tube is an expanded area to accept the released egg. This is adjacent to but not perfectly attached to the ovary. There is no "hermetic seal" between the two structures and the newly released egg can stray off course. An egg can actually veer so far astray that instead of entering the fallopian tube it winds up in the abdominal cavity. While this is certainly no place for a lonely ovum, if a sperm happens to also make it up and out of the fallopian tube and into the abdominal cavity, the two can pair up and set up shop right there. The resulting fertilized egg embeds itself into the lining of the abdominal cavity somewhere. This could be on the external surface of the intestines or it could be some place along the abdominal wall. Another site ectopic pregnancies have landed on is the *omentum*, an apron of fatty tissue draped over the intestines. Regardless of precisely where, the abdominal cavity is definitely no place to raise a fetus! And it is extremely dangerous for the mother. Nevertheless, all these abnormal non-uterine sites of implantation and subsequent embryonic growth serve as proof that it *cannot* be the uterus alone that bestows the embryo with shelter from immune attack. These ectopics are well away from the warmth and comfort of the uterus yet they are still considered "off limits" as far as the immune system is concerned.

As they are so potentially deadly to the mother, most ectopic pregnancies are terminated as soon as they are discovered. Nevertheless, the fact that these peculiar extra-uterine settings exist, and on rare occasion have ended as viable full-term pregnancies, make one wonder: Could a *man,* perhaps "get pregnant" and carry a fetus to full term? Safely, absolutely not. But in theory, maybe. The omentum, abdominal wall, and lining of the colon have all served as sites of ectopic embryonic implantation and development. Even though they pose serious risk of bleeding and death during fetal development and especially during surgical "delivery," there have been a few viable intra-abdominal ectopic fetuses in women.

So, since we have decided that it is not the uterus that confers the immune sanctuary to the embryo and fetus, maybe a hormonally prepared *male* could serve as a "surrogate mother." Could a fetus develop within a male intra-abdominal cavity? If hormones were provided to fully duplicate the hormonal milieu of pregnancy, might a man be able to harbor and nurture a fetus to full term? Although it might seem possible, skeptics cogently argue that if this *were* possible, given all the marginally ethical undertakings in the history of medical science, someone would have already done it. As there are no verifiable claims about male pregnancies, one might understandably relegate the conception (if you'll forgive my choice of words) to the realm of science fiction.

The organized, elaborate, and convincing "Man(?) of the Year" website that detailed the pregnancy of "Mr. Lee" was nothing more than a clever ruse.[3] This hoax fooled a lot of folks and got a lot of attention. One could even read the fabricated comments (which sadly sounded pretty much like what one would expect from genuine, unfiltered Internet commentary!) on both sides of the argument. Some would praise Mr. Lee's courage for going public and bringing new life into this world whereas others would preach fire and brimstone for this egregious act against God and nature. In the end however, like the movie *Junior* starring Arnold Schwarzenegger, Mr. Lee's pregnancy was pure fantasy.

Fantasy or not, this little aside about ectopic and male pregnancies was intended to show that it *cannot* be the womb alone that provides the immune sanctuary for the embryo. No matter where the developing embryo resides, it bears a bright, bold badge stating loud and clear, "HANDS OFF!" and the immune system seems to respect that warning.—

Just how "immune from immunity" is the embryo? Is this immune privilege really that profound? As I alluded to earlier, for the survival of the species, this immune safe haven must be extreme and complete. Just how extreme, you ask? Well, let's think about this. Back when I was first considering all this, some of my colleagues argued that it was the female uterus that bestowed the immune asylum. We have ruled this hypothesis out above. Then others argued that the immune system's lackadaisical attitude is because the fetus is half-self. They conceded that the embryo is half-father, but the fact that it is also *half-mother* is why the maternal immune system doesn't attack it. I cannot accept this. For if this were the explanation, how can surrogate motherhood be possible? In such cases there is *no* genetic relationship between the surrogate mother and the developing fetus. It is not half-self (or semi-allogeneic to use the technical term). Yet surrogate motherhood is possible. Thus, one must conclude that semi-allogeneity also cannot be the explanation.

In fact, the "hands off!" policy is intact not just when mother and fetus are completely unrelated genetically, but also seems in place when the mother and fetus are entirely different species! One would logically predict that the immune system would absolutely forbid any such thing, and I know that many of my colleagues were most skeptical of such possibility. Nevertheless, it can and has happened. One documented account of successful interspecies surrogate motherhood involved a full-term river buffalo calf being born from a swamp buffalo surrogate mother after transfer of an in vitro fertilized embryo.[4] Astonishingly, the two species don't even have the same chromosome count! The river buffalo has fifty total chromosomes whereas the swamp buffalo only has forty-eight. This certainly isn't the first and only successful account of such interspecies surrogate pregnancies. The immune laxity is amazing in such situations, but for successful full-term development a few tricks may be required. For instance, Spanish ibex embryos are aborted early when injected alone into the womb of a goat, but if a goat embryo is simultaneously injected, the ibex fetus is sometimes left alone and can reach full term.[5]

In an attempt to address the endangered status of giant pandas, researchers explored the potential of using a cat as a surrogate mother.[6] In this case, to overcome any immunological barriers, a cat embryo was also inserted into the surrogate cat's womb. (Actually the whole thing was even more complicated—the panda DNA was obtained from muscle cells and transferred into rabbit egg cells before ultimately being transferred to the cat uterus). Somewhat surprisingly, the trick worked. The cat's immune system, perhaps lulled into slumber by the adjacent feline embryo, did not recognize and eliminate the panda embryo. Unfortunately, this experiment ended badly as the mother cat died of pneumonia before delivery. Although ploys sometimes have to be used to deceive the immune system and avoid rejection, these tricks do often work and such stories make one wonder if extinct creatures may someday be resurrected by capitalizing on this immunological leniency toward embryos. If adequate amounts of intact mammoth DNA could be found and extracted from fossil specimens preserved in permafrost, and if that nuclear DNA could somehow be microinjected into an elephant egg, and if that egg could then be implanted into a surrogate mother elephant's womb, could we possibly revive the species? There are an awful lot of "ifs" there, but from an immunological point of view, this might not be utterly outlandish.[7]

This whole "paradox of pregnancy" has piqued the curiosity of scientists and physicians for well over half a century. Sir Peter Medawar, "the father of reproductive immunology," first systematically raised the question of how pregnancy so effectively evades maternal immunity in 1953.[8] Medawar is also regarded as the "father of transplantation" thanks to his pioneering research in tissue grafting, which served as the basis for later organ transplantation research. For his efforts he was awarded the 1960 Nobel Prize in Physiology or Medicine along with Sir Frank Macfarlane Burnet for their groundbreaking work in the understanding and applications of immunological tolerance. One particular study topic of Medawar's that proved to be of great clinical relevance was *acquired* immunological tolerance. This research was critical in the clinical management of immune rejection of extensive skin grafts after severe burns. Medawar's research refocused from simply coping with a fully developed immunity reaction to actively attempting to dampen the immune response itself. In this manner, suppression of immune rejection opened the door to successful organ transplants.

However, analogies between pregnancy and cancer significantly predate even Medawar. Something called the "embryonic rest hypothesis" of the origin of cancer was first proposed by French gynecologist Joseph Récamier in 1829.[9] (Incidentally, Récamier was the first to coin the word *metastasis* to describe the spread of cancer. Prior to his time, doctors thought that each additional tumor was arising anew in a cancer patient. Récamier ventured the hypothesis that these were not all brand new tumors but instead were bits and pieces of the original mother lode breaking free and settling elsewhere in the body.) The embryonic rest concept was later expanded by German pathologists Rudolf Virchow and Julius Cohnheim in the mid to late nineteenth century.[10] According to this intriguing hypothesis, "embryonic rests" (bits of embryo somehow left behind during the developmental process that failed to mature into fully formed tissues and organs) could someday come back to haunt an individual as malignant tumors. In another early variation on the theme, around the turn of the nineteenth century, Scottish embryologist John Beard published a paper in the *Lancet* on the "trophoblastic theory of cancer," which holds that the trophoblast, an early embryonic structure that later forms much of the placenta, as described in chapter 25, possesses many characteristics of cancer.[11] For example, both the trophoblast and cancer are invasive, induce the formation of new blood vessels, and produce similar specific markers such as human chorionic gonadotropin (HCG). And as we discussed in chapter 25, when things go horribly wrong with the trophoblast, as in gestational trophoblastic disease, it can actually directly develop into cancer.

Building on these ideas, Medawar surmised that the embryo's perplexing degree of freedom from maternal immune attack was due to one of three things: the uterus is an immunologically privileged site; the fetus itself is immunologically inert, meaning it has nothing that the mother's immune system can "see" yet; or the mother's immune system is out of commission during pregnancy.[12] We have discussed above that the first suggestion—that the uterus is an immunological sanctuary site—cannot be the sole explanation. The other two explanations have also been proven incorrect or incomplete. Another proposed explanation was that the placenta imposes an anatomical, or physical, barrier that cannot be breached by cells of the mother's immune system.[13] This assumes that the immune system is unaware of the enclosed embryo or, if it is aware, simply cannot get at it thanks to the physical barrier. But the fact is that there is an awful lot of immunological activity going on at the maternal-fetal interface (i.e., the placenta). This is evident right from the start. The presence of numerous immune cells at the implantation site of the blastula has been considered as a proof of a response by the maternal immune system to the embryo. The maternal immune system *is* acutely aware of this foreign tissue. However, instead of an offensive and aggressive assault on the embryo, these numerous immune cells are not attacking the developing embryo. Rather they are there to stand guard. Today, there is no doubt about it—*the placenta is intrinsically an immune regulatory organ!*

In an article from the Mayo Clinic, the authors succinctly summarized the modern perspective on the analogy between cancer and pregnancy.[14] To begin, both the early embryo and cancer proliferate rapidly. Both express similar molecules to enable this, such as those involved in the IGF pathway discussed in the context of Laron syndrome in chapter 19. Other molecules in common between the early rapidly growing embryo and cancer include *survivin* and *telomerase*, the latter of which was discussed in the context of HeLa cells in chapter 20.

Additionally, both cancer and the early embryo are highly invasive. As the embryo grows, it must invade into maternal tissues and anchor itself. In doing so it creates an active interface between itself and the maternal tissues in order to gain access to oxygen-rich and nourishing blood vessels. This interface is called the placenta. Invasion is key to a successful pregnancy and is a hallmark of cancer. Once again the cellular processes involved in the early phases of pregnancy are oddly familiar to an oncologist. Epithelial-to-mesenchymal transition, or EMT, involves many molecules and ultimately ends in a loss of cell-to-cell contact inhibition.[15] We have spoken about loss of contact inhibition as a proposed mechanism for the enigmatic cancer resistance of naked mole rats.

As the early embryonic cells that form the trophoblast invade further,

Author's Note:
Images with extended captions can be found in the appendix on page 295.

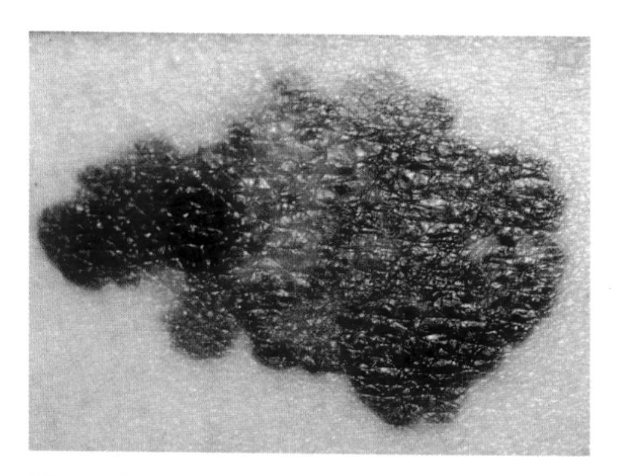

Figure 1:
A cutaneous melanoma.
Photo from the National Cancer Institute.

Figure 2:
William Coley (1862–1936) in 1892.
Photo from Wikimedia Commons.

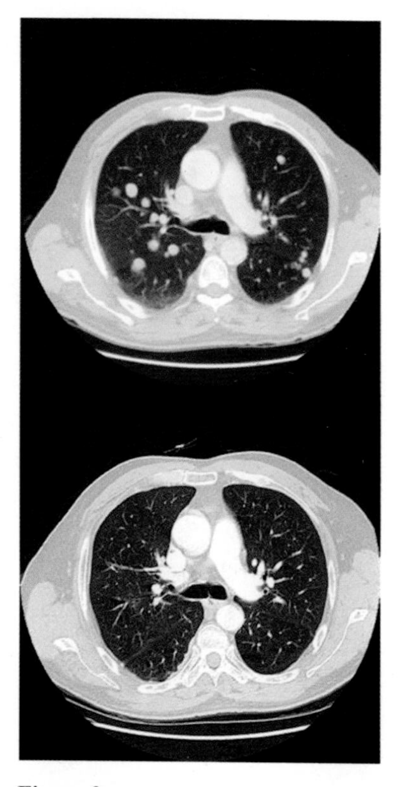

Figure 3:
An example of the abscopal phenomenon. *Photo from Michael Lock et al., "Abscopal Effects: Case Report and Emerging Opportunities,"* Cureus *7, no. 10 (October 7, 2015): e344, doi:10.7759/cureus.344.*

Figure 4:
Tasmanian devil in a defensive stance, at Tasmanian Devil Conservation Park, Tasman Peninsula. *Photo by Wayne McLean, from Wikimedia Commons.*

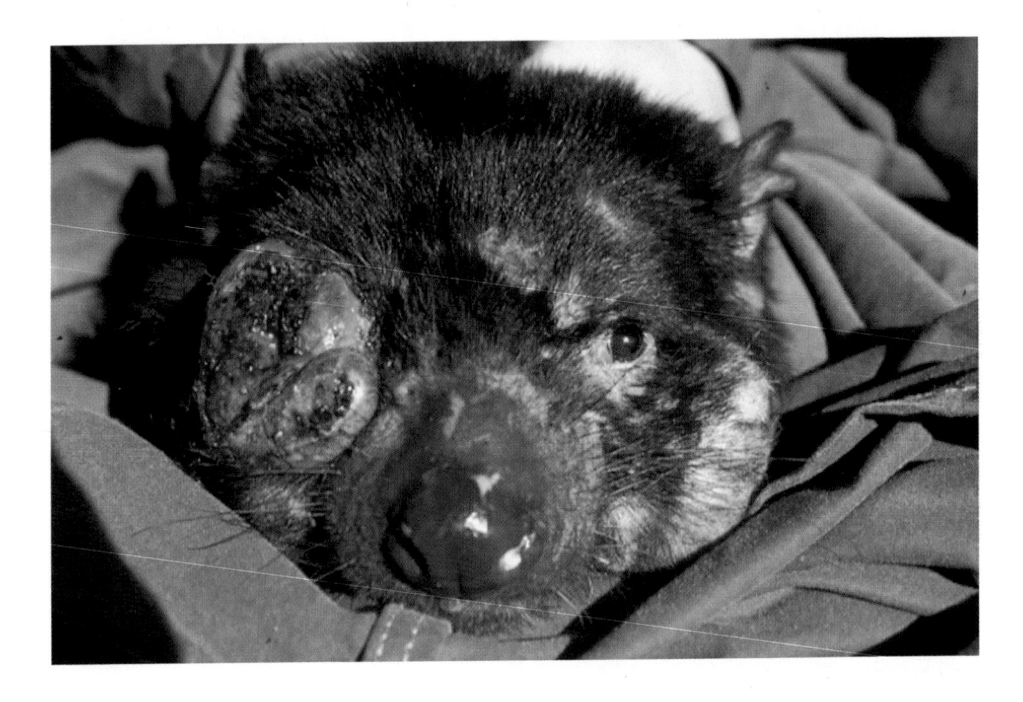

Figure 5:
Tasmanian devil with devil facial tumor disease. *Photo by Menna Jones, from Wikimedia Commons.*

Figure 6:
The last thylacine. *Photo from Wikimedia Commons.*

Figure 7:
Golden jackal.
Photo by Mariomassone, from Wikimedia Commons.

Figure 8:
Robin's pincushion, a tumor induced by a wasp. *Photo copyright Anne Burgess, from Geograph.org.uk.*

Figure 9:
Cuckoo wasp. *Photo by Alvesgaspar, from Wikipedia.*

Figure 10:
Artist's depiction of a gamma ray burst. *Photo by ESO / A. Roquette.*

Figure 11:
A soft-shell clam. *Photo by Kirsten Poulsen, from Wikimedia Commons.*

Figure 12:
"Santa," a great white shark with a tumor. *Used with permission from Sam Cahir / Predapix.*

Figure 13:
Naked mole rat in a zoo. *Photo by Roman Klementschitz, from Wikimedia Commons.*

Figure 14:
Palestine mole rat. *Photo by Bassem18, from Wikipedia.*

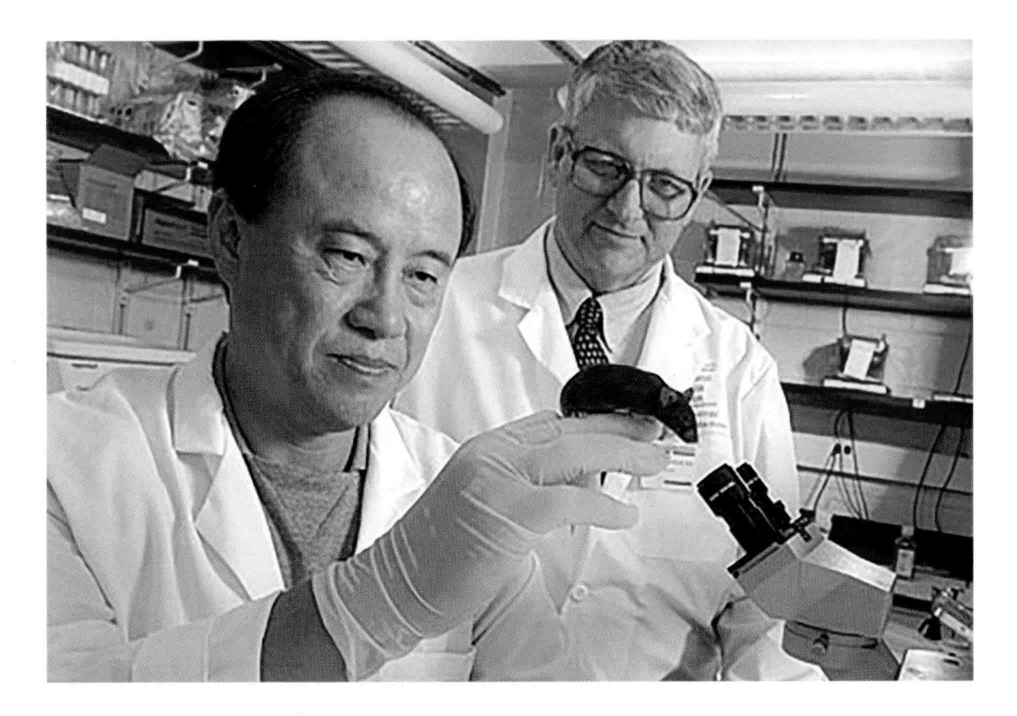

Figure 15:
Dr. Zheng Cui and Dr. Mark Willingham with "mighty mouse." *Image from* Popular Mechanics.

Figure 16:
Homo floresiensis, the "hobbit." *Photo by Ryan Somma, from Wikimedia Commons.*

Figure 17:
Drs. Arlan Rosenbloom and Guevara-Aguirre with some of "the little women of Loja." These Ecuadorean individuals with Laron syndrome exhibit a short stature but also an extremely low incidence of cancer. This protection from cancer is thanks to a cellular defect in the growth hormone receptor of patients with Laron syndrome, which leads to very low levels of insulin-like growth faction I (IGF-1) in their blood. Laron syndrome is a genetic trait that is passed along in an autosomal recessive fashion. *Photo courtesy of Dr. Arlan Rosenbloom.*

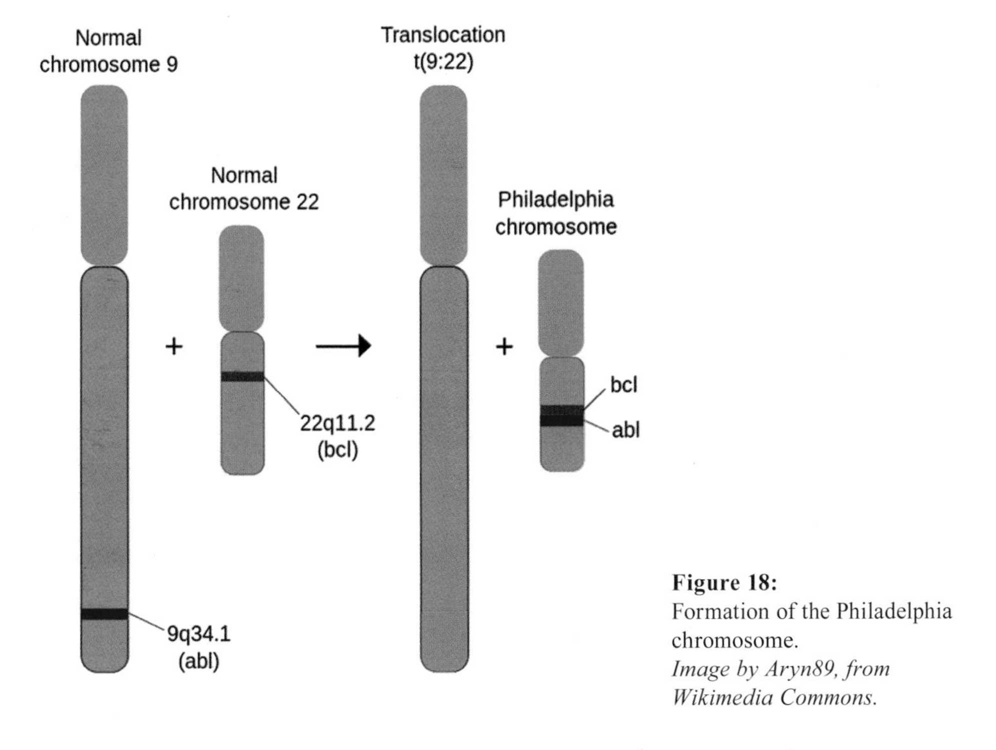

Figure 18:
Formation of the Philadelphia chromosome.
Image by Aryn89, from Wikimedia Commons.

Figure 19:
A Syrian hamster. *Photo by Peter Maas, from Wikimedia Commons.*

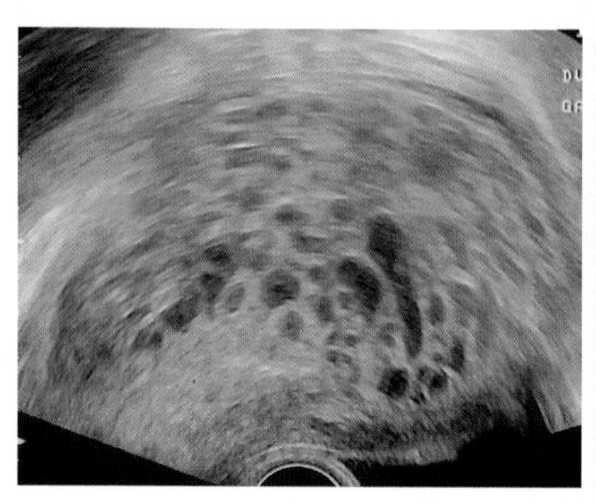

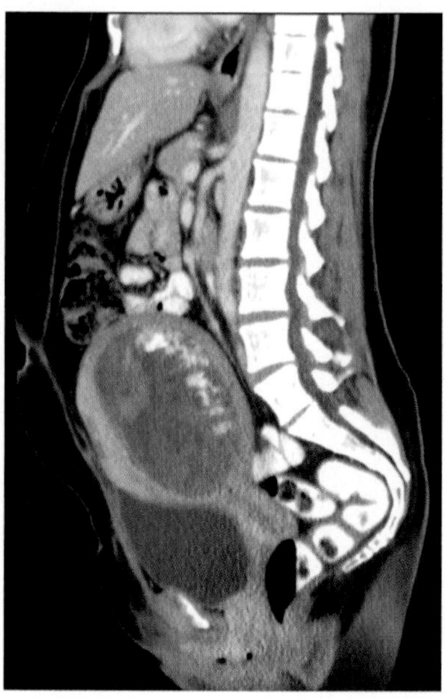

Figure 20- a & b:
Hydatidiform mole or molar pregnancy. A: CT image. *Image by Hellerhoff, from Wikimedia Commons.* B: Ultrasound image. *Image by Mikael Häggström, from Wikimedia Commons.*

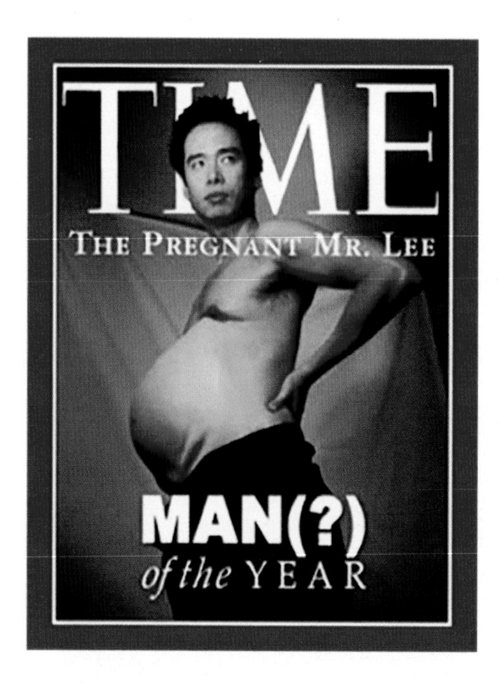

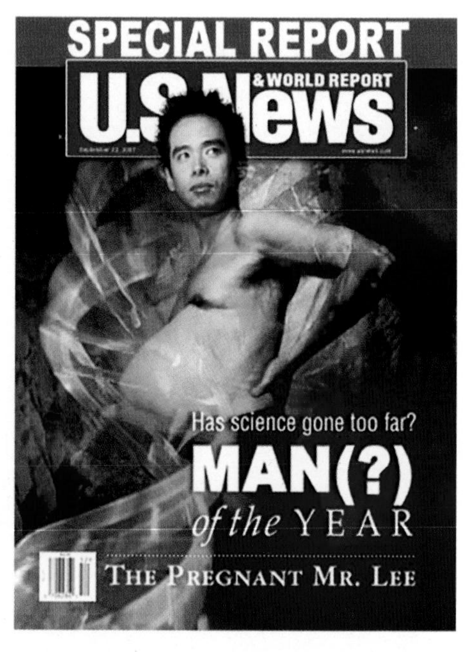

Figure 21- a & b:
"Mr. Lee," an ingenious Internet ruse. The website has been up since 1999, suggesting a rather prolonged pregnancy for poor Mr. Lee! *Images from http://www.malepregnancy.com.*

Figure 22:
An African elephant. *Photo by Godotl3, from Wikimedia Commons.*

Figure 23:
"Harpooning the Greenland Whale," 1876. *Image courtesy of the Freshwater and Marine Image Bank of the University of Washington.*

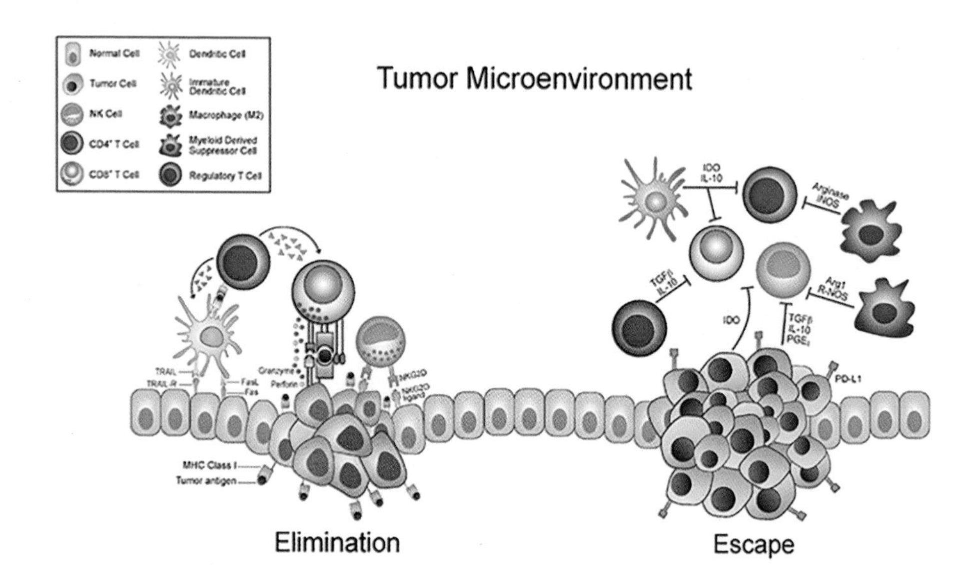

Figure 24:
Tumor microenvironment. *Image from A. M. Monjazeb et al., "Immunoediting and Antigen Loss: Overcoming the Achilles Heel of Immunotherapy with Antigen Non-Specific Therapies,"* Frontiers in Oncology *3, no.197 (2013). doi: 10:3389/fonc.2013.00197.*

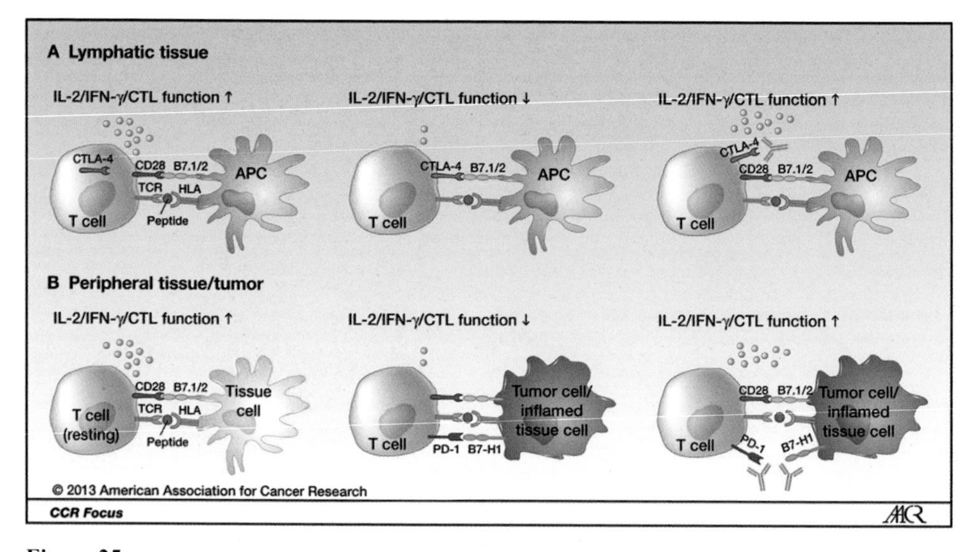

Figure 25:
T cells and tumors. *Image from Patrick A. Ott, F. Stephen Hodi, and Caroline Robert. "CTLA-4 and PD-1/PD-L1 Blockade: New Immunotherapeutic Modalities with Durable Clinical Benefit in Melanoma Patients,"* Clinical Cancer Research *19 (October 1, 2013): 5300.*

they gain an ability to directly form blood vessels. Normally, only blood vessel precursor cells have the ability to form blood vessels. This *vasculogenic mimicry*, or capacity for non-blood-vessel cells to turn into blood vessels, is seen in the early embryo as well as in certain cancers such as melanomas. A key molecule in this process is Mig-7.[16] Mig-7 is involved with invasion and vasculogenic mimicry in cancer and may also be expressed by trophoblast cells during creation of the placenta. Most interestingly, this is the *only* known expression of Mig-7 outside of cancer.

The creation of new nurturing blood vessels, or angiogenesis, is obviously critical for the growth of a developing fetus and, as we discussed in the context of shark cartilage, is a key component of cancer progression as well. As mentioned in chapter 14, central and familiar molecule in angiogenesis is VEGF, or vascular endothelial growth factor.[17] This molecule is targeted by the drug bevacizumab (Avastin) in cancer therapy. However, VEGF also is highly expressed on trophoblast cells and appears to play a crucial role in the creation of a firm connection to the maternal blood supply in embryos.

Although these and other parallels exist between the developing embryo and cancer, the most important and relevant commonality is that they both amazingly and effectively hide from the maternal immune system.[18] There is still much to be learned but finally we are gaining a basic understanding of some of the underlying mechanisms. Hopefully, if the methods whereby the semi-allogeneic fetus eludes the mother's immune system can be fully elucidated, new ways to both defeat cancer and protect fetuses in mothers prone to miscarriage might soon be discovered.

As stated earlier, in the past, the placenta was simply considered a passive mechanical barrier that physically protected the embryo and fetus from the mother's immune system. Today we know otherwise, and recent data has revealed that up to 40 percent of maternal cells in the placenta belong to the innate immune system.[19] The placenta is undoubtedly an immunologically active organ. As will be discussed later, new evidence now indicates that, much to our chagrin, *cancer, too, is an immunologically active entity*. The activity in both cases involves a set of highly potent guardians of the embryo or tumor. The most abundant immune cells in the placenta are a specialized form of cells called *uterine natural killer cells*, or uNKs. Interestingly, these uNKs are different from ordinary natural killer (NK) cells in that they do not express a molecule called CD16 on their surfaces. This loss of CD16 might be caused by high levels of TGF-β in the placental environment. As we shall see later, the *tumor microenvironment* is also loaded with TGF-β, and this could be the cancer's way of converting cancer-killing NK cells into cancer-guarding cells.[20] In the placenta, these uNK cells have a highly *protective* role and work through a variety of mechanisms. They are there to protect the fetus. It might be that an inadequate number of uNK cells contributes to multiple miscarriages and a pregnancy-associated condition called *preeclampsia*. Preeclampsia, or "toxemia of pregnancy," causes high blood pressure and too much protein in the urine (*nephrotic syndrome*). Preeclampsia accompanied by seizures is a very serious disorder called eclampsia. The exact cause of preeclampsia remains unclear, but many suspect that there is a malfunction wherein the immunological truce between mother and fetus is broken.[21]

In addition to the abundant uNK cells, there are many other representatives of the immune system present in force in the placenta, including macrophages, dendritic cells, and special T lymphocytes known as *regulatory T cells*, or *Tregs* (often pronounced, "tee-regs"). These regulatory T cells have only relatively recently been discovered and, as their name indicates, are imperious supervisors of other members of the immune system. These could be the hypothetical "royal guardians" I alluded to earlier in the chapter. For the sake of survival of the species, regulatory T cells must be able to tame and control the ever-eager attackers and keep them at bay. Women with a deficiency of regulatory T cells are prone to preeclampsia and spontaneous abortions.[22]

In recent years, regulatory T cells have also been found lurking around tumors. Here however, unlike in the placenta, they are not serving us well. As we shall see in an upcoming chapter, these regulatory T cells have been tricked by the tumor into protecting it as though it were a fetus. And as these are the elite regulators whose job it is to protect the fetus from all comers, these are fierce guardians!

Cytotoxic T lymphocytes are a chief component of our *adaptive* immune defense and boisterously participate in attacks against bacteria, viruses, and other unwelcomed intruders.[23] But in the vicinity of the embryo, these cytotoxic T cells are transformed into fatherly protectors of the fetus. Similarly, *macrophages* are elements of our *innate* immune systems that upon encountering intruders and infected cells, literally swallow them whole. But in the neighborhood of the fetus, these lions become lambs. Upon swallowing certain trophoblast cells, macrophages transition from *M1-polarization* to *M2-polarization*.[24] What this means is that these macrophages that were previously obsessed with killing foreign enemies are now elite members of the fetal guardian squad. And as members of this royal guard they secrete IL-10 and TGF-β—cytokines that further contribute to this new tolerant attitude among all who are present at the placenta.

This new tolerant attitude can be described as a *Th2 environment*, an immunological environment that preaches peace and tolerance. This Th2 environment replaces the previously hostile *Th1 environment* where any intruders were most unwelcomed and viciously attacked. Here in the Th2 world of the placenta, the foreign fetus is (most appropriately) treated as family. [25]Most *inappropriately* however, as we shall discuss later, the vicinity surrounding most malignant tumors is also a highly tolerant Th2 world. Cancers, rather than being ruthlessly attacked by cytotoxic T cells, macrophages, and the rest of the army, are instead treated like guests! Little baby guests! Molecules that are involved in the Th1 to Th2 transformation (or treachery, depending on the context), such as NF-κB, could someday serve as targets in cancer therapy.[26]

Perhaps the most striking and important point in our entire story is the extreme "Hands Off!" policy that both embryos *and cancers* enjoy when it comes to immunological assaults. Naturally, the embryo and fetus must be given absolute asylum from immunological attacks. But it does seem that cancers somehow "trick" our immune systems into granting them immunological asylum as well. It is almost as if the cancer innocently pleads, "Please don't kill me! I'm just a baby!" and our immune systems fall for the deception, hook, line, and sinker!

The cancer not only fools the immunological army into not attacking it, but also suckers the all-powerful immunological regulators into enforcing the *noli me tangere* policy. In essence, cancer, our bitter enemy, has recruited the protective guardian of our species. Our soldiers and our royal guard have forsaken us, to fight on cancer's side. This would certainly explain why cancer has proved to be a most formidable foe in the clinic.

Yet, all hope is not lost. It is most enlightening to remember that the permissive Th2 environment of pregnancy is not maintained forever. Late in the third trimester, there is a reversal, and the Th2 polarity is restored back to the normal Th1 condition.[27] This allows immune cells that were once eager to eliminate any foreign enemies, but restrained during pregnancy, to once again punish unwanted germs following the birth process. This reversion back to a state in which unwelcomed intruders are expelled is exactly what we want, and need, in cancer therapy.

The fact that many molecules common to both fetal protection and cancer protection are now being discovered is of great significance. Such research could prove invaluable in obstetrics, where it might lead to better ways of protecting fetuses and preventing miscarriages in women whose fetal protection system is imperfect. And of course, such research could open new doors in the cancer clinic. Inhibiting the recently identified "hypnotic molecules" could enable the immune system to wake up from its trance and do what it wants to do—destroy its enemy.[28]

Breaking the spell would permit the immune system to realize that it has not been protecting an innocent baby. Most embarrassingly, it had been duped into defending a tumor! Cancer immunotherapy might unleash an angry, vengeful, and open-eyed immune system on its mortal enemy. Maybe for the first time in history, the balance has been shifted.

These are some bold statements. What evidence do we have to support such assertions? In the next few chapters, I will review some of the basic science of oncology. We will first explore the cellular origins of cancer and then briefly touch upon the molecular details. Next we shall transition to *tumor immunobiology* and then finally return to the promise of *clinical cancer immunotherapy*. But it is very easy to get lost along the way and lose track of our ultimate goal. For me, as a clinician, the goal is a cure for cancer. As odd as it might sound, it may be highly instructive to remain cognizant of a pronounced dichotomy between *cancer, the curious biological phenomenon* and *cancer, the dread disease.* It is all too easy to get lost in the woods of cellular and molecular oncology and lose track of the clinical aim of defeating cancer the disease. Cancer biology has proven to be one of the most complex scientific disciplines on Earth. And paradoxically, gaining an increased understanding of the intricate details of cancer biology at the cellular and molecular levels has not consistently yielded as much in the way of clinical treatments as one would have imagined (at least not yet anyway). Nevertheless, it might be possible to take significant steps in addressing *cancer, the dread disease*, even if we have not yet mastered a comprehensive understanding of *cancer, the curious biological phenomenon.*

Chapter 28

COMPETITION: THE CAUSE OF THE CELLULAR DISEASE

Population, when unchecked, increases in a geometric ratio. Subsistence increases only in an arithmetic ratio. A slight acquaintance with numbers will show the immensity of the first power in comparison to the second.

—**Thomas Robert Malthus,**
An Essay on the Principles of Population

Cells ordinarily organize themselves into recognizable tissues, which in turn organize themselves into clearly identifiable organs. For example, hepatic cells organize themselves into discrete and identifiable liver tissue, and liver tissues in turn amass themselves into a clearly defined organ—the liver. Similarly, in a highly concerted fashion, various cells will arrange themselves into cardiac tissues, which organize themselves into a clearly defined organ—the heart. All the cells that make up these organs and tissues must interact with each other in an intricate and coordinated fashion to permit proper function. For example, cardiac muscle cells must work in perfect harmony to permit synchronized beating and pumping of blood. Disruption of this harmony, as in a cardiac arrhythmia, leads to uncoordinated, dysfunctional heart beating and possible death.

All cells in the human body are expected to collaborate with one another and respect each other's boundaries. Normal cells show respect for each other, not crowding each other or encroaching on one another's territories. For instance, kidney cells remain organized into kidney tissue and remain confined to the paired organs in the back just under the lowest ribs. Cooperating with each other and their neighbors, they remain content in their designated position and don't invade the nearby muscles, ribs, liver, or pancreas.

From this perspective, cancer is a *cellular disease*.[1] It is the inevitable final result of the gross violation of the one basic principle all cells should abide by: cooperation. Unlike the highly orchestrated and constructive behavior of normal cells, cancer cells ditch the concept of cooperation and choose to compete. They compete with each other, and they compete with normal cells. In any biological system, whether individual cells or organisms in a large population, where there is competition there is natural selection. And where there is natural selection there is evolution. In fact, NASA once went so far as defining life as "a self-sustaining chemical system capable of Darwinian evolution," underscoring the inextricable link between life, competition, and evolution.

In a population of predators competing for prey, over time there will emerge faster, stealthier, stronger, and craftier predators as they compete amongst themselves. To counter these upgrades, increasingly agile, more effectively camouflaged, more heavily armored prey will survive and predominate. Survival of the fittest. This outcome of constant competition seems intuitively obvious and is the basis for Darwinian evolution by natural selection.

Individual cells within tissues or organs are not supposed to compete. For the sake of the *entire organism*, they must all cooperate in a collaborative, peaceful, respectful manner. In this regard, a cancer cell is a rogue cell. Thus, cancer as a disease is characterized by a bastion of competitive, uncooperative, self-centered rogue cells. This is why cancer is considered a cellular disease. Cancer cells do *not* look out for the body as a whole. Instead, these subversive elements can undermine the whole organism through their selfish, competitive behavior. Rather than honoring each other's boundaries, a classic characteristic of cancer cells is their tendency to trespass. In fact, this tendency to invade can serve as a definition of *malignant*, as opposed to *benign*, tumors. Malignant tumors disrespect their neighbors and invade their property. While all tumors are, by definition overgrowths of cells, a *benign* tumor of the breast, for example, might remain in the breast as a small lump for many years, generally minding its own business and not causing trouble. In stark contrast, a *malignant* lung tumor, for instance, will have no qualms about crowding its normal neighboring lung tissue, obstructing airways and agonizingly eroding into adjacent ribs.

In order to constantly grow, the rapidly reproducing cancer cells must have access to a constant supply of nutrients. They must compete among themselves and with their normal neighbors for these limited resources. To accomplish their goal of incessant growth and domination, malignant tumors often evolve a strategy that confers another competitive advantage. They send out signals that entice new blood vessels to sprout and feed them. These resultant new blood vessels allow the cancer to grow at an alarming rate in some cases. Some malignancies exhibit disturbingly fast "doubling times" (the time periods required for a twofold increase in size), but there still is a big discrepancy between how fast cancer cells can grow in the lab versus in a person. Certain cancer cells might exhibit very rapid doubling times *in vitro*, but such rapid growth might not be actualized *in vivo* because the tumor literally outgrows its own blood supply. As a result, the tumor core dies and grows rotten from lack of oxygen and needed nutrients—it is necrotic. By inducing the growth of a new blood supply, the tumor perimeter, in contrast, grows without restraint. Unlike in the lab where the scientists can provide all the needed nourishment for continued tumor expansion, in a human body the tumor growth rate is constrained by this limitation in nutrients and oxygen. The limited resources lead to more intense competition and malignant masses inevitably outcompete and outgrow their normal, noncompeting neighbors. The malignant cells starve and suffocate their normal neighbors by stealing all their oxygen and nutrients. The single-minded, Malthusian lack of self-control exhibited by cancer cells means that in addition to outcompeting their normal neighbors for resources, they ferociously fight among themselves in classic survival of the fittest fashion, leading to weirder and weirder—and more dangerous—forms.

Another feature that eventually emerges thanks to the competition is the ability to migrate long distances. For instance, as the fierce internal competition within a camp of angry, uncooperative, and hungry colon cancer cells grows ever more intense, some cells might attain the ability to flee the local colon confines and set up shop in distant, more fertile territory. From a colon cancer's perspective, why remain cramped in the bowel when there is so much more lush land to expand into? Favored spots for colon cancer such as the liver, lungs, brain, and bones offer better digs.

This ability to travel to distant organs is one step beyond the disrespect shown by early malignant cells that simply invade adjacent tissues and organs. To leave home for good and start a new colony remotely requires the ability to enter the blood or lymphatic systems and journey. Normal cells would not survive such a journey; they soon die upon detachment from their home base. This phenomenon (called *anoikis*, from the Greek for "homeless") is not seen in malignant cells; the unusual ability to survive a long-distance trip is another key characteristic of cancer cells.

For most cancers, gaining this ability to *metastasize* represents the final evolutionary step.[2] This is the step that marks the culmination of many generations of constant competition for limited local resources. Now the cancer can spread wherever it wants to in order to find those sought-after nutrients that it formerly had to compete for in its original confinement. But unfortunately for the host, this is the step that often signifies the final phase of progression into a fatal cancer. Once a cancer can spread from place to place, it can wreak havoc anywhere. Ultimately the dispersing tumor cells will land in a region that is absolutely essential for continued survival, such as the brain. From that point on, the clock is ticking.

As mentioned in chapter 27, in the distant past, scientists and doctors thought that metastases represented individual new tumors spontaneously cropping up all over the body at various locations. It wasn't until 1829 that French physician Joseph Recamier figured out that the new tumors were not truly new at all but instead were bits and pieces (i.e., cells) from the original tumor breaking free and spreading via the circulatory system. Thus, this phenomenological description of cancer spread has been known for nearly two centuries. But what is *really* going on that switches cooperative normal cells into competitive, evolving, malignant cells? What is happening at the cellular and molecular levels?

In the 1970s, the decade when President Nixon declared the "War on Cancer," cancer was recognized as a cellular disease but not yet understood as a "molecular" disease. In fact, not only was it not understood, it was not even accepted; at that time there was some staunch opposition to the essential underlying premise that molecular biology could in any way improve our understanding of cancer. But as new discoveries were made, it became increasingly recognized that cancer, the cellular disease, was ultimately driven by (or at least accompanied by) molecular changes, or *mutations*, in the DNA. Cancer was finally recognized as a cellular disease *and* a molecular disease.[3]

When I was a Yale graduate student in the 1980s, major strides were being taken in the understanding of the molecular biology of cancer. Rather than just a simple, qualitative portrayal of "cells gone bad" in a certain organ, cancer was finally being characterized at the DNA level as a transformation from a normal appearance and behavior (phenotype) to a malignant cellular phenotype. The intricate molecular genetic details were finally being uncovered. Initially, the discovery of *oncogenes* and *tumor suppressor genes* promised to, at long last, establish a clear and complete understanding of this ancient disease. It appeared we were on the cusp of truly grasping just what cancer was at the molecular level. Specific mutations in key growth-controlling genes such as *SRC*, *RAS*, and *MYC* promised to explain just what went wrong in cancer cells, or at least that is what I was led to believe back then. How naïve we all were.

Although it was a time of enthusiastic excitement and optimism, cancer biologists were about to find that there was far, far more to this gargantuan iceberg than just the tiny tip that was recently uncovered.

Chapter 29

A STANDARD MODEL OF MOLECULAR ONCOLOGY

When conveying the concepts of cancer biology, I occasionally find it helpful to draw analogies to particle physics. Ostensibly the two fields might appear to be as disparate as two disciplines can possibly be. But in fact there are surprising parallels both scientifically and historically. And that history dates back over two thousand years.

To begin with, as you may be aware, the name "cancer" was coined by "the father of Western medicine," Hippocrates of Kos. Hippocrates is credited with transforming medicine into a genuine discipline, since before his time disease was shrouded in superstition and attributed to angry acts of the gods. While Hippocrates and colleagues were busy creating the Corpus (which includes the famous Hippocratic Oath, which in some form or another, is still taken upon graduation by most physicians), a contemporary of Hippocrates, Democritus, was contemplating matter. Democritus was fathoming the idea of breaking things in half, over and over and over again, until finally there was a tiny fragment that could no longer be split. He named this smallest fragment the "atom," or that which cannot be broken. But far from always being the stately and stoic philosopher, Democritus was prone to unexplained and uncontrollable bouts of laughter. (Today we might term such outbursts *gelastic fits* if they were caused by a brain disorder, such as a hammartoma of the tuber cinereum). Thus, when leaders feared that Democritus might be going insane, they ordered him to see a doctor—and coincidentally that doctor was none other than Hippocrates. Hippocrates soon realized that there was nothing wrong with Democritus at all (who allegedly might have been laughing at the stupidity of his fellow man). Hippocrates's diagnosis: "a happy disposition," and history occasionally refers to Democritus as "the laughing philosopher." Thus, one might argue that particle physics and cancer medicine were twin siblings born at the same time about 2,500 years ago. And now it is time for their reunion.

Around the turn of the twentieth century, the atom was apparently quite simple. Hantaro Nagaoka's beautiful "planetary" model of electrons orbiting the nucleus like planets orbiting the sun was simple and aesthetic.[1] Today, among countless other uses, this iconic model is seen on television's *The Big Bang Theory* and on the emblems used by the US Nuclear Regulatory Commission, the IAEA (International Atomic Energy Agency), and CERN (the Conseil Europeen pour la Recherche Nucleaire, or European Organization for Nuclear Research—the physics laboratory that recently confirmed the existence of the Higg's boson). Nevertheless, despite its simplicity, beauty, and elegance, this model is far from reality.

Instead of the science-fiction-inspiring "solar system" model (in which there could be microinhabitants on the third electron from the nucleus analogous to inhabitants residing on the third rock from the sun, and then nanoinhabitants on the third electrons in atoms of that world, *ad infinitum*), a more realistic description of the atom is a relatively large cloud of "electron density" surrounding a tiny nucleus. The size discrepancy between the electron cloud and the nucleus has been likened to a fly in a huge cathedral and serves as the basis for the name of the book by Brian Cathcart that eloquently recounted key aspects of the dawn of the nuclear era.[2]

The dismantling of this beautiful but simple atom was just the beginning. The nucleus itself was found to be composed of smaller components—individual protons and neutrons (collectively known as *nucleons*). In the mid-twentieth century, following the advent of the "atom smashers," or particle accelerators, the electron, proton, and neutron were joined by a cast of previously unsuspected characters such as pions, kaons, and muons. Before long, a veritable particle zoo came into being with hundreds of new players.

Instead of the elegant solar system concept of the atom, the atom and the subatomic world had gotten quite messy, crowded, and unwieldy. Far from the simple world of the electron, proton, and neutron, there were soon hundreds of uncategorized new particles of various masses with no coherent connections. Something was desperately needed to bring order to the chaos.

Fortunately, the *Standard Model of Particle Physics* emerged and provided some order.[3] The Standard Model offered a classification scheme that helped to tame the particle zoo by organizing the hundreds of particles into neat categories. For instance, particles could be classified according to their responsiveness to the various forces of nature. (The four fundamental forces or interactions are gravity, electromagnetism, the strong nuclear force, and the weak nuclear force.) Particles susceptible to the *strong nuclear force*, such as protons and neutrons, were collectively called *hadrons* (from the Greek *hadros* meaning "thick" or "strong"). Hadrons in turn could be divided into two subcategories: *mesons* (from the Greek *mesos* for "medium") and *baryons* (from the Greek *barys* for "heavy"). Readers might be familiar with the growing field of bariatrics, the branch of medicine that shares the same etymology as baryons and deals with the causes, prevention, and management of overweight patients. In the Standard Model, baryons are those hadrons made of three *quarks*, whereas mesons are hadrons composed of a pair of quarks, specifically a quark and antiquark pair. In this fashion, at least there was now some semblance of order emerging.

In the way of brief review, there are six quarks lumped into three families: up and down; strange and charmed; and bottom and top (or alternatively, beauty and truth), in order of increasing mass. Analogously, the electron is now known to be a member of the family of subatomic particles known as *leptons* (from the Greek *leptos* for "light," "thin," or "small"). Like the six quarks, there are six leptons, and they, too, are lumped into three families: electron and electron neutrino; muon and muon neutrino; and tao lepton (or tauon) and tao neutrino.

The most familiar types of baryons are the proton (made of two up quarks with one down quark) and the neutron (composed of two down quarks and one up quark). Since the down quark is slightly heavier than the up quark, this partly explains why neutrons are somewhat more massive than protons. Because hadrons must have no net "color," baryons are composed of three quarks of three different colors (red, blue, and green) for "color neutrality." Similarly, mesons are pairs of quarks and antiquarks with a given color and "anti-color" such as red and anti-red. In this manner, a given proton might consist of a blue up quark, a red up quark, and a green down quark to balance the colors. A pi-meson (pion) might consist of a blue up quark and an anti-blue down quark for color neutrality. Of course, quarks do not have true color in the literal sense of the word. The term "color" here is simply a name chosen to represent certain quantum states, which, according to quantum mechanical rules, must be distinguishable.

In contrast to the leptons, which have whole number electrical charges (e.g., the electron has a charge of -1), somewhat unexpectedly, quarks do not have whole number electrical charges but instead have fractional charges. For example the up quark, strange quark, and top quark all have charges of +2/3, whereas the down quark, the charm quark, and bottom quark have charges of -1/3. Therefore, the proton, which is composed of two up quarks of charge +2/3 each and one down quark of charge -1/3, has an overall charge of +1. The neutron, which, as its name suggests, is electrically neutral, is composed of two down quarks of charge -1/3 each and one up quark of charge +2/3.

Another method of classifying particles is by their intrinsic angular momentum, or *spin*. Simply, particles with integral spin (e.g., 0, 1, 2, etc.) are *fermions*, whereas particles with half-integer spin (e.g., 1/2, 3/2, 5/2, etc.) are *bosons*. Fermions and bosons behave quite differently. For example, fermions, such as electrons, obey the *Pauli exclusion principle* and remain aloof (or more technically, occupy different quantum states), whereas bosons, such as photons, don't mind being cramped together into the same quantum state. The latter phenomenon allows lasers to operate. It turns out that the known mediators of the fundamental forces are bosons. Thus, the electromagnetic force is mediated by photons; the weak nuclear force is mediated by the W-, W+, Z bosons; and the strong nuclear forces are mediated by gluons. Matter itself, on the other hand, is composed of fermions. Thus, the quarks and leptons and all the composite building blocks of the known universe are fermions.

While just about everyone is fascinated with fundamental particles (I think . . .), I suspect that I personally have a greater-than-average fondness. For a brief while, I had the privilege of treating cancer patients with high-energy neutrons ("fast" neutrons) at Fermi National Accelerator Laboratory, or Fermilab, when the NIU Institute for Neutron Therapy at Fermilab was operational. More recently, I have had the opportunity to switch to another form of hadron therapy, proton beam radiation therapy, through my current institution (Loyola University Stritch School of Medicine), which has a contract with the nearby Northwestern Medicine Chicago Proton Center. Thus, my academic avocation is now actually vocational, and the Standard Model truly represents the "pharmacology" of what I regularly treat cancer patients with.

In any case, the previously confusing morass of unclassified, subatomic particles discovered in the mid-twentieth century eventually yielded to an orderly organizational scheme, which, although some might say is inelegant (Why are there six quarks divided into three families for instance?) is far better than hundreds of random "fundamental" particles. In a parallel fashion, over the past several decades, the original small handful of cancer-causing genes (*oncogenes*) such as *SRC*, *MYC*, and *RAS* have been joined by dozens more. Many dozens. Additionally, well over a hundred *tumor suppressor genes*, such as *RB*, *TP53*, *APC*, and *BRCA1 and BRCA2*, have been identified. In many ways the quagmire of cancer-related genes was reminiscent of the particle zoo. A "Standard Model of molecular oncology" was desperately needed.

The point of this short chapter was not to do a gross injustice to the Standard Model, but simply to point out that particle classification, based on parameters such as composition or fundamental force responsiveness, greatly facilitated a better understanding of particle physics. In a similar vein, the "oncogene zoo" can likewise be tamed. Classification based on certain parameters has greatly facilitated a better understanding of molecular oncology.

Chapter 30

RUNAWAY TRAIN!

It all began in 1910 when a woman brought what appeared to be "a strong, young barred Plymouth Rock hen of light color and pure blood" to what was then the Rockefeller Institute for inspection.[1] The examining physician, Dr. Peyton Rous, noticed a "somewhat movable . . . irregularly spherical" mass in the right breast.[2] Little did anyone know back then that Rous was about to discover something that would someday change the world of cancer biology.

Dr. Rous was born Francis Peyton Rous in Baltimore, Maryland, in 1879. His mother's family were Huguenots who settled in Virginia after the Edict of Nantes. Rous's maternal grandfather, anticipating chaos associated with the impending Civil War, bought land in Texas and moved his family there after the war ended. Rous's father, a Baltimorean of English descent, married his mother while visiting Texas. The couple moved back to Baltimore, where he became an exporter of grain to Europe. Rous's father died early, leaving his mother with three small children and only a meager means of support. Despite all this, Rous's mother was determined to provide the best possible education for her children and elected not to return to the security of Texas and her family. Somehow or another, Rous's mother most definitely did afford him that education, and Rous put that education to good use.

One notable oddity about the tumor-bearing hen was that this was not the only afflicted chicken. Several Plymouth Rock hens from the same farm had similarly peculiar tumors—there appeared to be some sort of epidemic going on. It turned out that Rous was far ahead of his time—perhaps too far. Maybe it was because he was such a young pathologist who had only recently graduated from Johns Hopkins School of Medicine. Or more likely, the world was just not yet ready for the discovery that cancer could be an infectious disease caused by a virus.[3]

Rous ground up tumors from the afflicted chickens and filtered the material through a mesh with pores so small that not even microscopic bacteria could slip through. Nevertheless the "cell-free extract" could have contained even tinier infectious agents, namely viruses. When Rous injected this extract into healthy chickens, he observed it was able to cause cancer. This work was the first to clearly show that cancer, under certain circumstances, has characteristics of an infectious disease. Rous reasonably concluded that this extract contained a cancer-causing virus. Lamentably for Rous, the scientific community was not as rational. Scientific and medical colleagues firmly adhered to the tenet that cancer had an *unknown* cause. Therefore, whatever Rous was studying, it *couldn't be cancer* because he *knew* its cause! Eventually the scientific world caught up with him and appreciated his work. In 1966, over fifty years after his pioneering research that identified the first cancer-causing gene (i.e., an *oncogene*), Peyton Rous was finally awarded the Nobel Prize in Physiology or Medicine. He was eighty-seven years old at the time. I think he holds the record for longest interval between scientific discovery and award of a Nobel Prize.

It wasn't until decades after his original research when other cancers were also suspected of being caused by viruses that Rous's work began to attract any attention. Today we know that many animal cancers are virally induced, and in such cases, retroviruses are often the guilty parties. In humans however, relatively few cancers are caused by viruses and of those that are, even fewer are caused by retroviruses. Nevertheless, retroviruses do cause a few human diseases, including HIV/AIDS, adult T cell leukemia/lymphoma, and tropical spastic paraparesis. But in animal cancers, retroviruses often play a very prominent and direct role.

Recall that retroviruses are made of RNA rather than DNA. This in itself is not unusual. Influenza, for instance, is an RNA virus. However, as mentioned in chapter 13, retroviruses are different in that their RNA genomes are "reverse transcribed" back into DNA, which then integrates itself into the host's chromosomal DNA. Upon integration into the host's DNA, the DNA copy of the retrovirus's genes is called a *provirus*.

The chicken-cancer-causing retrovirus that Peyton Rous worked with is now known as the *Rous sarcoma virus* in his honor.[4] In the 1970s, Peter Vogt isolated mutant variants of the Rous sarcoma virus that could still infect chickens but did *not* cause cancer. Since the Rous virus only contains four total genes in its entire genome, this was a monumental discovery in a tiny virus. One of the four genes encoded the reverse transcriptase enzyme, the second encoded the capsid structural proteins, the third encoded viral envelope proteins and the fourth . . . well, the fourth gene was the one missing in the mutants that could no longer induce cancer. In this manner, Vogt's work enabled Michael Bishop and Harold Varmus to subsequently identify the specific DNA sequences of the cancer-causing gene in the Rous sarcoma virus. Eventually the cancer-causing gene was isolated and named v-src because it was a viral gene that induces sarcomas.

It was initially hoped that this would be *the* key to understanding cancer. If we could just figure out what the cancer-causing v-src gene did, we might at long last understand cancer. It was soon discovered that v-src coded for a *tyrosine kinase* enzyme. (I will explain more about kinases below.) For a brief instant, it appeared that tyrosine kinase activity might possibly hold the key to comprehending cancer.

Soon afterward, however, it was learned that *normal* genes closely resembling v-src were found in a variety of animals—including humans.[5] Given the ubiquity of this gene in animals ranging from fish to humans it was concluded that this v-src look-alike must have some critical function or else it would have been lost long ago. In other words, it was an evolutionarily conserved gene. But since the v-src gene causes cancer, just what were the normal v-src look-alikes doing, and what was so important about them that just about every living creature has some form of them?

In the early years of such study, the normal cellular counterparts to v-src were given the name *c-src* for "*c*ellular." But in short order several other viral oncogenes were discovered and shockingly, they *all* seemed to have normal cellular counterparts. It was concluded that the viral oncogenes were mutated versions of the normal cellular genes and the "c" prefix was eventually dropped. The normal cellular genes were then called *proto-oncogenes* since upon mutation, they could cause cancer. In this way, v-src is the *v*iral cancer-causing oncogene, whereas c-src (now just known as *SRC,* using standard all-capitals and *italic* for human genes) is the normal *c*ellular version of the gene (i.e., the proto-oncogene). *SRC* held some then unknown key function, but did not cause cancer under ordinary circumstances.

We now understand that the *SRC* proto-oncogene codes for an enzyme in the protein kinase family. Generally speaking, a *kinase* is an enzyme that adds *phosphate groups* to other molecules. To use the standard biochemical terminology, kinase enzymes *phosphorylate* other molecules. Thus, *protein-phosphorylating* enzymes are called "protein kinases." More specifically still, the *SRC* gene product is a *tyrosine kinase*, which means it adds phosphates exclusively to the amino acid *tyrosine* in proteins. Kinases, and especially tyrosine kinases, are often involved in *molecular signaling path-*

ways that control cellular growth, survival, and reproduction. When things go wrong with key players in these signaling pathways, cellular growth goes uncontrolled—in other words, a tumor forms. It turns out that many of the key players in cellular reproduction signaling pathways happen to be tyrosine kinases such as *SRC*. (Although not always adhered to, the rule is that human gene names are all in capitals and italicized, whereas the names of their protein products are not italicized but the first letter is capitalized: *ONC* vs Onc. I have tried to stick to this nomenclature in this book.) Under normal circumstances, the Src tyrosine kinase is inactive until called upon, whereas the mutant version carried by the Rous sarcoma virus is always "on"—in other words, it is *constitutively activated*. Since the Src tyrosine kinase is part of a cellular signaling pathway for growth and reproduction, unregulated cellular proliferation ensues, producing a tumor.[6]

But just when it started appearing clear and simple, other cancer-causing viruses began popping up—and several did not work in the same fashion as the Rous sarcoma virus. But once again, it was the trusty chicken who led the way. Some cancer-causing retroviruses appeared to work fast, often within days of injection. These *acutely transforming* retroviruses had to be distinguished from slower-acting retroviruses, which might take months or years to induce tumors. A key distinction between the acutely acting and the slow-acting oncogenic retroviruses was the presence or absence of an oncogene such as v-src. Fast-acting viruses possessed an oncogene. But then how might a slow-acting retrovirus with no oncogene cause cancer at all? The answer was through another mechanism called *insertional mutagenesis*.[7]

Among the first studied examples of insertional mutagenesis was the chicken retrovirus called avian leukosis virus (AVL). After first converting its RNA genome into DNA via reverse transcriptase, AVL insets its proviral DNA into random locations in the host chicken cell's chromosomal DNA. So far this is no different from any other retrovirus. However, under certain circumstances, the AVL proviral DNA serendipitously lands in a lucky spot (actually a very *unlucky* spot for the chicken). This is when it lands near a normal cellular gene called *MYC*. When the viral DNA inserts itself in the vicinity of *MYC*, it misguidedly excites the *MYC* gene. The hyperactive *MYC* gene then transcribes a lot of Myc protein, which stimulates cell proliferation. The *MYC* gene and the Myc proteins themselves are perfectly normal; it is their excessive activity that is abnormal and drives the cancer.

How does the AVL's proviral DNA actually turn on the *MYC* gene? At both ends of retroviruses are sequences of DNA that tend to repeat the same nucleotides over and over and are therefore called *long terminal repeat* (LTR) sequences. These LTR sequences aid in integrating the proviral DNA into the host's DNA and also aid in activating gene transcription. The LTRs need only activate the nearby *viral* genes, but they seem to be a bit more efficient than necessary. They tend to also activate *any* genes they happen to lie next to. Thus, if the AVL proviral DNA happens to haphazardly land near a proto-oncogene, that proto-oncogene would be unnecessarily activated and would spur on cell growth and reproduction when it should not. In other words, a tumor forms. It just so happens that the AVL proviral DNA occasionally inserts itself near the proto-oncogene *MYC* and the super-efficient LTR sequences stir the sleeping giant.

However, *MYC* is involved in cancer production through several other means besides slow-acting viral insertional mutagenesis. There is yet another cancer-causing retrovirus of chickens (poor chickens!) called *avian myelocytomatosis virus* (AMV) that harbors its own malignant *mutant* version of the *MYC* gene (i.e., it possesses a viral oncogene). In fact, it is because of this association with the bone marrow myelocytomatosis malignancy that *MYC* gets its name. In this way, AMV is just like the Rous sarcoma virus—it contains a direct and fast-acting cancer-causing gene; the big difference is that instead of v-src, this time it is v-myc. Upon entry into an ordinary chicken cell the virus's v-myc stimulates undesirable cellular proliferation: a tumor.

Additionally, *MYC* is implicated in a *human* cancer, Burkitt's lymphoma. In this case, a virus known as *Epstein-Barr virus* (EBV) is the trigger. In most healthy people, EBV causes little trouble, but in people where the immune system is disrupted (e.g., by HIV infection or malaria) EBV-infected cells proliferate with abandon. In antibody-producing lymphocytes there are several genes that are exceedingly active, including those involved in antibody production on chromosomes 2, 14, and 22. In humans, the *MYC* gene resides on chromosome number 8. In rare instances, a piece of one chromosome can be transferred to another chromosome in an aberrant process called *translocation*. If, in an antibody-producing lymphocyte, the part of chromosome 8 that harbors the *MYC* gene, gets translocated to the naturally hyperactive promoters of antibody-producing genes on chromosomes 2, 14, or 22, *MYC* becomes overproduced and cellular proliferation goes unrestrained. Once again the result is a tumor.

Exactly how the EBV infection promotes the translocation process that leads to Burkitt's is unclear at present.[8] It might be that EBV plus malaria, HIV, or another "cofactor" leads to incessant proliferation of cells that are already shuffling genes around to create antibody molecules, and the confluence of circumstances terminates in an unfortunate molecular outcome.

If *SRC* and *MYC* were all there was to it, the cancer conundrum would have been deciphered a long time ago. It was soon apparent, however, that these two had plenty of company. It is now recognized that there are many viral oncogenes that cause animal cancers. These viral oncogenes, or v-oncs, have normal cellular counterparts (c-oncs, or proto-oncogenes) that carry out essential functions in healthy cells. The difference is that the viral versions are uncontrolled and always "on," whereas the normal versions are tightly controlled and typically "off" except under certain circumstances. The bottom line is that in addition to *SRC* and *MYC*, the list of proto-oncogenes that could somehow be derailed and lead to cancer soon expanded and also included *SIS*, *RAS*, *ALK*, *FOS*, *ABL*, *RAF*, *ERB*, *RET*, *JUN* . . . okay, I'm going to stop now. You get it. The list goes on and on.

Discouragingly, this was just a *first* step in understanding the molecular and genetic basis of cancer. There was another, entirely different pathway being explored that uncovered even more cancer-causing genes en route. This pathway was the study of *hereditary cancer syndromes*.

In 1971, Alfred Knudson began analyzing pedigrees of families with a tumor called *retinoblastoma*. This rare pediatric cancer of the light-detecting retinal cells in the back of the eye was once invariably fatal but now has a survival rate of nearly 90 percent.[9] The study of cancer survivors and their families revealed that there are two different patterns—a hereditary form that accounts for around 40 percent of cases and a sporadic, nonfamilial form that accounts for 60 percent of cases. Children with the familial form tended to get the cancer earlier, have multiple tumors, and have tumors in both eyes (bilaterally), while those with the sporadic form almost always have just a single tumor.

Recall that (except for sperm and eggs) every cell has two versions of each chromosome—one from the mother, one from the father. Having two versions of each chromosome means the cell is *diploid* (coming from the Greek *diplos*, or double). In healthy humans, the diploid chromosome count is forty-six, or twenty-three pairs. Since there are two copies of each chromosome, there are two copies of any given gene. In this context, the genes are sometimes referred to as *alleles*, so that each cell has a maternal allele and a paternal allele of any given gene. The alleles may be identical if both the paternal and maternal genes are the same; a person with identical copies is *homozygous* for that gene. On the other hand, the maternal and paternal alleles might be slightly different (e.g., one for brown eyes and one for blue eyes) and such an individual would be *heterozygous* for that gene.

Upon analysis of the particular pattern found in retinoblastoma, Knudson formulated his "two-hit hypothesis" of cancer induction.[10] Basically, this model holds that *two* mutations must occur in a specific, critical cancer-related gene within a single retinal cell. In other words, both the maternal allele and the paternal allele must be affected. Since the average mutation rate in human cells is only about one mutation per million genes per cell division, the odds of two separate mutations striking both alleles of a key cancer-related gene in a single cell are quite remote. Thus, the frequency of sporadic retinoblastoma is only around one in twenty thousand children. In dramatic distinction, among families with the inherited form of retinoblastoma, nearly half of the children are afflicted with this cancer. According to Knudson's model, children with the familial form are born with one "hit" already. They already have a mutation that inactivates either the maternal or paternal version of the retinoblastoma gene in every one of their retinal cells. Given the background mutation rate of one per million genes per cell division, combined with the fact that there are over ten million retinal cells, one can see that if a child inherits a first genetic hit in every retinal cell, the odds are quite high that he or she will develop the malignancy in at least one retinal cell. If the child does indeed have the first hit in every cell, there is about a 90 percent chance that he or she will sustain a second hit somewhere in the retina potentially leading to a retinoblastoma tumor. This accounts for the observed frequency of nearly 50 percent of children in affected families coming down with the disease.[11] The roughly fifty-fifty odds of inheriting the genetic defect from the affected parent's chromosomes comes from the fact that a child will inherit only one member of each chromosome pair from his/her mother and one from the father. Recall that sperm and eggs only have a *single* copy of each chromosome (i.e., a *haploid* count) rather than the normal *diploid* count. Thus, there is a 50 percent chance that the defective chromosome might be passed on if either the mother or the father harbors the defective gene. This frequency is consistent with a so-called *autosomal dominant* pattern of inheritance.

Cellular *cytogenetic* studies (examination of the chromosomes) revealed that in the familial form of retinoblastoma, there was an inherited defect (specifically a deletion) in one of the two members of chromosome 13.[12] This inherited deletion was present in not just the tumor cells but *in every single cell* in the afflicted individual's body. In the *malignant* cells this defect in one member of chromosome 13 was joined by another happenstance deletion in *the other* member of the chromosome 13 pair. In contrast, but still supporting the concept, in the sporadic form of retinoblastoma the malignant cells also harbored deletions in both members of chromosome 13, but this was found *only* in the tumor cells. In either situation, whether hereditary or sporadic, a single deletion in chromosome 13 was apparently insufficient to produce cancer. *Both* alleles had to be defective. Formation of the cancer required a second mutation in chromosome 13—a second hit—during the multiple rounds of cellular division involved in development of the retina. The key gene on chromosome 13 was eventually isolated and named *RB* for its role in retinoblastoma.

RB turned out to be the first identified *tumor suppressor gene.*[13] Although it and other tumor suppressor genes are involved in tumor production, their primary physiological function is to restrain cellular proliferation. As such, when both the maternal and the paternal versions of the gene are defective, cellular growth goes unchecked and . . . surprise! A tumor.

Another well-studied familial cancer syndrome is named the Li-Fraumeni syndrome after the two physician-scientists (Frederick Li and Joseph Fraumeni Jr.) who first recognized it in 1969.[14] In this case, the gene in question was found on chromosome 17, and the cancers associated with the syndrome were not just pediatric tumors confined to a single organ but rather ranged far and wide both in age and anatomical distribution. For example, family members with the Li-Fraumeni syndrome might be afflicted with brain tumors, osteosarcomas (a malignancy of bone), rhabdomyosarcomas (a malignancy of muscle), or leukemias in childhood; plus breast, lung, colorectal and adrenal cancers, lymphomas, melanomas, or a variety of other malignancies in adulthood. Overall, the chances of developing some sort of cancer are nearly 90 percent. The molecular defect in this cancer predisposition syndrome involves a protein called p53 (named because of its molecular weight of 53 kilodaltons), which is the product of the *TP53* gene. (For those interested, the actual molecular weight is actually closer to forty-four kilodaltons but because of the high number of proline residues, it migrates slowly through electrophoresis gels and appears more massive.) As in the case of retinoblastoma, both copies of the gene must be defective for a cancer to form. But given that each and every cell (again except sperm and eggs) in a person with the Li-Fraumeni syndrome already has one defective copy of their two *TP53* genes, the odds favor development of a second mutation in at least one other cell somewhere, someday.

As mentioned in chapter 12 (when we digressed about the low cancer rates in elephants, which might be attributable to amplification of their *TP53* genes), the p53 protein carries out numerous key protective roles that help fight off cancer; p53 has even been nicknamed "the guardian of the genome."[15] Upon injury to a cell's DNA, p53 steps into action to activate enzymes that repair the damage. It also halts the cell cycle so that cells with the wounded DNA don't propagate the damage before it is mended. If the DNA is damaged beyond repair, p53 initiates a march toward cellular death (programmed cell death, or apoptosis). All these duties serve to inhibit the formation of cancer. Being responsible for all these critical functions, it is no wonder that the loss of p53 function within a cell leads to bad news.

It should therefore come as no surprise that although the Li-Fraumeni syndrome itself is extremely rare, cancers in otherwise ordinary people quite frequently exhibit damage in their *TP53* genes. Chemical carcinogens such as the *polycyclic aromatic hydrocarbons* (or PAHs, which will be familiar to those interested in astrobiology) found in cigarette smoke and other agents frequently cause problems via damage to the *TP53* gene. In fact, about half of all lung cancers possess *TP53* mutations that are characteristic (i.e., "molecular fingerprints") of exposure to PAHs.[16] Mutations in *TP53* are also identified in nearly half of sporadic melanomas, lymphomas, sarcomas, and cancers of the bladder, colon, liver, stomach, and breast among many more.

Tumor suppressor genes such as *RB* and *TP53* do pretty much what their name would suggest they do. Under normal conditions, these genes produce proteins that control cell growth and reproduction and thereby suppress tumor formation. They are the restrainers, or "brakes," on the cellular division process. Not surprisingly, when these brakes are lost or impaired, cellular reproduction proceeds unconstrained. (In other words, a tumor appears.) Loss of function of tumor suppressor genes can be viewed as a necessary, but not sufficient, molecular step in the creation of a truly malignant cell. Incidentally, this loss of brake function can occur either by tumor suppressor gene mutation or *epigenetic changes* to the DNA. Epigenetic changes, such as adding methyl groups to cytosines in the DNA, can shut down function every bit as effectively well as a classic mutation can, and these epigenetic changes are passed along to progeny cells to boot. (Epigenetic modifications are not limited to adding methyl groups to cytosine residues on the DNA. Other less familiar chemical modifications to the DNA molecule including acetylation, ubiquitylation, phosphorylation, and sumolyation also qualify. As long as the DNA base sequence itself is not altered, such chemical modifications are not called mutations but instead are in the realm of epigenetic modifications.)

But just as with the oncogene situation, there is a lot more to the cancer-suppression story than just *TP53* and *RB*. Other hereditary cancer syndromes were identified such as familial adenomatous polyposis, which is caused by the *APC* tumor suppressor gene and results in thousands of polyps in the colon. Over time, at least one of these benign polyps is likely to degenerate into a malignant polyp. Then there is *hereditary breast and ovarian cancer syndrome*, which recently gained media attention through movie and television stars Angelina Jolie[17] and Christina Applegate.[18] Both celebrities greatly increased public awareness of hereditary cancer syndromes through their public disclosures of their mutations of the breast-cancer-associated *BRCA* genes and subsequent bilateral mastectomies. Curiously, for reasons presently unknown, women with mutations in *BRCA1* or *BRCA2* born after 1940 seem to have a slightly higher risk of developing breast cancer than women born before 1940.[19]

The *CDH1* tumor suppressor gene is linked to hereditary diffuse gastric cancer.[20] A careful review of the autopsy report confirmed that Napoleon Bonaparte died of stomach cancer in 1821 at the age of fifty-two. It has also been suggested that he might have had a hereditary stomach cancer. On his deathbed, Napoleon was said to utter "pyloris . . . my father's pyloris," alluding to the fact that his father died of cancer of the stomach (the pyloris). But allegedly so did his grandfather, his brother, and three of his sisters.[21] Although most researchers today feel that Napoleon did *not* have *CDH1*-associated hereditary stomach cancer, if the alleged family history is true, I think there is indeed a very strong case for some sort of hereditary cancer syndrome.

If all we had was this trivial handful of oncogenes and tumor suppressor genes to cope with, things might have been manageable. But of course, nothing in biology (especially oncology) is ever so simple. In short order *RB*, *TP53*, *APC*, *BRCA1/2*, and *CDH1* were joined by *PTEN*, *NF1*,

WT, *VHL*, *ATM*, *XPA*, *FANCA*, *BLM*, *MSH2*, *MSH6*, *MLH1*. Okay, again it's time to stop. By the turn of the century there were dozens of oncogenes and well over a hundred tumor suppressor genes. Any hope that we would find *the* gene that caused cancer and therefore *a* cure for cancer was ruthlessly dismantled.

Obviously, the previous hope for the discovery of a single genetic change that caused cancer was lost. But given the new cards we were dealt, what sense could be made of the muddle? Could the morass of tumor suppressor genes be organized? The answer is yes.

One way to classify tumor suppressor genes is by deciding whether their primary function is to repair the DNA or to inhibit cellular proliferation. It appears that most tumor suppressor genes do neatly fit into one or the other of these two broad categories, which are now called "caretakers" and "gatekeepers" respectively.[22] For example, the so-called DNA mismatch repair genes such as *MLH1*, *MSH2*, and *MSH6* are responsible for correcting DNA bases that are improperly matched (recall the Watson-Crick pairs: A pairs with T, while G pairs with C). When there is a Watson-Crick mismatch, the mismatch repair genes are called into action to clean up the mess before it gets passed on to the next generation as a permanent mutation. When there are inherited mutations in these DNA mismatch repair genes, colorectal and other cancers are more prevalent, and the condition is called *Lynch syndrome* (or hereditary non-polyposis colon cancer [HNPCC] syndrome).[23] Therefore these tumor suppressor genes are classic caretakers.

On the other hand, the tumor suppresser gene *RB* acts to hold the cell back from dividing. The Rb protein is a molecular brake of sorts and inhibits cellular proliferation. Rb protein blocks the *cell cycle*, or, in other words, Rb action prohibits cells from multiplying. *RB* thus is a classic gatekeeper. Loss of *RB* function therefore leads to unrestrained reproduction—the direct formation of a tumor.

In a similar way, the various oncogenes can be categorized by their particular function into six main groups: *growth factors*, *growth factor receptors*, *membrane G-proteins*, *tyrosine kinases*, *transcription factors*, and *cell cycle/cell death regulators*.[24] Without going into the details, basically all these oncogenes, when inappropriately activated, serve to send cellular division into overdrive. In other words, a tumor forms.

In spite of the simplification, when trying to explain all this to students, especially biology newcomers, such classifications still often fail to get the message across. Thus I resort to my own version of an overused but nevertheless still very effective analogy—cancer as a runaway train. Let's start out with an ordinary and functional family sedan that serves to drive us back and forth to work or pick up the kids from school or soccer practice. This would represent the normal cell, which cooperates with all its neighbors and gets the job done. It has a gas pedal to accelerate it and has two sets of brakes (front and back wheels) to slow it down. And let's assume it has a decent driver who obeys the traffic rules. Overactive oncogenes, which quicken cellular reproduction and tumor growth, can be thought of as the gas pedal. When stuck, the gas pedal forces the car to accelerate beyond control; the cells go into overdrive. Usually, if the brakes still work, the car can be slowed and stopped. Tumor suppressor genes can be thought of as brakes. When these brakes malfunction, the family sedan can no longer stop; the cells no longer cease dividing. Nevertheless, even without brakes, a car might still be controllable if the gas pedal is not pressed and other mechanisms such as the emergency brakes or transmission are operating. The car can still stop without disaster; the cells might not develop into a full-blown tumor.

Using this analogy, we can readily see that a *combination* of malfunctioning oncogenes *and* tumor suppressor genes is needed to convert an ordinary cell into an unruly, uncontrollable cancer cell.[25] A combination of various oncogenes and tumor suppressors is indeed what is actually seen in cancer. Very rarely is cancer caused by only a single genetic mutation; an amalgamation of malfunctions in oncogenes and tumor suppressor genes is required.

A car that has a stuck accelerator pedal and no brakes is a quite apt analogy to early tumor growth. But a truly malignant cancer is far more than just a group of quickly growing cells. Recall that real cancers can invade other tissues, spread far and wide, and are immortal. Each of these progressive steps toward worsening malignancy has their own characteristic genetic mutations.

Returning to our car analogy, even if the accelerator pedal is stuck and the brakes are lost, the car will eventually run out of gas. Cancers work their way around this with ease. Remember that one of the early adaptive steps in tumor formation is *angiogenesis*, the formation of new blood vessels to nourish and sustain the ever-growing tumor. Certain oncogenes act by promoting this abnormal angiogenesis. These improperly triggered oncogenes effectively provide unlimited nutrients, oxygen, and waste removal service for the mushrooming tumor and can be thought of as an unlimited fuel supply—an infinitely large fuel tank in our car analogy. Our family sedan has now certainly morphed into something wicked and wild—no brakes, jammed accelerator pedal, and unlimited fuel. What else could go wrong?! A lot.

Cancer cells, unlike ordinary cells do not obey stop signals. Normal cells will eventually reach maturity, as say, a liver cell, blood cell, or brain cell, and thereafter cease growing and dividing. Not so with cancer cells. They retain their immature state and never "grow up," or *differentiate*.[26] This is why, although it is usually easy to distinguish *normal* cells from different organs (e.g., liver cells look different from lung cells, which look different from muscle cells, etc.), with cancer cells it is not always as easy. Often the pathologist must be told in advance which organ a tumor comes from since cancer cells are less readily distinguishable from one subtype to another under the microscope. Also, normal cells that are damaged beyond repair are taken out of action through programmed cell death, or apoptosis. Cancer cells, upon loss of additional tumor suppressor functions, have disdain for this rule, too. Our sedan with a stuck gas pedal, no brakes, and unlimited fuel now also has complete disregard for traffic lights, and shows no signs of ever slowing down or stopping. This is now a dangerous situation! And things just keep getting worse. Recall that cancer cells disrespect their neighbors and try to crowd them out. Eventually they gain the ability to not just crowd but aggressively invade. Specific tumor suppressor genes are responsible for curtailing this tendency toward invasion; when these genes malfunction, tumor cells go on the rampage. Our former family sedan has now gained "off-road capabilities" and is storming through neighborhoods, backyards, and school grounds with reckless abandon.

How is our formerly friendly family car gaining all these unfamiliar and dangerous new capabilities? Well, one additional feature of cancer cells is their tendency to rapidly evolve. They do this through an unprecedented *mutation rate*. This rapid mutation rate enables them to adapt to their new circumstances and evolve at an amazing clip. This *genetic instability* permits cancers to quickly gain new undesirable features along the way. The condition is sometimes referred to as a "mutator phenotype" since it allows new mutations to quickly crop up to fulfill nearly anything the tumor needs.[27] Loss of certain caretaker tumor suppressor genes creates the mutator phenotype, which in turn facilitates the formation of new and even more dangerous cell types over time. With time, our formerly calm little car turns itself into an off-road-ready SUV, a battle tank, or maybe even a runaway train.

And as if that weren't enough, there is one more thing. Recall that ordinary cells are mortal—they have a limited ability to replicate. Inappropriate activation of the telomerase gene in malignant cells endows eternal life, as we have seen in the HeLa cell line. Thus, our once friendly and functional family sedan has transformed itself into a runaway train with no brakes, a stuck gas pedal, an infinite gas tank, and an ability to travel off road, and it is piloted by an insane, hell-bent, and immortal engineer who has gone totally insane, stolen the brake handle, and the train just won't stop going. Not good. Not good at all.

Now that we have a better understanding of the molecular mechanisms of cancer, how can this knowledge be put to practical use in the clinic? Well, although most cancers are caused by a constellation of calamities in oncogenes and tumor suppressor genes, there are rare instances in which there really is a single principal driver mutation, and if that mutation can be overcome, a cure might be attainable. The prime example, and a wonderful success story, is chronic myelogenous leukemia, or CML.

Kareem Abdul-Jabbar is perhaps the greatest NBA basketball player in history. Abdul-Jabbar was a record six-time NBA most valuable player and to this day, still holds the title as all-time leading scorer, largely thanks to his unstoppable "sky hook." He played roles in several movies and television shows and was an accomplished martial artist (he was one of the very few who studied Jeet Kune Do directly under Bruce Lee). In 2008, Kareem was diagnosed with CML.[28] Had he been diagnosed with this malignancy only a decade earlier, his life expectancy would have been only three or four years. However, thanks to modern molecular medicine, he has survived far beyond this. Hopefully he and many others with CML are for all intents and purposes cured of their cancers. In 2011, Kareem Abdul-Jabbar received one of his greatest accolades ever—the Double Helix Medal for increasing awareness and funding for further cancer research.

CML is caused by a chromosomal translocation, or a swapping of pieces of two chromosomes, in white blood cells. The specific translocation involves swapping parts of chromosomes 9 and 22, which generates a weird hybrid chromosome called the *Philadelphia chromosome*.[29] The name comes from a custom started back in the 1950s in which every genetic abnormality was to be named after the city in which it was discovered. Since the first observation of the abnormal chromosome was made by David A. Hungerford of Fox Chase Cancer Center and Peter C. Nowell of the University of Pennsylvania in the city of Philadelphia, the chromosomal abnormality became known as the Philadelphia chromosome. This was the first chromosome abnormality consistently associated with a malignancy. The end result of this chromosomal translocation is inappropriate activation of a novel oncogene called *BCR-ABL*. It turns out that the gene product of this abnormal oncogene is an overactive tyrosine kinase stuck in the "on" position—but this tyrosine kinase can be inhibited by a specific drug called imatinib, or Gleevac. For patients who do not respond to, or cannot tolerate imatinib, two slightly different drugs, dasitinib and nilotinib, might more nimbly access the active site of the Bcr-abl tyrosine kinase enzyme and produce the desired effect. The use of these tyrosine kinase inhibitors has revolutionized the care of chronic myelogenous leukemia patients. It now appears that, given the large number of individuals who have survived at least five years since their diagnosis of CML, this former death sentence is now a curable condition. More details about the Philadelphia chromosome, CML, and the new drugs used to cure it can be found in Jessica Wapner's *The Philadelphia Chromosome: A Genetic Mystery, a Lethal Cancer, and the Improbable Invention of a Lifesaving Treatment*.[30]

As I mentioned earlier, however, the vast majority of cancers do not have a *single* dominant driver mutation but instead have *multiple* malfunctions in hyperactive oncogenes and inactive tumor suppressor genes. In fact, it appears that most cancers actually have dozens of these malfunctions, and many, such as lung cancer and melanoma, typically have well in excess of a hundred. Therefore, in order to cure common cancers, we will have to delve deeper still.

Chapter 31

ORDER OUT OF CHAOS

Young man, if I could remember the names of all these particles I would have been a botanist!

—Enrico Fermi to a young Leon Lederman when asked about the latest new subatomic particle

Obviously, the quagmire of tumor suppressor genes and oncogenes desperately needed some sort of simplification—a classification scheme of sorts. One start was to determine if the abnormality was due to a *gain-of-function* (GOF) in a gene that was normally quiescent or if the abnormality was a *loss of function* (LOF) in genes that were normally operational. If it were the latter, an LOF problem, then *both* alleles of the gene in question had to be lost for the tumor to materialize, whereas if it were the former, a gain of function, then *only one* of the two genes had to be set into overdrive.

In 2000, cancer researchers Douglas Hanahan and Robert Weinberg proposed a cohesive framework aimed at integrating all of the known cellular and molecular features of cancer (in a manner a wee bit more scientifically rigorous than my car-into-runaway-train analogy!) They called their framework "The Hallmarks of Cancer."[1] The hallmarks of cancer provided a gestalt of cancer, with all its intricate molecular and cellular characteristics. In essence, the complicated chaos of tumor suppressor genes, oncogenes, and other unique cellular characteristics of cancer could be reduced to a set of six basic principles:

The first principle is *self-sufficiency in growth signals*. Cancer cells do not need "go" signals from outside; they stimulate their own growth.

Second, unlike normal cells, cancer cells ignore inhibitory signals that should halt their growth. This is known as *insensitivity to anti-growth signals*.

Third, they defy programmed cell death and refuse to differentiate into mature, mortal cells. In other words they *evade apoptosis*.

Next, they can multiply indefinitely—they possess *unlimited replicative capacity*. Or more simply, malignant tumors are immortal.

The fifth hallmark is *sustained angiogenesis*, the ability to stimulate the growth of blood vessels to supply themselves with nourishment.

Finally, tumors *invade* adjacent tissues and *metastasize* to distant sites.

All of the above-mentioned hallmarks can be viewed in light of specific gain of function mutations in certain oncogenes or loss of function in tumor suppressor genes. This landmark paper published in the journal *Cell* remains one of the most highly cited works in all of the cancer-related scientific literature.[2]

To illustrate the power of such a network of cancer characteristics, here is a quick aside about the third hallmark above, that cancer cells refuse to differentiate into mature, mortal cells: *Acute promyelocytic leukemia* (APL, or PML) is a unique subtype of *acute myelogenous leukemia*. First identified in 1957, PML is one of the fastest growing and most rapidly fatal of all leukemias.[3] Today, PML is one of the most treatable forms of leukemia, with a ten-year survival rate exceeding 75 percent. Like many cancers, especially leukemias and lymphomas, this malignancy is characterized by yet another specific chromosomal translocation—this time involving chromosomes 15 and 17 and is abbreviated t(15;17). In nearly all cases, this chromosomal translocation involves the *retinoic acid receptor* and a gene aptly called *PML*, for promyelocytic leukemia

But what is unique and significant about PML is its responsiveness to a drug called *all-trans retinoic acid* (ATRA, or tretinoin). Unlike traditional chemotherapy, ATRA does not generally kill rapidly growing cells. ATRA works by inducing *terminal differentiation* of the leukemic cells; it is a "differentiating agent." In other words, it causes the immature and immortal malignant cells to "grow up" and become mature but mortal normal cells. Following differentiation into mature cells, like all mature blood cells, the previously malignant cells eventually die on their own. Thanks to this differentiating agent, more than 90 percent of patients with newly diagnosed PML achieve complete remissions, and about 75 percent can be cured by the combination of ATRA and chemotherapy.

ATRA is a *retinoid*, or vitamin A derivative. Students of chemistry might recall that vitamin A itself is the alcohol form (*retinol*) and as the name suggests, ATRA, or all-trans retinoic acid, is the carboxylic acid version, with its double-bonds in *trans* configuration. Other readers might recognize the alternative name *tretinoin* as the topical skin creams called

Retin-A or Renova. In this form, it is a popular prescription acne medication and is also used for reducing fine wrinkles and age spots. A *cis* isomer, isotretinoin (or 13-*cis* retinoic acid) was once marketed as Accutane by Hoffmann-La Roche as a drug primarily used to treat severe cystic acne and related conditions. It also found some use in the prevention of nonmelanoma skin cancer in high-risk individuals and was tested in preventing second cancers from developing in those already with a past history of lung or head and neck cancer. Isotretinoin has also been used to treat a very rare and often fatal skin disease, *harlequin-type ichthyosis*, as well as a related serious skin disorder called *lamellar ichthyosis*. However, the drug can cause serious birth defects when used by pregnant women (or even if they are accidentally exposed); in technical terms, it is *teratogenic*.[4] (A teratogen is any agent that disrupts the development of an embryo or fetus; exposure to teratogens during pregnancy can lead to birth defects.) Accutane has been discontinued by Hoffmann-La Roche although generic forms are available in other countries.[5] Generic isotretinoin may still be available in the United States through a very strict prescription system.

Returning to our hallmarks of cancer, Weinberg and Hanahan proposed some new cancer characteristics that were formally incorporated as emerging hallmarks in 2011.[6] The first is the peculiar preference for "abnormal" metabolic pathways. As mentioned in chapter 11, many cancer cells perplexingly prefer "primitive" metabolic pathways to generate energy instead of the more efficient aerobic respiration. Given their intense energy demands, it remains odd and poorly understood why they would choose to use the relatively inefficient glycolytic pathway when aerobic respiration is fully available to them. This phenomenon, the Warburg effect,[7] has been known since the early twentieth century, but only recently has it recaptured attention because of new possible therapies that exploit it. In essence, any feature that is consistently different between normal and malignant cells could potentially be therapeutically exploited.

Another characteristic of cancer that was recently inducted as a hallmark is *genetic instability*. Cancer cells are exceptionally prone to mutations. It is well known that malignant cells often possess not just one, but several chromosomal aberrations along with numerous invisible DNA mutations. It appears that loss of function of key caretaker tumor suppressor genes is the underlying mechanism of this cancer *hypermutability*. The bottom line is that the DNA of cancer cells tends to mutate far more readily than the DNA of healthy cells. This trait allows cancers to constantly and quickly adapt to their environment. This of course is quite bad for us when we are adding chemotherapy and radiation to that environment, only to watch the cancers rapidly evolve mechanisms to overcome our best efforts to stop them. The fact that cancer cells exhibit hypermutability is the underlying premise behind the *Goldie-Coldman hypothesis*. Simply stated, the Goldie-Coldman hypothesis[8] is the idea that cancer cells can quickly grow resistant to just *one* prescribed chemotherapy agent, but they are much less likely to gain resistance to multiple agents *given at the same time*. The former situation would require only one quick mutation to acquire resistance, whereas in the latter case, multiple, *simultaneously acquired* mutations to provide resistance to *all* of the administered agents would be required. Such a multitude of simultaneously acquired mutations is mathematically highly improbable. Therefore, *combination chemotherapy plus radiation therapy* is now frequently employed as a method to circumvent hypermutability and the rapid development of resistance.

The next new hallmark is *chronic inflammation*. Inflammation was known to physicians as far back as the ancient Greeks who noted that *calor* (heat), *rubror* (redness), *dolor* (pain), and *tumor* (swelling) often went together. But inflammation should be divided into acute and chronic forms. *Acute* inflammation is a critical component of immunity and signifies the attack of our immune system against a pathogen. Acute inflammation is thus a necessary and welcomed component of overcoming an infection. Chronic inflammation on the other hand is no one's friend. No one's friend except cancer, that is.

Countless studies have now documented the strange role that localized chronic inflammation plays in inducing, and sustaining many types of cancer in both animals and humans.[9] For example, marmosets have a high incidence of *colitis* (a general term for inflammation of the colon), and they have an accompanying high incidence of colon cancer. In humans, *inflammatory bowel disease* (which includes *Crohn disease* and *ulcerative colitis*) can be associated with colorectal cancers and other cancers of the GI tract.

Chronic bacterial, viral, and parasitic infections, or persistent physical agents such as non-digestible particles (e.g., asbestos) and chemical irritants can lead to chronic inflammation.

Coal, silica dust, and asbestos are indigestible and cannot be jettisoned by the immune system. The constant battle between the immune system and these foreign bodies leads to chronic inflammation, and certain cancers such as *mesothelioma* may result. Similarly, the chronic inflammation associated with some older-model breast implants has, in rare instances, led to anaplastic large cell lymphoma. The inflammation linked to autoimmune diseases of the thyroid (Hashimoto thyroiditis), intestines (gluten-sensitive celiac sprue), and of the salivary glands (Sjogren syndrome) may in some situations cause lymphoma in these organs. Parasites such as the *liver flukes*, *Opisthorchis*, and *Clonorchis* settle into the bile ducts, causing chronic inflammation there (*cholangitis*) and possibly *cholangiocarcinoma*, a malignant tumor of the bile ducts. Parasitic infections with *blood flukes* (*Schistosoma)* are prevalent primarily in developing nations and are hard to address since reinfection from contaminated water supplies is common. Chronic infection with these parasites (i.e., *schistosomiasis*) induces inflammation in various organs, occasionally causing cancers in the bladder, liver, and possibly the rectum, along with lymphoma of the spleen. Lymphoma of the eye region (orbit) has been linked to the chronic inflammation associated with an infection called "parrot fever," or *psittacosis*. Psittacosis, which was once believed to only be caused by contact with birds of the parrot family (i.e., psittacine birds), is a lung disease contracted by inhalation of air-borne dust contaminated with bird droppings. On rare occasions, it affects other organs. DNA from the causative bacteria (*Chlamydia psittaci*, now called *Chlamyophila psittaci*) has been found in orbital lymphomas (i.e., lymphoma in the area around the eye) in some patients. In such cases, treatment of the infectious organism with doxycycline may cause this lymphoma to regress. Certain malignancies of the stomach can be traced to chronic inflammation caused by bacterial infection with *Helicobacter pylori*.

The final hallmark of cancer—and the one most relevant to our main theme—is an *ability to evade immunity*. As we have stated several times throughout this book, cancers seem to possess an uncanny and most vexing ability to hide themselves from our immune systems.

Although cancers do somehow effectively conceal themselves from our immune systems, they *have* to. Throughout our story we have repeatedly emphasized the countless ways in which cancer cells are exceedingly different from normal cells. Thanks to the fact that they are just *so* bizarre and *so* vastly different from ordinary cells, unless there was some *active* mechanism to hide them, cancer cells would be easily picked out of a lineup by our immune systems and recognized as the obvious perpetrators. I and many others suspect that this obvious difference could prove to be cancer's Achilles's heel.

Our goal is to exploit this Achilles's heel and enable the immune system (either through some sort of natural awakening, gentle prodding, or powerful pharmaceutical assistance) to recognize and react to the truly enormous differences between malignant and normal cells.

Chapter 32

IMMUNE THEORY OF CANCER

Much of what we have been discussing in the past few chapters is consistent with the so-called *mutation theory of cancer*. Cancers, according to this theory, are caused by mutation after mutation after mutation, piled on top of each other, until we finally arrive at a phantasmagorical, completely malignant, and deadly phenotype.[1] However, if that was all there was to it, that increasingly bizarre cell with dozens upon dozens of mutations would be so grotesque and remote from anything called "normal" that the immune system should immediately recognize and eliminate it. Yet, as we have seen, the immune system typically does not. Now the question is why not?

At this point we reach a fork in the road. The most popular explanation for the invisibility from our immune systems is that the never-ending chain of mutations ultimately produces a malignant cell possessing a mechanism to evade immunity. According to the mutation theory, the cancer cells continuously evolve until they serendipitously stumble upon a cell shape or form that the immune system simply cannot recognize. When that phenotype eventually evolves, it will be "chosen" through the natural selection process, since the cells not possessing this immune-evading form will be culled out. Over time, that type of cancer cell will predominate.

Although most cancer biologists favor the mutation theory, there are some glaring deficiencies. For instance, if the mutation theory were entirely correct it would imply that we would *all* eventually get cancer if we lived long enough. Given enough time, chance alone would dictate that inevitably, some cell, somewhere in the body would in due course have accumulated the necessary repertoire of mutations to initiate that fatal chain reaction. It is a mathematical inevitability. But let's double check that math.

Assuming that all living cells (or at least all those cells that divide) have an equal chance of someday becoming malignant, extremely large animals (with many more cells than we have) should also inevitably get cancer simply through mathematical probability. We saw in chapter 12 that dinosaurs, despite their immense size, did not appear to have higher cancer rates than modern animals do. One might criticize this reasoning since dinosaurs had reptilian (or avian) physiologies and are therefore not perfectly analogous to the human situation. On the other hand, as we have seen in chapter 30, chickens and other birds certainly have their fair share of cancers. Granted, the dinosaur analogy may be flawed not only because they were nonmammalian but also because they lived at a time when Earth's environment was vastly different.

Therefore a better analogy would be a mammalian cousin. How about the blue whale? Blue whales can reach weights perhaps three thousand times heavier than a human being (by some estimates, blue whales might reach weights as high as 420,000 pounds!). Although it can be difficult to estimate the lifespan of blue whales and other baleen whales (since they don't have teeth, which easily record the animals age), it is believed that blue whales can live around eighty years—not very different from humans. Now, if we do some simple (and admittedly grossly oversimplified) arithmetic, a human and a big blue whale have perhaps a three-thousand-fold difference in the number of cells. Thus, according to the mutation theory of cancer, we would have to live about three thousand times longer (or roughly three hundred thousand years) in order to have the same number of cell divisions, and therefore the same probability of cancer, as a blue whale! You might criticize my reasoning since the cellular division rates of people and whales might be different and also because much of a whale's mass is blubber and supportive structure that is less cancer-prone than actively dividing tissue. But even if my math is off—by let's say, a couple of orders of magnitude—we would still have to live about *three thousand years* in order to have a similar number of cell divisions and the same probability of cancer!

Based on their mass, number of cells, and number of cellular divisions over a lifetime, whales and elephants should have a *much* higher risk of developing cancer than do humans or mice—but they do not. Elephants, as noted in chapter 12, noticeably deviate from this prediction, and scientists have offered explanations for this difference, including the increased number of TP53 genes that might offer cancer protection. In general (and in clear conflict with theory), all mammals, regardless of mass, tend to have relatively similar rates of cancer[2] (ranging from 20 percent to about

45 percent, with some glaring exceptions such as elephants, naked mole rats, and blind mole rats discussed in earlier chapters). Therefore, I am quite skeptical of the mutation theory as the only answer for cancer.

I am certainly not the only one to recognize the paradox. This sharp contradiction between data and theory—that is, the absence of a strong relationship between body size and cancer risk, is known as *Peto's paradox*,[3] named after epidemiologist Richard Peto of Oxford University. Peto, in the 1970s, was the first to appreciate that across species, the rates of cancer do not convincingly correlate with body mass, despite the predictions of the mutation theory.

To explain the paradox, some have speculated that larger animals possess some ill-defined protective mechanisms that smaller animals do not. (Although I must admit, the recent discovery of additional copies of TP53 genes in elephants could be a legitimate additional contribution. However, as I speculated in chapter 12, the strong immune system of elephants, even in old age, might be the real reason for their resistance to cancer.) Alternatively, maybe the very rough comparisons based on body mass are imperfect since we are restricting our analysis to only three dimensions. We have not included the all-important temporal component.[4] Proponents of the mutation theory could argue that *time* is needed and is crucially important for the malignant mutations to accrue. The probability of someday coming down with cancer is not simply related to the *number* of cells per se; those cells must have ample time to divide and accumulate the needed mutations. But this argument also has its limitations.

First, we know that animals with relatively *short* lives (such as pet dogs and cats) all too often succumb to cancer, despite only being with us for a decade or two. Somewhat to my surprise, about 50 percent of dogs living beyond the age of ten years will develop cancer.[5] This is not unlike the situation of humans in their later years; cancer is generally a disease of older people. The types of cancers that dogs develop are different, however. As in humans, lymphoma is a common malignancy of dogs, but dogs also tend to get *mast cell tumors*, which are quite rare in humans, along with *osteosarcomas*, which tend to be pediatric tumors in people. Soft tissue sarcomas are also better represented in dogs than in humans. Dogs have a higher share of *hemangiosarcomas*, *histiocytic sarcomas*, and thymomas. Regardless of the specifics, if the mutation theory were the sole explanation for cancer, dogs and other animals with short lifespans should not have enough time to develop cancer. The number of mutations required for the evolution of a malignant neoplasm simply shouldn't be possible in the limited length of time allotted. Conversely, according to the mutation theory, if animals lived long enough, they *all* eventually should get cancer. The record shows however, animals with exceptionally long lives *do not* exhibit higher cancer rates.[6] In fact, it appears to be quite the opposite; longevity seems to correlate *inversely* with cancer occurrence. Jalopy's case from the preface notwithstanding, Galapagos tortoises, which can live over 150 years, do not routinely succumb to malignancy. Furthermore, long lived animals such as parrots, koi, and elephants also do not inevitably die of cancer.[7]

As we saw in chapter 13, clams can get cancer. However, outside of the markedly bizarre contagious hemic neoplasia epidemic, clams as a rule do not often get cancer. Yet, they should since clams are among the very longest-lived animals we know of. In fact, some clams can live for centuries. For instance, one specimen of the ocean quahog (*Arctica islandica*) was dated at a whopping 507 years old (through annual growth rings and carbon-14 radiometric methods). This venerable clam was named "Ming" since the Ming dynasty of China was still in power when the clam was born in the year 1499. Nevertheless, when Ming died in 2006, it was not due to cancer. The team of scientists who gathered Ming and friends did not immediately recognize his Guinness Book potential and tossed the venerable clam into the deep freezer, abruptly and unceremoniously ending his 507 years on the planet.[8]

Critics will understandably point to obvious differences between human and molluscan physiology and assert that long-lived clams who do not come down with cancer proves absolutely nothing. Thus, it might be worthwhile to look to long-lived *mammals* again. Focusing again on whales, the fact that these long-lived mammals are also *gigantic* makes them perfect test subjects for the theory. In contrast to clams, as mammals, they have generally comparable physiology to humans. But thanks to their size and lifetimes, their hundred- or thousand-fold more cells will go through many more replication cycles than an average human. Therefore, according to theory, whales are essentially guaranteed to develop cancer. But yet again, in contrast to predictions, these behemoths do not invariably come down with cancer.

Flying in the face of the mutation theory, an exceptionally long-lived mammal is the bowhead whale, which is second in size only to the blue whale. Recent estimates of the bowhead whale life span places them at the very top of the list among mammals, with some pod members estimated between 135 and 172 years in age and one great-great-great grandpa possibly 211 years old. These estimates have been made based on characteristic changes in the lens of the eye, which reliably correlate with age, along with the identification of old harpoon tips that failed to reel in their quarry still embedded in living whales. Those stone and ivory harpoon tips went out of style back in the 1880s! If the mutation theory of cancer were entirely true, in addition to harpoon tips going back well over a century, these leviathans should also harbor massive tumors nearly as old.

In people, cancer is a disease that increases in frequency with age. However, after a certain age is attained the odds of dying of cancer actually decrease. In other words, very old people and very old animals have somewhat reduced chances of ever succumbing to cancer. Nevertheless, rather than abandoning the theory, some scientists have posited that there must be some inherent protective mechanism in very large and long-lived animals.[9] Since theory predicts that long-lived, massive beasts should all succumb to cancer, but they don't, molecular biologist are presently scouring the genome of elephants and whales for the "secret recipe" that protects them from developing cancer.

Additionally (much to my chagrin since I was once a staunch supporter), clinical studies of *antioxidants* have failed to prevent cancer. Antioxidants should reduce DNA mutations, and the reduction in mutation rate should be reflected in a reduction of cancer.

Contrary to predictions however, several large, prospective clinical trials have shown the opposite—use of antioxidants paradoxically *increased* the risk of cancer. For instance, I and the entire cancer community, were stunned and dismayed when the ATBC (alpha-tocopherol/beta-carotene) cancer prevention trial involving over twenty-nine thousand male smokers was stopped ahead of schedule because preliminary results showed that subjects taking the beta-carotene were developing *more* lung cancer than the control group.[10] Beta-carotene is a powerful antioxidant and the single oxygen quencher present in carrots and other orange fruits and vegetables. Many people believed that beta-carotene, either alone or in combination with another potent antioxidant, vitamin E (alpha-tocopherol), would reduce the high rates of lung cancer in men who smoked heavily. Disappointingly, not only were the men paradoxically *more* at risk of developing lung cancer, they were also more at risk of dying of cancer compared to men in the placebo group. Following these discouraging results, the CARET (Carotene and Retinol Efficacy Trial) study was also terminated early for the same reasons.[11] In this clinical trial of beta-carotene and vitamin A (retinol), rather than showing the anticipated reduction in lung cancer, individuals with a history of smoking or asbestos exposure actually demonstrated an *increase* in lung cancer and overall mortality.

Selenium is an essential micronutrient (meaning it is required in only minute dietary quantities) that is a component of several antioxidant enzymes. Early epidemiological studies indicated that selenium deficiency was associated with various cancers including prostate cancer. Thus it was hoped that supplemental selenium, either with or without vitamin E, would reduce the incidence of prostate cancer. But like the other cancer prevention trials, the SELECT (Selenium and Vitamin E Cancer Prevention Trial) study (involving over thirty-five thousand men) was also terminated ahead of schedule when data showed a 17 percent *increase* in prostate cancer in those taking both selenium and vitamin E.[12] This was a statistically significant increase, indicating it was unlikely to be due to chance alone. For men randomized to the groups taking either selenium alone or vitamin E alone, there was also more prostate cancer but the increases were smaller and not statistically significant, indicating that they could be due to chance. Although there are plenty of other explanations for the unexpected and disappointing results, one interpretation is that these clinical trials contradict the mutation theory of cancer.

As I mentioned near the start of the chapter, there is a fork in the road when it comes to ideas about the origin of cancer and one of the hallmarks (evasion of the immune system). Competing with the mutation theory is a completely different hypothesis, which might alternatively explain the etiology of malignant neoplasms.

It remains possible that cancer is caused not primarily by the accrual of mutations but instead because of a failure of the immune system.[13]

According to this hypothesis, the multitude of mutations may be a necessary *but not sufficient* step in the evolution of a cancer. In classic cancer biology, we talk about *initiators* and *promoters*. In the laboratory, initiators, which damage the DNA (i.e., induce mutations) are necessary but not sufficient. To create a tumor, initiators must be followed by application of another agent, namely a promoter. Promoters do not necessarily damage the DNA. In most classical lab experiments, promoters simply accelerate the reproduction of cells. In this manner, the DNA damage caused by the initiator is not properly corrected because the promoter is forcing cell division to proceed before the DNA is fully repaired. In this manner, the mutations caused by the initiator become permanent and the accelerated cell division induced by the promoter is the last straw. While these are fairly basic tumor biology concepts, it is not customary to think of immune deficiency as a promoter. However, to do so is perfectly logical and consistent with tumor biology. Although immune deficiency is not accelerating cell division like a classical promoter, it is nonetheless "promoting" the DNA damage inflicted by an initiator in vivo by not removing cells harboring such damage.

Perhaps under *ordinary* circumstances a grotesque, heavily mutated cell *is* actually "visible" and *is* quickly taken out of commission by our immune systems. Maybe we all have new "cancer-could-be cells" that crop up regularly through exposure to cigarette smoke, ultraviolet light, and other ubiquitous mutagens during our normal activities. But on a daily basis, our immune systems recognize these damaged cells and purge them before they can amass more mutations and become cancers. Thus, mutations alone do not a cancer make.

This purported detection and eradication of incipient cancers was alluded to earlier as the *immunosurveillance hypothesis*.[14] In its modern iteration, this hypothesis (the *immuno-editing* model), still incorporates immunosurveillance as an essential component (but immunosurveillance is now called the "elimination phase" since the immune system is detecting and eliminating burgeoning tumor cells). The elimination phase can be complete (meaning the cancer has been completely eliminated) or it might be incomplete, as when only a fraction of the tumor cells are actually killed. In this scenario, the remaining malignant cells might be held in a state of "equilibrium." As we saw in the case of transplanted cancers (chapter 22), this *equilibrium phase* might go on for quite some time, perhaps indefinitely. If the cancer cells that are held in equilibrium are quiescent, all should be fine as long as the immune system never lets its guard down. On the other hand, it could very well be that the equilibrium phase is a very *dynamic* equilibrium, with an ongoing tug-of-war between the immune system and the cancer cells. The cancer cells might be incessantly mutating and perpetually evolving—and the immune system tirelessly keeps pace. In other words, the immune system is also evolving to match the multitude of nefarious new mutations the malignant cells come up with. In this manner, both the neoplastic cells and the immune system are engaged in eternal battle. This is why the whole process is called "immuno-editing"—the unending war constantly edits, or refines, the relationship between the combatants. As long as the immune system remains at the top of its game and stays one step ahead of the cancer, all is fine. And as I mentioned in chapter 22, it appears that the immune system can indeed remain a step ahead of its foe for decades, if not forever.

In this amaranthine war, however, tumor cells will continue to evolve, accumulating changes that continually increase their strength and their ability to deceive the immune system. In fact, the immuno-editing process almost *forces* this to happen since rogue cells that are discernible will be culled, whereas those that are less conspicuous will get away—and further hone their deception skills. What probably happens at the cellular and molecular level is that the tumor cells modulate their expression of tumor-specific antigens and perhaps MHC molecules during the late equilibrium phase. As this process continues, the immune system exerts an evolutionary *selective pressure* by removing "visible" cells and thereby inadvertently selecting for those with characteristics that enable them to hide. Eventually, this natural selection process might lead to the evolution of a clone of cancer cells that the immune system simply cannot detect and destroy. This is the *escape phase*. During the escape phase, the immune system is no longer capable of constraining the malignant cells, and a progressively enlarging tumor results. The previously imprisoned traitorous cells have finally outfoxed their guards and can now go on the rampage.

Well, that is the theory at least.

At this time it is uncertain if "clinically significant" cancers have, through the mutation theory, evolved the ability to hide from the immune system or if through the immune theory, the immune system has faltered and allowed the cancer to escape. (By "clinically significant cancer," I mean a cancer that could someday prove fatal. This would be a malignancy that one could die *from* rather than die *with*. Presently there is heated controversy about management of early-stage, low-grade cancers of the prostate and certain precancerous conditions of the breast. Do all these conditions require aggressive treatment? Certainly not. But determining with 100 percent certainty in a given patient whether or not their particular tumor might someday progress into a dangerous cancer is most challenging.)

We do know that if the immune system is suppressed, clinically significant cancers can evolve, and the likely conclusion in that scenario would be that the cancer has escaped immune surveillance. We have seen that suppression of the immune system dramatically increases cancer risk in the organ transplant clinic and in the immunosuppressed HIV/AIDS population.[15] These immunosuppressed individuals have a several-fold increased risk of getting cancer. Although the types of cancers in this immunosuppressed population are predominantly virus associated, there is also an increase in cancers not known to be virally caused.

Given these observations, it makes us again wonder (or worry) if cancer is caused in otherwise healthy people through a gradual waning of immune capabilities. Do ordinarily healthy people eventually experience a deterioration of cancer immunity over time, which in turn allows neoplastic cells to evade the immunity that once was fully capable of eliminating them? If this situation materializes, mutant cells will no longer regularly be eradicated or be kept in check by the immune system. Instead, mutant cells would be allowed to accrue even more mutations until eventually a truly dangerous form materializes.

Returning to a very simple analogy, perhaps the diminution of immunity to cancer is similar to the waning of immunity to viral conditions such as chicken pox. As chicken pox immunity fades with age, the previously controlled virus reemerges in the form of shingles. Presently there is a lot of attention surrounding shingles vaccinations, and these apparently do help prevent shingles. Will we someday similarly get a "booster shot" that will ward off cancer just as a shingles or tetanus shot reinvigorates our previous immunity?

We do know that cancer is generally a disease of older individuals. The incidence of cancer is considerably higher among people in the sixty-to-eighty-year-old age bracket than it is in those in the twenty-to-forty-year-old age bracket. Immunological metrics, such as the *lymphocyte stimulation index*, confirm a weakening of the immune system with age. It is also known that people who make it to age ninety and beyond are somewhat less likely to ultimately succumb to cancer as their final cause of death.[16] Perhaps such people simply have immune systems that better withstand the test of time. We have seen repeatedly throughout this book that animals that live longer (such as mole rats) tend to have lower cancer rates. Longevity and cancer resistance appear linked in a more intricate manner than the simple, old idea that, "if one doesn't get cancer, one might live a long time." Maybe the gene that confers longevity is the gene that confers cancer-resistance. And while the amplified *TP53* in elephants could be a candidate, there may be other mechanisms—and maybe those additional mechanisms work via the immune system.

Dogs that reach a ripe old age are less likely to ultimately succumb to cancer. Just why dogs are cancer-prone at all, given their relatively brief life spans, remains uncertain, but there are a number of hypotheses, such as tick and flea killers, a host of environmental toxins, and the typical diet given to pet dogs. However, could it be that, as in humans, the immune system of the aging dog simply starts stumbling? If true, this faltering immunity could allow small cancers that were quashed in the previous decade to emerge, grow, and cause trouble during the second decade. Additionally, if this is true, our veterinary colleagues might be the first to figure out something effective (in the canine population) that could be extrapolated to humans.

But if it were this simple then all we would have to do is stimulate or strengthen the immune system in some form or fashion and cancers would never materialize. It is probably a lot more complicated than this. If all we had to do was beef up our immune responses, we would have cured cancer a decade or so ago. Interferons, interleukins, vaccines . . . such immune-stimulating agents have been used in the cancer clinic for a while. But unfortunately they have met only limited success.[17] It seems unlikely that simple stimulation of the immune system—whether by exercise, low-dose radiation exposure, better nutrition, or drugs—is going to prove fully adequate in addressing an aggressive cancer. We have seen in earlier chapters that both the fetus and malignant tumors surround themselves with a squadron of supreme guardians and regulators that will never be vanquished by the single-minded attacking soldiers, no matter how strong these soldiers become. Past methods to "strengthen the immune system" did so with equal opportunity—they strengthened the cancer-killing soldiers but likewise strengthened the royal guards. And in this manner, the cancer always remained protected.

To summarize, in this chapter we have encountered serious challenges to the mutation theory of cancer. These challenges make it unlikely that the theory is complete. It does appear that the immune theory of cancer holds some merit. But as far as defeating cancer in the clinic, merely boosting the immune system as a whole augments not only the cancer-killing soldiers but also bolsters the cancer-protecting regulator's power over the cancer-killers. A major paradigm shift is thus required.

Fortunately, a major paradigm shift in cancer immunotherapy is underway.

Chapter 33

WHAT CAN COWS TEACH US ABOUT CONQUERING CANCER?

VACCINE

Syllabification: *vac·cine*

Pronunciation: *\vak-ˈsēn*

Definition: a substance used to simulate or increase immunity to a particular disease

Etymology: Late eighteenth century from Latin *vaccinus;* ***from*** *vaca,* ***"cow"***

Although everyone knows what a vaccine is, not many of us are familiar with the origin of the word.

The story all begins with Edward Jenner and the creation of the smallpox vaccine in 1798. Since this was the first successful vaccine to be developed, Jenner is rightly considered to be the "father of vaccination." But how did Jenner come up with his great idea? The answer is: from cows.

For years, while smallpox was ravaging the human population, it was noticed that some women seemed immune to smallpox even during the most devastating outbreaks. What was keeping them from succumbing to this scourge? It appeared that the common denominator was that these ladies served as milkmaids. As an occupational hazard, these young ladies often contracted cowpox, a relatively mild disease that manifested with a low-grade fever, slight headaches, and a few days of not feeling well. A case of this disease, however, seemed to provide lifelong immunity to smallpox, the far deadlier malady. Oh, by the way, the scientific name for the cowpox virus is *Vaccinia*.

Smallpox, or *variola*, has been known by a variety of names throughout history (for example "the pox" or "the red plague"), but the name "smallpox" was probably first used in Britain in the fifteenth century to distinguish it from the "great pox"—what we would today call syphilis. In its "major" form (there was also a "variola minor" with a much lower mortality rate), this disease was indeed a major killer. Very major. Smallpox was estimated to take a full four hundred thousand European lives per year around the end of the eighteenth century. It was a highly contagious disease with a terrifyingly high mortality rate of 20–60 percent. In children the death rates were even worse at around 80 percent. Smallpox might have taken up to five hundred million lives in the earlier part of the twentieth century. During the later twentieth century, most thankfully, smallpox was rendered extinct in the natural human population.[1] The last naturally occurring case was diagnosed on October 26, 1977, and the World Health Organization (WHO) formally announced the eradication of smallpox in 1979.[2] In June 2011 came the announcement of the eradication of a second infectious disease. The United Nations Food and Agriculture Organization officially proclaimed the eradication of *rinderpest*.[3] Rinderpest, or cattle plague, was once an infectious viral disease of cows and other artiodactyls (even-toed ungulates) with horribly high death rates (essentially 100 percent in immunologically naïve populations). Animals would develop high fevers, mouth ulcers, nasal discharge, diarrhea, and often die of dehydration within two weeks. Fortunately, the world was made rid of rinderpest in 2001. Presently, smallpox and rinderpest are the only two infectious diseases officially declared dead.

The path to smallpox eradication goes back a long way, with immunization methods dating back to fifteenth-century China and late-eighteenth-century Africa and Middle East. There were several variations on the main theme of exposing individuals to material taken from a patient with smallpox. The most popular method was to rub a modest amount of powdered smallpox scabs or fluid from smallpox pustules into small scratches made on the skin. The aim, of course, was to induce permanent immunity through a very mild case of smallpox.[4] Introduced into England and North America in the 1720s, the process was called *variolation* because of the alternative name for smallpox, variola. As one might imagine, such smallpox inoculations came with significant risks and on occasion the anticipated mild cases were anything but mild. Thus in modern medicine, the use of genuine *inoculations* (that is use of fully potent, live pathogens as in variolation) has fallen out of favor. On the other hand, *vaccination*, the use of weakened/killed pathogens or administration of a variant that doesn't cause major disease but does induce immunity (such as the cowpox and smallpox situation) has evolved and emerged as the preferred approach. Incidentally, vaccines can be effective even if they do not contain any real virus, alive, dead, or in-between. Two cancer-causing viruses, hepatitis B virus (HBV) and human papilloma virus (HPV), now have effective vaccines solely composed of *protein parts* of the virus without any viable nucleic acid infectious material at all.[5]

Although Jenner is credited for the first successful vaccine development, he had a few predecessors who paved the pathway. You would think that a pithy title like, "Cowpox and Its Ability to Prevent Smallpox" would have garnered some attention, but John Fewster's 1765 publication attracted little more than dust. In the interval just before Jenner's breakthrough, there were several accounts of primitive but successful vaccination efforts. For example, farmer Benjamin Jesty vaccinated his wife and two children with cowpox (and probably successfully induced their immunity to smallpox) during a 1774 smallpox epidemic.[6]

But based on the milkmaid observations, along with anecdotal reports that farmers and others who regularly worked with cattle also were often spared during smallpox outbreaks, Edward Jenner was the first to formally propose that prior infection with cowpox could prevent smallpox. Jenner then formulated a method of validating this hypothesis and implementing a systematic program. In May 1796, Jenner first tested his idea through "vaccination" of his gardener's son, eight-year-old James Phipps. He obtained pus from cowpox blisters on the hands of Sarah Nelmes, a milkmaid who had an active case of cowpox. Jenner then vaccinated James on both arms, which led to a slight fever and some malaise but no serious symptoms. Afterward, Jenner challenged James with genuine smallpox, which was the standard method of attempted immunization at that time. However, James did not develop the anticipated mild case of smallpox. In fact, he showed no signs of smallpox at all. Jenner variolated James again, but, as before, James showed no sign of infection. The vaccination was successful. Young James Phipps was now immune to smallpox. Two years later, Jenner published a series, the world's first scientific work on vaccination—a procedure for which we owe cows a great deal. (Incidentally, Blossom the cow was the bovine who gave Sarah Nelmes her cowpox. In her honor, Blossom's hide now hangs on the wall of history at St. George's medical-school library).

Although originally restricted to the application of *Vaccinia*, the cowpox, the term "vaccination" now generally applies to all efforts at artificially inducing immunity against any infectious disease through administration of immunogenic material. In fact, as I have been presenting throughout this book, maybe the term can now be expanded further to include induction of immunity not just to infectious disease, but also against malignant disease.

Since this book is not about smallpox, how does this digression about cows and vaccination relate to our main topic, cancer? Readers may be familiar with the bacterial species *Mycobacterium tuberculosis*, the cause of tuberculosis. Tuberculosis has been known throughout history by a variety of names including "the white plague" and consumption. Today, most in the medical fields call it by the abbreviated name, TB. In any case, it turns out that the quest for a cure for tuberculosis in cows circuitously led to an effective immunotherapy for bladder cancer in people.[7]

Tuberculosis was, and sadly still is, a dread disease—and with the emergence of multidrug-resistant strains, TB is staging a serious and scary comeback. In the nineteenth century, TB was a major cause of morbidity and mortality in both man and beast. In the late nineteenth century, in major urban areas, an estimated 80 percent of the population was exposed to TB, and among those with active tuberculosis, about 80 percent eventually died of the disease. At the dawn of the twentieth century, in France it was estimated that three people out of every one thousand would die of tuberculosis each year. Frightening indeed. Although caused by a closely related but not identical *Mycobacterium* species (*M. bovis*), tuberculosis is also a serious scourge of various animals of commercial importance, including cows. To address this, beginning in 1908, microbiologist Albert Calmette and veterinarian Camille Guérin began a lifelong collaboration at the Pasteur Institute in Lille, France.[8]

First, they isolated a virulent strain of *M. bovis* from the udder of a cow with a bad case of bovine tuberculosis. With a strong tendency to clump together, their bacterial culture proved a bit unwieldy in the lab, so they modified the culture medium. They learned that by cultivating the bacterial strain in a medium of cow bile, potatoes, and glycerin, they could not only calm the clumping problem but could also reduce the *pathogenicity* (the potential to produce disease) of the bacteria. Over time, the strain gradually demonstrated diminished virulence, and in 1915, they administered an early version of this attenuated *M. bovis* strain to cows and showed that this vaccination provided protection against TB. After many years of refinement (and 230 passages in the lab), in 1921 they came up with a genetically stable, nonvirulent version, which was subsequently named in their honor. Thus was born the *bacille Calmette-Guérin* (or BCG) strain of *Mycobactrium bovis*.

Contemporaneous with William Coley and somewhat reminiscent of his toxin story (described in chapter 8), some researchers noticed that tuberculosis seemed to have anticancer effects. An autopsy series conducted at the Johns Hopkins Hospital showed an inverse relationship between cancer and tuberculosis. When people who somehow survived cancer were compared with those who had died of cancer, it was evident that the cancer survivors had higher rates of active or prior tuberculosis than those who had succumbed. This relationship held even when the cohorts were matched by age, sex, and race. Conversely, the study showed lower rates of cancer in people who ultimately died of TB than in similarly matched individuals. Dr. Raymond Pearl, the author of the autopsy series,[9] concluded in 1929 that some sort of mutual antagonism must exist between TB and cancer, but like Coley and other early pioneers of tumor immunology and cancer immunotherapy, he was unable to provide a detailed mechanism. Nevertheless, a putative link between tuberculosis and cancer was made, and that link remains the foundation for what remains to this day the most effective treatment for early-stage bladder cancer.

Pearl's observations that patients with tuberculosis had lower rates of cancer ultimately led to investigations into the use of BCG as a form of cancer immunotherapy. In the 1950s, Lloyd J. Old[10] at the Sloan Kettering Cancer Center found that mice injected with live BCG demonstrated resistance to certain experimental cancers. The mechanism appeared to be one of generally amplified immunological activity. Following the BCG injections, activated macrophages possessed the ability to recognize, attack, and eliminate cancerous cells and destroy mouse tumors. This provided the first direct proof of BCG's anticancer efficacy. (Incidentally, Old went on to further fame in the immunological world through his 1975 discovery of a cytokine frequently mentioned in this book, tumor necrosis factor alpha, or TNFα). While some Italian investigators initially explored *M. tuberculosis* itself as a cancer chemotherapeutic in animal models, it wasn't until the early 1970s when research at the NCI (National Cancer Institute)[11] clearly demonstrated the tumor-inhibiting properties of BCG upon injection into animals. In a seminal paper, the authors wrote: "Complete tumor inhibition was observed when infection with living BCG occurred at the site of tumor inoculation. This inhibition . . . was mediated by . . . immunologic response to the infecting organisms."[12]

Following up on this highly encouraging animal data, beginning in 1969, French clinical tests of BCG therapy for acute lymphoblastic leukemia began, and some work was done in the United States against melanoma in 1970. In 1975, the American team of researchers reported the first successful treatment of a metastatic melanoma lesion in the bladder.[13] They treated this by instilling live BCG organisms right into the bladder cavity itself.

Based on these data, a Canadian team of doctors at Queen's University in Kingston, Ontario, led by Alvaro Morales began instilling live BCG right into the bladders of patients with early-stage bladder cancers.[14] In a small (very small) trial of patients who had frequent bladder cancer recurrences despite prior standard treatments, all seven patients who were available for follow-up showed excellent response to the BCG therapy. Follow-up bladder biopsies displayed intense immunological reactions with no residual tumor. Their results, published in 1976, marked the first phase of genuine evidence-based cancer immunotherapy. In a bold move (considering it was based on only seven cases!) the NCI funded two large, prospective, randomized clinical trials comparing BCG against surgery alone. These trials confirmed Morales's results, and in 1990 the FDA approved the use of *intra-vesicle* (that is, directly into the bladder) BCG therapy for early-stage, superficial bladder cancer and precancerous bladder lesions (also known as bladder carcinoma in-situ, or CIS).[15] Thanks to these solid data, BCG supplanted cystectomy (complete surgical removal of the bladder) as the treatment of choice for bladder CIS in the mid-1980s. As mentioned before, to this day, *minimal* bladder surgery (that is, an internal scraping of the tumor off the bladder wall, rather than a full cystectomy) plus BCG immunotherapy remains the most effective treatment for early stage bladder cancers.

In fact, one can argue that BCG therapy for early-stage bladder cancer is presently the most successful immunotherapy in human cancer medicine. BCG appears to induce a potent anticancer response involving T cells, macrophages, and various chemical cytokines that kills or stymies cancer cells in a highly orchestrated fashion. The cellular and molecular details about how it really works are still unclear, but when instilled into a bladder with early-stage cancer, BCG eradicates small tumors and precancerous lesions, reduces recurrences, delays progression to invasive or metastatic tumors, and most important, improves survival. While several other agents have been tried, nothing has bested BCG for early-stage bladder cancer and bladder CIS.

In the bladder, the live BCG bacteria are gobbled up by macrophages. BCG organisms may also enter bladder cancer cells themselves, and macrophages will engulf these bacteria-laden cancer cells. Next, within the macrophages, the cancer cell proteins are broken down into smaller peptide fragments, which are combined with MHC antigens and displayed on the cell surface. This ultimately tags any cancer cells owning these proteins and peptides for destruction. It appears that the actual killing of tumor cells is done by *cytotoxic T cells*, lymphocytes that specifically recognize the coexpression of abnormal MHC antigens (on the cancer cells) and foreign BCG antigens. Importantly but not surprisingly, the response to BCG is diminished if the patient is immunosuppressed.

Now that we have some "proof of principle" that cancer immunotherapy can clearly be of use in certain situations, let's dive into the details and determine if other cancers may someday also yield to similar immune-based strategies.

Chapter 34

TOUGH MOTHERS AND JUVENILE DELINQUENTS

Undesirables such as pathogenic bacteria and "abnormal cells" are treated harshly by our immune systems. Importantly, virally infected cells and cancer cells are on this list. To start, certain cellular members of the immune system engulf, chop up, and then show minced pieces of such uninvited visitors to other cells of the immune system. The diced up pieces of the unwelcomed invaders or abnormal cells are small sections of proteins called peptides. In the context of immunology, these peptides are called *antigens*. The main job of some immune cells is to display or present these cut up pieces of pathogens (antigens) to other members of the immune system. For these reasons, the cells that display antigens to other cells can be called "professional *antigen-presenting cells*,"[1] or APCs. Mature *dendritic cells* are one example of such professional antigen presenting cells. As such, they *properly* display pieces of invading pathogens or abnormal cells (including virus-infected and cancerous cells) on their surfaces so that other immune cells such as T cells might then recognize these antigens. The key word here is *properly*, since the antigens must be displayed in a very specific fashion for the T cells (and other immune cell recipients) to recognize and respond to the presented antigens.

Following proper presentation, the T cells will then be able to recognize these antigens on their own through their unique T cell receptors. In a way, the antigen presentation process can be thought of as "school" for T cells. Upon graduation, the well-trained T cells will be capable of identifying invading germs, virally infected cells, or malignant cells. Each T cell receptor, or TCR, is slightly different and recognizes a specific unique antigen in a lock-and-key fashion. Potentially, the human body can generate a staggering 10^{18} (one quintillion) different TCRs.[2] That's a lot of lock-and-key combinations!

In immunology, *Signal 1*[3] refers to the process whereby a dendritic cell engulfs and dismantles a cancer cell, shreds that cancer cell's surface proteins into peptides, and then presents those peptides as antigens to naïve (i.e., "unschooled") T cells. Following Signal 1, not surprisingly, is *Signal 2*—the "activation" of those T cells who have the proper lock-and-key antigen/receptor fit. These activated T cells now have the potential to identify, attack, and destroy any invading or abnormal cells that possess the targeted antigen. In order to successfully achieve this goal, one final signal called *Signal 3* is required. Signal 3 is basically supplying "fuel" in the form of *cytokines* to enable the T cells to multiply, accumulate, and persist. In the absence of such cytokine fuel, T cells would ordinarily die young before completing their missions.

In review, it appears that the real action starts with the professional antigen presenting cells, such as dendritic cells. And since dendritic cells play such a key role, it should come as no surprise that the body doles out harsh punishment for any dendritic cells that don't mature and carry out their appointed duties—they are killed! The immune system is one tough mother!

Well, at least under normal circumstances that is.

There is one perverse location, where all the old rules are broken and mayhem rules. This chaotic setting, where anarchy lives, is the *tumor microenvironment*.[4] The tumor microenvironment was only relatively recently discovered. But only more recently has it been appreciated just how truly bizarre and backward everything is there. Rather than being punished by death for not maturing and carrying out their responsibilities, immature dendritic cells are *welcomed and rewarded* in the tumor microenvironment! Here, not only are immature, inept dendritic cells not exterminated, these juvenile delinquents are nurtured and encouraged. Just like true juvenile delinquents, these immature dendritic cells go out of their way to interfere with the tumor-eliminating aims of the mature and capable adults.

However, recent research has revealed that the strange new world of the tumor microenvironment is not in reality a place of complete chaos. Rather, there does appear to be some unambiguous order and a distinct set of new laws. There is clearly a new sheriff in town—but he is not wearing a white hat.

Chapter 35

AS CRAZY AS THE QUANTUM CAFÉ

Et tu, Brute?

—**William Shakespeare, *Julius Caesar***

In earlier chapters we talked about cancers being "invisible" to the immune system, perhaps by using a cloaking device of sorts to conceal their identity and avoiding attack. We also suggested that tumors might have stolen the "secret of invisibility" from the embryo and fetus, pleading with the immune system not to kill the innocent baby. It is now clear that these hypothetical concepts do have a factual biological basis; cancer does indeed effectively hamstring the immune system. And today we are unraveling the mysteries at the cellular and molecular level.

The key to this immunological corruption lies within the mysterious *tumor microenvironment*, which corrupts and cripples the immune system so that normal anticancer immune actions are stymied. This bizarre region comes complete with a generous supply of needed nutrients through the creation of its own new blood vessels—angiogenesis. As mentioned earlier, tumor angiogenesis stems from mutant oncogenes, and angiogenesis is in fact one of the hallmarks of cancer. The development of new blood vessels (also known as neovasculature) is mandatory if the tumor is to continue its upward growth trajectory. These new and rapidly growing blood vessels can serve as a therapeutic target in their own right, with several anticancer agents such as bevacizumab (Avastin) in clinical use to slow the growth of tumor neovasculature.

But the tumor microenvironment is a lot more wicked, thornier, and weirder than just a ramped-up blood supply. The treacherous ways that tumors subvert the immune system from "attacker" to "defender" is most vexing. The tumor microenvironment sometimes grants a cancer complete sanctuary from immune attack. In the previous chapter we explored one mechanism—the recruitment of juvenile delinquents in the form of immature dendritic cells to help shield the cancer from attack. It turns out, however, that the tumor microenvironment receives a lot more help than just from the immature, double-crossing, dendritic cells.

And as Walt Kelly's Pogo once said, "We have met the enemy and he is us."[1]

Macrophages are main members of the *innate immune system* and as such are one of our natural defenses against invading pathogens. They are often seen infiltrating tumors. One might presume that this is a good thing; the macrophages must be fighting the cancer cells just as they would be fighting a bacterial invasion, right? Wrong! The tumor-associated macrophages, or TAMs, are under the hypnotic spell of the tumor microenvironment. Somehow, the tumor microenvironment brainwashes the macrophages into *guarding* the cancer cells instead of attacking them.[2] This switch in behavior is called an M1 to M2 conversion. Whereas M1 macrophages would normally participate gleefully in the elimination of malignant cells, M2 macrophages surround and protect those malignant cells as if they were the crown jewels. But these M2 macrophages are among the weakest and least sophisticated of the cancer protectors found in the microenvironment. Several other former allies have been recruited to sabotage the mission.

Cancer cells manufacture certain chemical compounds called cytokines that subvert the attack. Under ordinary circumstances cytokines are key components in the immune attack on uninvited invaders. For example, certain cytokines such as interferon gamma (INFγ) and tumor necrosis factor alpha (TNFα) are often released in the process of attacking and eliminating virally infected cells. They could and should be instrumental in attacking and eliminating cancer cells as well. But once again, in the tumor microenvironment the attack is off—and the shields are up. Instead of cancer-*fighting* chemical compounds, the tumor microenvironment is flooded with cancer-*defending* cytokines. For instance, interleukin10 (IL-10) and transforming growth factor beta (TGFβ)[3] are elaborated by many cancers and these immune-suppressing cytokines are instrumental in the M1 to M2 macrophage transformation.

But IL-10 and TGFβ do a lot more than just affect macrophages. They, along with several other suppressive cytokines, also convert *helper* T cells into cancer-tolerant *regulatory* T cells. Both of these T cell subtypes possess a surface marker called *CD4* and are thus called *CD4+ cells*, but they function in completely opposite fashions. Put very simply, among other functions, helper T cells assist cytotoxic T cells carry out their responsibilities (cytotoxic T cells carry a different surface marker called CD8 and are therefore *CD8+ cells*). These CD8+ cytotoxic T cells are the cells that do much of the actual killing of unwanted pathogens and abnormal cells, including cancer cells. Helper T cells help cytotoxic T cells do their thing, assisting in their mission to recognize, bind to, and destroy any uninvited intruders. In this way, cytotoxic T cells are a principle means by which the body wards off attack from bacteria, fungi, parasites, and viruses. Virally infected cells that cannot be salvaged are sacrificed by the immune system through cytotoxic T cells. In a similar fashion, abnormal cells such as cancer cells are eliminated by cytotoxic T cells.

Or at least they should be.

However, once again in the tumor microenvironment things are not as they seem. Instead of the CD4+ T cells becoming helpers that aid the cytotoxic T cells in killing the malignant enemy, the immune-suppressing cytokines secreted in the tumor microenvironment transform the CD4+ cells into a class of potent "police" T cells that constrain the anticancer crusade. These policemen are called *regulatory T cells*, or *Tregs*, and were introduced in chapter 27. Under ordinary conditions, Tregs function to limit the rowdiness of any immune attack. They control inflammation. As mentioned, acute (i.e., short-term) inflammation is instrumental in fighting off infections. However, after the infection has been cleared, inflammation is no longer needed. Chronic (prolonged and unneeded) inflammation is detrimental and can damage tissues. This is where Tregs come in. They quell overexuberant immune attacks and prevent auto-immunity, one of the most important functions the immune system has.[4] But what are the Tregs doing in the tumor microenvironment? Well, it turns out that here, like so many (supposed) allies, the Tregs are under the spell of the cancer. As if in a trance, they aid and abet the tumor; they have become vicious guardians. And these guardians are tough!

As suggested earlier, it might be anticipated that the guardians of the embryo and the fetus would prove to be a most powerful set of defenders—maybe *THE* most powerful of them all. Charged with survival of the species, the guardians must be stronger than the attackers. For a species to survive, these enforcers of the "Hands Off!" policy must have supreme strength and must be capable of overpowering any rebellion. And so it is in the tumor microenvironment. Here, the Treg cells quite effectively quash any and all anticancer activities.

The tumor microenvironment seems to have no limits in its resourcefulness when it comes to outwitting and outlasting the immune system. The list of cells susceptible to the mesmerizing charms of cancer in the microenvironment is staggering. We mentioned in the previous chapter that a key initial player in the immune process—dendritic cells—have become defectors who have chosen to aid the enemy. This happens when, instead of maturing into capable cancer conquerors, they remain in a juvenile state and fail to carry out their assignments. But not only do they fail to finish their assigned homework, they actively interfere with normal anticancer activities of the mature adults in the vicinity. In addition to these dendritic cell defectors, helper T cells, which normally assist cytotoxic T cells annihilate adversaries, have joined the dark side. It appears that helper T cells may come in several varieties. The category of helpers that help cytoxic T cells conquer their enemies are "true" helpers in that they are true to our mission of defeating the enemy and are sometimes called TH1 cells. But just as macrophages have their M2 sinister counterparts, and mature dendritic cells have their immature foils, so, too, do TH1 cells have their archenemies—TH2 cells. Unlike TH1 cells, TH2 cells *inhibit* actions of cytotoxic T cells. In other words, they protect the tumor.

Speaking of inhibiting T cells, the tumor microenvironment has another set of immune cells in abundance called *myeloid-derived suppressor cells* (MDSCs).[5] The name "myeloid-derived" simply means they are derived from the bone marrow, just as all immune cells are. So, what are they doing in the tumor microenvironment? Are they helping the T cells fight the cancer? Not a chance. Instead they synthesize and squirt out enzymes such as nitric oxide synthase and arginase I, which kill T cells! Local production of these enzymes impedes the proliferative potential of T cells and can trigger programmed cell death (apoptosis) in cancer-fighting T cells. The nitric oxide synthase enzymes induce oxidative stress in the tumor microenvironment, which blocks proper formation and function of T cell receptors, so that even if they wanted to do their jobs, the T cells wouldn't be capable.

A pillar of innate immunity, the *neutrophil*, is perhaps the most dynamic of all white blood cells when it comes to confronting and killing germs. Neutrophils are capable of engulfing enemy bacteria and bathing them in a sterilizing stew of hydrogen peroxide, superoxide radicals, and other potent antibacterial oxidizing agents. They are like our own internal bleaching system. Neutrophils are often found in abundance in the tumor microenvironment of certain cancers. So are the neutrophils there to kill the cancer? No, of course not! Neutrophils, under these circumstances, are anything but our friends. Here, beguiled by the tumor microenvironment, they depress the action of anticancer T cells, stimulate growth of tumor neovasculature, and enhance the abilities of tumors to invade beyond their initial surroundings.[6] Furthermore, *ordinary* neutrophils last only about a day or so before they retire and die, whereas in the tumor microenvironment (where they are up to no good), they persist for a disappointingly long time.

Unlike the adaptive immune system (which must first be trained before it can seek and destroy cancer), the innate immune system sees cancer as its natural born enemy. Another key member of the innate immunity team is the NK (natural killer) cell. NK cells do not directly attack and eradicate invading pathogens but instead eliminate badly compromised host cells. Although this usually means virally infected cells under ordinary circumstances, tumor cells also fall into this category, the way NK cells identify unhealthy host cells in need of elimination by their reduced levels of MHC markers. As mentioned, all normal cells in the body express specific MHC I antigens on their surface. The particular MHC I antigens serve as recognition flags that signify to the immune system that they are "self." Any cells displaying the wrong MCH flags will be identified by the adaptive immune system as foreign and attacked. Likewise, any cells devoid of MHC markers will be identified as outsiders and attacked by the innate immune system (especially by NK cells). Virally infected cells reduce their MHC expression, thereby alerting NK cells that something is seriously amiss. What is amiss is that their MHC is missing—the so-called "missing self" situation.[7] Cancer cells often try to evade adaptive immunity by hiding their MHC markers and thereby making themselves invisible to T cells. In so doing, however, these clever cancer cells should be identified by their missing self and eliminated by NK cells. In fact, NK cells *do* efficiently identify and remove circulating tumor cells if they exhibit no MHC surface markers. Nevertheless, the same cancer cells, when ensconced in the tumor microenvironment, are somehow shielded from NK cell attack.

How is this happening? Well, it appears that in addition to detecting MHC surface markers, NK cells distinguish healthy cells from sick cells by measuring the net amounts of "activating" versus "inhibiting" signals coming from the targeted cells. Thus, activation signals plus a reduction in expression of MHC class I molecules earmark virally infected or malignant cells for extermination. Once again, however, in the tumor microenvironment all bets are off. The net balance of activating versus inhibiting signals emanating from malignant cells (and the bewitched immune cells) stonewalls any NK-cell-led anticancer campaign despite the grossly abnormal reduction in MHC expression.

And we are not done yet. *Cancer-associated fibroblasts*[8] were once believed to only serve a structural or physical supportive function within tumors. In normal tissues, fibroblasts are metabolically relatively quiescent and are one of the main structural cells, a sort of building block of the body. So, are the cancer-associated fibroblasts merely functioning as bricks and mortar for the growing tumor? Negative! In the tumor microenvironment, these characteristically quiet building blocks are quite metabolically interactive—they secrete a host of cytokines that stimulate angiogenesis and can also secrete TGF-β (which, as we mentioned earlier, is an immunity-subduing cytokine that aids in the M1 to M2 macrophage conversion and in the helper T to Treg conversion). TGFβ might also be instrumental in something called EMT (epithelial to mesenchymal transition) a process that might help cancer cells metastasize.[9]

At this point I feel quite badly for *B cells*. Having come this far into our story and only now bringing up B cells makes me fear I have given them short shift. In most immunology books, B lymphocytes take center stage. In our epic, however, B cells have been overshadowed by their more illustrious lymphocyte cousins, the T cells. This is because certain T cells (the CD8+ cytotoxic T cells) have the glorious distinction of directly killing cancer cells. In contrast, B cells are the lymphocytes with the mundane job of manufacturing antibodies (immunoglobulins), which play a more prominent role in fighting off infectious diseases. Vaccinations work by motivating B cells to create large quantities of specific antibodies that ward off specific infections. But B cells play a very prominent role in immunology from any perspective. In fact, one common way of describing the immune system is to distinguish two very distinct arms of immunity, namely, *cellular immunity*, which includes T cells, and *humoral immunity*, which is antibody-based.[10] Antibodies derived from B cells are active against toxins and freely wandering pathogens in the bloodstream and other bodily fluids. Since antibodies are found in the "humors" (an old name for body fluids), the antibody-mediated protection afforded by B cells is called "humoral immunity." T cells, in contrast, do not produce antibodies but instead directly attack invaders. Thus, T cells are a component of cell-mediated immunity.

Although the earliest prenatal site of B cell differentiation is the fetal liver, in mammals the primary location for B cell development is in the bone marrow. However, the "B" in B cell does not come from bone marrow but rather comes from the *bursa of Fabricius*, the organ in which B cells mature in birds. The bursa is an out pouch of the *cloaca*, or common urinary, gastrointestinal, and genital opening in certain animals (and was briefly described in chapter 4). In any case, in our tale, B cells may have been given the short end of the stick. How can I atone for this omission?

Well, B cells are found in abundance in the tumor microenvironment. Maybe I can make it up to them and restore their status if they are gallantly fighting off the cancer. Perhaps they deserve a full chapter for themselves! Alright, are the B cells in the tumor microenvironment cancer-slaying, unsung heroes there to fight the good fight? Nah.

Although B cells normally inspire powerful immune responses through antibody production and by presenting antigens to T cells, there is a flip side to the B cell coin. Just as T cells have their regulatory T cells, B cells, too, have their *regulatory B cells*.[11] These regulatory B cells, or *Bregs* (pronounced "bee regs"), include "B10 cells," which are so-named because of their production of the anti-inflammatory cytokine interleukin-10. Under ordinary circumstances, B10 cells seem to have important roles in controlling autoimmune conditions such as inflammatory bowel disease. However, cancer is anything but ordinary. Although the details remain murky, in the tumor microenvironment it appears that B10 cells promote tumor survival and growth by inhibiting cytotoxic T lymphocytes and by promoting the conversion of CD4+ T cells into Tregs. Sadly for the B cell saga, they will not be thrust into the limelight as heroes at the eleventh hour.

The list of former friends and allies found guilty of treason goes on and on. Although immune cells are often found in abundance within tumors, things are not as they appear. Our army of protection—our immune system–is not there fighting cancer. In the tumor microenvironment, these immune cells are cancer's BFF. This place—the tumor microenvironment—is as crazy as Brian Greene's Quantum Cafe! In the microscopic world of quantum mechanics, the day-to-day rules we are so familiar with in the macroscopic world just do not apply. Particles behave like waves, and waves exhibit particle-like properties. One cannot truly know a particle's position and momentum simultaneously; one cannot predict many aspects of quantum mechanics with certainty but instead can only assign probabilities to a certain outcome; entangled particles light-years apart respond instantaneously to each other's condition. Common sense goes out the window. I never thought anything could be as weird as the quantum world. This was until I learned about the tumor microenvironment.

A painful new theme is patently evident: Cancer, through an array of perfidious cells and counterproductive cytokines, has surrounded itself with the most powerful, protective police squad imaginable in the tumor microenvironment. Our normally capable cancer-fighting soldiers really don't stand any chance against these turncoat cancer-defenders.

Before moving on, however, it should be pointed out that although the tumor microenvironment might seem as mad as a hatter, such a weird world would in fact be perfectly normal—if it were in the placenta defending a fetus. Similarly, these seemingly traitorous cancer-defending cells would be perfectly welcomed if they were doing what they ordinarily do: preventing autoimmune reactions. In years to come, I predict that a great synergy will unfold as we learn more and more about the underlying cellular and molecular mechanisms behind autoimmune diseases and cancer, for there are significant parallels. For instance, in autoimmunity, the same coterie of cells that we so despise in the tumor microenvironment (for protecting the cancer), are again failing us. But unlike in cancer where the Tregs and company are alive and well but defending the tumor, in autoimmune diseases, such as lupus and rheumatoid arthritis, these regulators—that are supposed to be preventing the overly zealous soldiers from attacking our own tissues—have gone to sleep on the job, and in so doing, are permitting rampant hostility against our own bodies. If only they were so lazy in the tumor microenvironment.

In summary, the laboratory has provided confirmation of our theory. Cancer has indeed stolen the secret of the embryo. The cloaking device is real in the tumor microenvironment. Okay, so what are we going to do about it?

Chapter 36

CONNECTED DOTS: LOOKING BACK AND GLIMPSING THE FUTURE

When I first began this project, many folks, including my own colleagues, were quite skeptical about cancer immunotherapy and its prospects for effectively addressing cancer. I was forewarned that immunotherapy had already been tried and met with little success. Why would the recent revival be any different? Has anything changed?

I would say the answer is a definite, YES! For me, what has changed is the fresh new way of looking at the entire concept of cancer immunotherapy.

Looking back, what has brought me to this new perspective is the conglomeration of clues gleaned from our global expedition. After encountering an amazing and inexplicable experience in the form of the abscopal effect in the cancer center, we travelled to Tasmania, and then off to Africa, and then over to Ecuador; while having fun with time travel from the Paleozoic to the Paleolithic.

Along the way we compared and contrasted the uniformly fatal Tasmanian devil contagious cancer with the dog's normally nonlethal and naturally retreating contagious cancer. We have noted the similarity of the former with advanced human cancers and drawn analogies between the later with the abscopal effect.

We have also seen that cancer, in rare circumstances, can be transferred from one person to another, which defies conventional wisdom. Unlike clams, humans and other vertebrates possess adaptive immunity and the sophisticated MHC system—which might have evolved expressly to forbid such transfer. Yet, such transfer has happened.

We have examined the profound immune deficiency in certain organ recipients and in HIV/AIDS patients. Their immunological insufficiency leaves them relatively prone to cancer, supporting our assertion that the immune system plays a critical role in cancer prevention.

An amazing discovery was made in the transplant clinic: surgeons have transplanted organs from donors *decades* after they were "cured" of cancer—only to learn that those healthy-appearing organs sometimes concealed hidden cargo of cancerous cells. These cancer cells were probably being ably held in check in the donor's body for years! The donor's competent immune systems were apparently capable of controlling that metastatic cancer for extended periods of time. Could they have done so forever? Could *our* immune systems do the same if or when we, too, are challenged with malignant cancer?

We have seen the poorly understood abscopal effect wherein treatment with radiation therapy, cryotherapy, or some other *local* therapy induces a *distant* and body-wide response. Could this be a manifestation of the immune system awakening and recognizing a foreign cancer and either eliminating it outright or, alternatively, "sending it to jail" indefinitely via what we might today call immunological equilibrium? Either way, I would argue that the answer is yes.

We have stumbled upon cancer-free critters such as naked and blind mole rats that use a variety of molecular methods to remain cancer-free. Next, we went on to meet Dr. Cui's mighty mouse, the amazing mouse whose immune system would not allow it to get cancer. But maybe more importantly, we have seen how transfusions of his white blood cells were capable of both preventing and curing cancer in other mice. We have speculated that there could be *people* out there with the same super power—cancer immunity. Can we find them? And through blood transfusions, could they possibly cure us all (regardless of whether we have a clear and full understanding of the underlying biological mechanisms)?

We have also explored the recently discovered weird world of the tumor microenvironment, a place where all is not as it seems. Here, although there is often robust immunological activity, the immune system is not present to kill the cancer. Rather, it is fervently protecting the cancer. The whole situation seems so bizarre—until we recognize that this scenario is strangely reminiscent of another far more familiar situation—protection of the embryo from a mother's immune system. During our visit to the obstetrician's office we observed that the embryo/fetus very effectively conceals itself from immune attack. Embryos and fetuses shroud themselves with a veil of invisibility, and so do tumors. Could the mechanisms be one and the same? We observed that late in pregnancy and after delivery, a mother's acceptant and forgiving immune system reverts back to its original intolerant self. Can we similarly revert the permissive tumor microenvironment back into a place where neoplasms are no longer welcomed?

We certainly have taken a long and winding expedition, but now that we have come full circle, can we put our tour to good use in the cancer clinic? Can we connect the dots from these sundry clues and tie all these scattered bits of information together to find a better treatment for cancer?

As we have learned, cancer seems to subvert the immune system in its vicinity. Instead of killing the cancer, vast arrays of immune cells in the tumor microenvironment are actually safeguarding the cancer like an overprotective parent. I have facetiously called these "royal guardians," but thanks to our visit to the obstetrician's office, we now know the real names of these spellbound tumor protectors. And in recent years, we have uncovered many of the biochemical and molecular mechanisms that cancer uses to hypnotize these cells into doing their bidding. All this takes us one step closer to breaking the spell. Can we, like Odysseus, figure out how to resist and reverse Circe's spell and take our immune systems back?

We have seen from past experience that simply "strengthening" the immune system typically does not produce the desired cancer-curing effect. First-generation cancer immunotherapy agents from decades back, such as interferon-α and interleukin-2, do augment the army of tumor-killing soldiers but also proportionately reinforce the previously unrecognized tumor guardians. This is perhaps why cancer immunotherapy did not fare well during the first go-around.[1] A very different strategy was required.

One such strategy is via checkpoint inhibition.

Actively thriving tumors are often infiltrated (or infested, depending on one's perspective) with lymphocytes. We would hope that the tumor-infiltrating lymphocytes, or TILs, are there to exert a powerful anti-cancer effect. But while these TILs do have the *potential* to attack cancer (as demonstrated in the laboratory), they remain asleep in the tumor microenvironment. The tumor microenvironment lulls these warriors into lassitude through soporifics such as TGF-β, IL-10, VEGF, IL-6, and many more. Additionally, police cells such as Tregs and MDSCs vigorously enforce a "thou shall not touch" policy regarding the tumor. Another recently revealed trick that tumor cells use is to display "halt signals" on their surfaces. In this fashion, the T cell warriors are fooled into believing the grossly abnormal cancer is normal tissue, or self, and they abide by their prime directive: "thou shalt not attack self."

These "halt signals" go under the name of *immunological checkpoints*, and disabling these checkpoints represent a promising new avenue of cancer therapy. [2]

Only recently have the concepts of immunological checkpoints been appreciated and understood. If we think about this, the whole idea now just seems so intuitive that it is somewhat surprising that the notion wasn't conceived many years back. Following a vicious but victorious battle when every last virally infected and compromised cell is purged, the still-armed and ready T cell army must know when to stand down. There must be some signal to call off the attack. This is where immunological checkpoints play their normal and essential physiological role. Activation of immunological checkpoints limits the extent of inflammation and curtails the T cell attack, which if unrestrained would cause serious damage. Checkpoints serve as brakes on the immune response. If the immune system is not kept under extremely tight control, excessive and ongoing inflammation or serious autoimmune diseases will result.

In other words, immunological checkpoints prevent persistent inflammation and avert autoimmune disease. Not surprisingly, immunological checkpoints are also quite active in cancer.

One example of immunological checkpoint activity involves the interaction between T cells and antigen-presenting cells such as dendritic cells. "Activation" of a naïve T cell requires two distinct cell-to-cell interactions.[3] First, the T cell receptors on the T cell surface must bind to MHC molecules and antigens (displayed as a unit) on the surface of dendritic cells. Only T cells bearing receptors that precisely match the size and shape of the antigen/MHC unit on the surface of the dendritic cell will be eligible for full activation. But this step alone is not sufficient to fully motivate a T cell. Another step, *co-stimulation*, is required. A well-studied example of T cell co-stimulation is via another receptor on the surface of T cells called *CD28*. CD28 binds to co-stimulatory molecules on dendritic cells called *B7.1* (also known as CD80) and *B7.2* (CD86). To mature and become fully active, a juvenile T cell must engage *both* its T cell receptor (by binding to the MHC/antigen unit on the dendritic cell) *and* CD28 (by binding to B7.1/B7.2 on the dendritic cell). Only when both of these are fully engaged will a naïve T cell be able to take that next step in the activation process. But there is more to this story.

While these activating and coactivating receptors must be engaged in order to commence confrontation with the enemy, there are other receptors on T cells that *postpone or suspend* activity. These *inhibitory* receptors[4] (such as *CTLA-4* for "cytotoxic T-lymphocyte antigen 4") also can bind to the very same B7.1 and B7.2 molecules on the dendritic cells. In fact, CD28 and CTLA-4 *compete* with one another for binding to B7.1 and B7.2—and CTLA-4 wins. CTLA-4 binds to B7.1 and B7.2 ten times tighter than CD28 does. Upon binding of CTLA-4 on T cells with B7.1 or B7.2 on dendritic cells, the whole system is dampened; T cell activity is suppressed. Although CTLA4 is expressed on *cytotoxic* T cells (thus its name), its biological role is perhaps more significant on *helper* T cells. Engagement of the CTLA-4 receptor on helper T cells tones down helper T cell activity and *enhances* the immuno*suppressive* activity of regulatory T cells. The end result is that activation of a T cell is invariably coupled with inhibition of that T cell. The two sides of the scale (co-stimulation and co-inhibition) are delicately balanced. In this manner, inflammation will ordinarily cease after the goal of infection-eradication has been achieved. Nature *needs* these precisely balanced checkpoints to ward off autoimmune disease and limit inflammation during infection.

This highly regulated and finely tuned function involving co-stimulation and co-inhibition is the immunological checkpoint and the site of interaction between the cells involved is sometimes called an *immunological synapse* in analogy to neuronal synapses between dendrites and axons.

Tumors skew the balance of activity at these immunological synapses in their favor. Some tumors have figured out how to defend themselves by exploiting immunological checkpoints and synapses. It is now appreciated that one very powerful mechanism that tumors use to evade immunological detection and destruction is through activation of these checkpoints. In so doing, they put blinders on attacking T cells. In fact, this could be a *major* means through which tumors evade immunity. Checkpoints inhibit the activation of cancer-killing T cells . . . and cancers activate these checkpoints![5]

In cancer, where an immunologically suppressive environment seems to dominate, what would happen if the "brakes" (CTLA-4) were somehow disconnected. Would this allow the immune attack against the tumor to proceed as planned? As it turns out there is now a drug, ipilimumab that does exactly this.[6] Ipilimumab is a monoclonal antibody directed against CTLA-4. Basically, what this drug does is inhibit the inhibitor. It impedes the action of CTLA-4—and since CTLA-4 serves as a brake on the T cell activation process, disengaging this brake allows the process to proceed unrestrained. In this manner, ipilimumab enables T cells to mature, become active, and attack the cancers they were formerly held back from.

CTLA-4, however, is certainly not the only immunological checkpoint. Several others exist and some operate at different locations and phases of the immune response. CTLA-4 primarily regulates T cells during the *priming phase* of T cell activation, which occurs in the lymph nodes. CTLA-4 is thought to function as a *general* immunological shutdown and broadly inhibits T cell activity. Another well-studied checkpoint involves PD-1 (short for programmed cell death protein 1), which dampens T cell activity during the *effector phase* of the immune response.[7] (The activation of T cells right there on the battlefield between pathogens and immune cells is called the effector phase.) Like CTLA-4, PD-1 does act partly in the nearby lymph nodes (i.e., during the priming phase) but it also acts right at the site of infection (i.e., during the effector phase). However, unlike CTLA-4, which is a general immunological inhibitor that helps prevent body-wide autoimmunity, the PD-1 checkpoint is more anatomically restricted. One normal function of the PD-1 checkpoint is to impede T cell activity in *peripheral tissues* (i.e., tissues outside the lymph nodes) in order to limit collateral tissue damage during an immune response to an infection. PD-1 checkpoint activity also probably plays a crucial role in promoting tolerance of "self" and thereby reducing autoimmune conditions.

Since PD-1 checkpoint action occurs right at the site of infection (or in the case of cancer, right at the tumor) engagement of the PD-1 checkpoint can shut down T cell activity in the tumor microenvironment. It turns out that this is definitely one of the numerous nefarious ways neoplasms are shielded in the tumor microenvironment. And detailed knowledge of our enemy's playbook may provide a means of defeating our foe.

The PD-1 checkpoint pathway is a bit different from the CTLA-4 checkpoint mechanism. PD-1 molecules are normally expressed on the surfaces of activated T cells, as well as on B cells and macrophages. When PD-1 is linked to its receptors (PD-L1 or PD-L2, which stand for programmed death ligands 1 and 2), T cell activity is curtailed. (Incidentally, in biochemistry and molecular biology, a "ligand" is a molecule that binds to a receptor.) One ligand for PD-1 (PD-L1), is found in abundance on most normal cells and serves to ward off attacks on normal cells by T cells engaged in battle with infectious agents. The other ligand, PD-L2, has a somewhat more restricted expression and is limited to immune cells and a few organs, such as the lungs and colon. While such termination signals are desirable in preventing autoimmune diseases such as lupus, loss of T cell activity in the tumor microenvironment would be a major handicap. And cancer cells do have a most contemptible method of triggering this checkpoint and activating the tumor-protecting PD-1 pathway. They impersonate immune cells by expressing PD-L1 and PD-L2 on their surfaces. This shuts T cells down; in essence, cancer cells stymie T cells by directly hitting their "off-switches." When PD-L1 or PD-L2 on the surfaces of tumor cells in the tumor microenvironment link up with PD-1 receptors on T cells, those T cells befriend the tumor. Such consorting with the enemy has been clearly documented in certain cancers of the lungs and ovaries and on some lymphomas and melanomas. In all likelihood, many other cancers play the same mean trick on our T cells.

But as with CTLA-4, if we could design a drug that interferes with cancer's ability to turn off T cells through the PD-1 pathway, we might have something. Fortunately we now do. New drugs such as nivolumab and pembrolizumab are monoclonal antibodies that bind to PD-1 and block the activation of the PD-1 pathway. Both were FDA approved in 2014[8] (well after I started this book) for use in the United States against melanoma. In 2015, the FDA approved use of nivolumab against previously treated and progressive non-small-cell lung cancer.[9] Another antibody aimed at interfering with the PD-1 checkpoint pathway, pidilizumab, presently is under study. Newly developed antibodies directed against the ligands for PD-1 have not even been given formal names yet but go by company code names such as MDX-1105, BMS-936559, MEDI4736, MSB0010718C, and MPDL3280A (actually this last agent has recently been given the name atelolizumab).

Another drug undergoing clinical testing at this time is tremilimumab, which like ipilimumab is an antibody against CTLA-4. And there are many more immunological checkpoints besides just the two examples of CTLA-4 and PD-1. Among the co-stimulatory checkpoints, are CD28, OX40, GITR, CD137, CD27, and HVEM. Can potent anticancer effects be evoked by strengthening such interactions? And besides the CTLA-4 and PD-1 inhibitory checkpoints, there are TIM-3, BTLA, VISTA, and LAG-3. Might releasing these brakes improve cancer outcomes?

So far we have only been focusing on *T cell* checkpoints. As mentioned in the previous chapter, there are many other players in the tumor microenvironment. And while awakening T cells is all the rage at the moment, in years to come, new methods of breaking the spell on other bewitched cells in the tumor microenvironment should further augment our anticancer crusade. For example, *natural killer* (NK) cells, like T cells, have both stimulatory and inhibitory receptors on them. KIRs (killer immunoglobulin-like receptors) can be activated (by ITAMs, or immunoreceptor tyrosine-based activation motifs) or inhibited (by ITIMs, or immunoreceptor tyrosine-based inhibitor motifs). As in the T cell checkpoints, the balance between activation and antagonism must be carefully weighed—and in general, inhibition wins out over activation. Once again, the tumor microenvironment takes advantage of this phenomenon and shields cancer cells from attack (even when they have reduced MHC expression). Obviously, our clinical goal is to obviate inhibition and simultaneously provide activating signals in the tumor microenvironment to encourage natural killer cells to go on a cancer-killing spree. And as in the case of T cell checkpoints, there is much more than just the one example of KIRs. Other checkpoints on natural killer cells include the *NKG2* pathways. Prostate cancer cells, for example, down-regulate (hide) their MHC class I molecules, which, while effectively concealing themselves from adaptive immunity by T cells, should make them vulnerable to natural killer cell mediated innate immunity—or at least that's how it *should* work. The way prostate cancer cells work around this is by shedding decoy NKG2D molecules. These decoy ligands bind to NKG2D receptors and slow down natural killer cell activity. As in the case with T cell checkpoint inhibition, we are devising ways to overcome NK cell checkpoints.[10] One example is *lirilumab*, a monoclonal antibody directed against KIRs on NK cells. Blocking these inhibitors should facilitate anticancer killing activities by NK cells. Regardless of the specifics, many new cancer-concealing molecules are being discovered and many new drugs are being designed to counter their cloaking abilities. Slowly but surely the "emperor of all maladies" is being rerobed. And once that tumor has all its identifying clothes back on and our immune systems can clearly see the cancer that has been fooling it for so long, that angry immune system just might dole out justice with a vengeance. One gets the exciting feeling that we are just uncovering the tip of a huge iceberg.

This academic excitement must of course be tempered by realities encountered in the clinic. Clinical trial results are coming in. At first blush they might not appear all that impressive. Two standard metrics, the percentage of patients who respond, and the median survival, are slightly, but not dramatically, different from what has been historically achieved with conventional treatments. However, the real difference lies among those who do respond. With conventional therapies for many advanced cancers, there might be a high response rate to chemotherapy, but the duration of that response might be disappointingly brief. In other words, median survival might be slightly longer than without treatment but there is no real "cure" rate to speak of. With the checkpoint inhibitors, there do seem to be some genuine long-term survivors. The proportion of people with melanoma, for instance, who are still alive at two years, five years (and more) is significantly higher with the checkpoint inhibitors. These individuals appear to be experiencing encouragingly long remissions. And the clinical indication for these new drugs is rapidly evolving. For instance, just this past year the results of CheckMate 067, a clinical trial comparing ipilimumab versus nivolumab versus a combination of the two for patients with melanoma, have come in.[11] Nivolumab appeared clearly better than ipilimumab alone, and the combination might be even better than nivolumab alone. Similarly, the KEYNOTE-006 study showed that pembrolizumab was better than ipilimumab in terms of both efficacy and side effects for melanoma patients.[12] We must of course remain mindful of the side effects of these new agents. Unlike the familiar hair loss, nausea and vomiting, and declining blood cell counts of conventional chemotherapy, modern cancer immunotherapy comes with a new set of relatively unfamiliar adverse effects such as autoimmune problems involving the thyroid, pituitary, GI system, lungs, and liver, among many more. Nevertheless, there is definitely room for cautious optimism.

We should not forget the lessons learned from prior generations of cancer immunotherapy. While it is true that early immunotherapy failed to live up to the hype, today, many of us believe that this was because the first iteration of cancer immunotherapy focused only on *strengthening* the immune system. We were unaware of the fact that such general strengthening would simultaneously and counterproductively bolster the *inhibitory* capacities of the immune system. We did not understand the inhibitory functions of the immune system back then. Today, there is still much we likely do not understand. This first generation of checkpoint inhibition is certainly encouraging but is unlikely to be *THE* cure for cancer; at least not as single agents used alone. What would *combinations* of immune-*strengthening* agents such as interferon-α or interleukin-2 plus checkpoint inhibition produce? I have not yet even mentioned the numerous evolving immunotherapy approaches such as the FDA-approved sipuleucel-T.[13] In this form of cancer vaccine, dendritic cells are removed from a prostate cancer patient's body and sent to the lab. There, they are "trained" to attack prostate cancers by exposing them to PAP, or prostatic acid phosphatase (which is overexpressed on malignant prostate cells), and expanded by incubation with a cytokine (specifically, granulocyte-macrophage colony stimulating factor, or GM-CSF), which stimulates their growth. Next, these trained and amplified dendritic cells are reintroduced into the patient with the hope of provoking an anticancer attack with renewed fury. The process is repeated a total of three times. This treatment has provided some benefit but is certainly not considered *the* cure. We have to wonder if we could magnify anticancer effects through potentially synergistic combinations of various immunological mechanisms.

Where might other immune-modulating therapies such as classical anticancer vaccines, tumor-directed antibodies, cell-based cancer immunotherapy, and very low-dose total-body irradiation fit into the new equation? Can we somehow integrate high-dose, localized radiation therapy into modern cancer immunotherapy to produce long-term abscopal effects on a regular basis? Will currently unyielding cancers finally concede defeat once we learn the right recipe?

Years of skepticism about cancer immunotherapy are finally fading. After this voyage in time and geography, exploring everything from anthropology to zoology, I personally have become a firm believer. Although I myself was once quite skeptical about any potential of the immune system to defeat cancer, my own clinical experience plus this prolonged "thought experiment" has dramatically changed my view. The puzzle pieces finally fit. It all makes sense now.

Many former cynics are experiencing a change of heart. *Science* magazine listed cancer immunotherapy as the 2013 "Breakthrough of the Year." Professor Jim Allison (CTLA-4 pioneer, chair of the Department of Immunology at the University of Texas MD Anderson Cancer Center, and winner of the 2014 Breakthrough Prize in Life Sciences), mentioned that even James Watson, Nobel Prize-winning codiscoverer of the DNA double helix (and former cancer immunotherapy skeptic) once told him, "This is going to do it."[14]

The revolution is on!

ACKNOWLEDGMENTS

There are far too many to whom I am indebted to list here, and I apologize in advance to all those I have missed.

I would like to begin by expressing my thanks to gifted writer and published author Ron Piana for the many creative conversations and for providing me with the courage to initiate this project.

I am especially indebted to "Big Professor" Christian Debranin for his extraordinarily sharp clinical and scientific insights along with his numerous stimulating conversations, emails, and thought-provoking questions. I'd like to thank the many friends and colleagues I met through Fermilab, including Alvin Tollestrup, Mark Palmer, Tom Kroc, Mary Anne Cummings, Carol Johnstone, Bruce C. Brown, G. P. Yeh, Fritz DeJongh, and many more who offered ideas and advice along the way.

Particular thanks go to Nobel Laureate Leon Lederman for his encouragement and endorsement of Prometheus Books. Leon (along with Christopher Hill) is coauthor of *Beyond the God Particle*, which was published in 2013 by Prometheus Books. When others dared not take a chance on an unknown, unpublished author, Prometheus was willing to consider the project, and I am forever in debt to Leon for the introduction. In the same vein, I am most grateful to Steven L. Mitchell for his willingness to consider my manuscript and return my initial call years ago. If not for him, and the voluminous e-mails and telephone conversations along the way, this project would not have ever gotten started, let alone come to fruition.

I'd like to acknowledge the numerous e-mail exchanges with members of SARI (Scientists for Accurate Radiation Information), including but certainly not limited to Rod Adams, Wade Allison, Jerry Cuttler, Mohan Doss, Bob Hargraves, Mark Miller, Noy Rithidech, Bill Sacks, Bobby R. Scott, Jeffry Siegel, and Yehoshua Socol.

Special thanks go to all my colleagues and coworkers in Radiation Oncology at Loyola University Medical Center and the Edward Hines Jr. VA Hospital for putting up with me during this lengthy project. I especially thank my chairman, Dr. William Small, for his extraordinary patience and understanding during this lengthy ordeal. I wish to also express my appreciation to my many friends and colleagues at ACRO, ASTRO, RSNA, ACR, ASCO, and SBMT who provided valuable feedback along the way; especially Drs. A. J. Mundt and Babek Kateb for their ideas, input, and advice on this project.

I wish to also thank my many veterinary colleagues for their patience and kind input including Drs. Lynette Greenwood, Greg Almond, John Hynes, Tom Beckett, and several others.

A very special thank you goes out to Jeff Limmer for all the stimulating and inspiring discussion about the various topics examined in the book. Without his input, I doubt I would have pursued the project. Similarly, I am especially thankful to Rock Mackie for his encouragement and the countless and intellectually stimulating conversations over the years that have provoked the in-depth exploration of much of the material covered in this book. Also, I wish to express my sincere special appreciation to Dr. Steven Finkelstein and Carmen Bigles for their ideas, interest, and support for this project over the years. Another thank you goes to friends and professional colleagues Drs. Eduardo Fernandez, David Gius, Minesh Mehta, Bill Rate, Gayle Woloschak, and Sarah Thurman for putting up with all my blabbering on this subject over the years.

Particular thanks go out to Dr. Arlan L. Rosenbloom, adjunct distinguished service professor emeritus of pediatrics at the University of Florida College of Medicine and world authority on Laron syndrome, for his careful criticism and corrections to a chapter that his own important research figures heavily in. That chapter is now far better and more accurate thanks to his input. I am similarly indebted to Dr. Zeng Cui for his highly appreciated information and feedback on the cancer-resistant mice discussed in one of the chapters in this volume.

I wish to also thank the staff at Wild Life Sydney Zoo in Australia for the valuable information on Tasmanian devils and devil facial tumor disease during my visit.

A unique thank you goes out to the Field Museum in Chicago and the American Museum of Natural History in New York for the excellent collections and information that fostered my imagination and slaked my thirst for knowledge. I actually spent a good deal of time at these marvelous institutions contemplating and constructing the book. Similarly, I should thank NIU physics chairman Larry Lurio for allowing me to teach astrobiology through the Northern Illinois University Honors Program and his department. That experience provided fertile ground for much of the unusual material covered in this present issue.

I also wish to say thank you to my longtime friend and professional colleague Michael Wilburn for the innumerable discussions on molecular biology, biophysics, and immunology dating back to our days at Yale. I would like to also express my appreciation to Stephanie Weiss and Hugh McLaughlin for the countless conversations and sometimes lively debates over the past several years that have substantially remodeled this effort. Ditto for the witty, instructive, and delightful dinnertime repartee with Drs. Manny Suarez and Ed Johnson in St. Vincent, which served as a platform on which to construct much more.

I wish to acknowledge George Johnson, author of *The Cancer Chronicles*, whose stimulating book and occasional e-mails further inspired me during the creation of this volume.

I am thankful to Said Daibes and many more of my friends and colleagues at the Nuclear Regulatory Commission for the instructive conversations we have had on immunology and nuclear physics.

I would like to thank Tom Donovan for his continued friendship, support, and clarity of ideas. He provided me with the needed perseverance to see this project through.

I thank my patients and the staffs at the various institutions where I have practiced who have asked poignant and thought-provoking questions and taught me to be a better physician. In Wisconsin: Wisconsin Rapids, Wausau, Madison, Marshfield, Antigo, Manitowoc, Freeport, Appleton, Rhinelander; in Shreveport, Louisiana; and in Illinois in Peoria, Pekin, and the greater Chicago area, including the NIU institute for Neutron Therapy at Fermilab. I am also quite grateful to my educators and all the patients I learned from at the Johns Hopkins Hospital in Baltimore where I did my clinical radiation oncology training.

I am extremely grateful to my sister Anna and my brother-in-law Alec for putting up with me, as well as their much needed encouragement, assistance, and constructive criticisms in molding this book from the very start. I am appreciative of all the intelligent input provided by my brother Bill and the careful reviews provided by my sister-in-law Natasha. I wish to express my love and appreciation to my dad (who for the past six years has been responding beautifully to his treatments at Memorial Sloan Kettering Cancer Center) for his review of the manuscript and the numerous corrections he made and to my mom for the encouragement and necessary nourishment during some highly intensive episodes of writing.

I am in deepest debt to my wife, Teri, for the innumerable hours put into this project. In many ways this final book is really her effort. I thank her for the invaluable, insightful, and inspiring input in the creative process, and I am especially appreciative of her intense efforts during the editing process. Without her love and support, this project would never have been completed.

Appendix

IMAGES WITH EXTENDED CAPTIONS

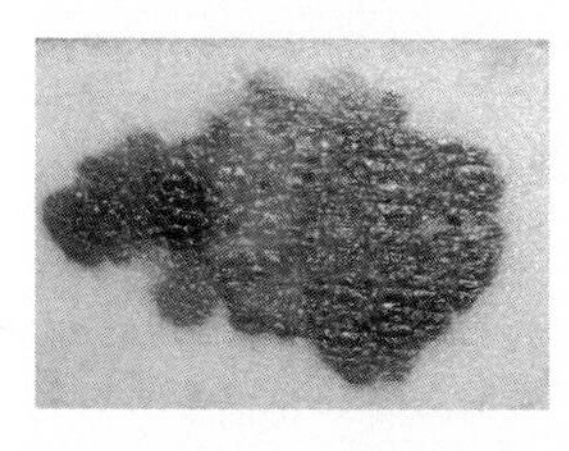

Figure 1: This dark and irregular skin lesion or mole could be melanoma—a dangerous form of skin cancer. One might recall the "ABCDE's" of melanoma: (A) **a**symmetry in the lesion; (B) a **b**order that is irregular or ragged; (C) a recent change in **c**olor or the presence of various different colors within the lesion; (D) a lesion **d**iameter that has changed, particularly if it is larger than the diameter of a typical pencil eraser; and (E) a lesion that is changing or **e**volving over time. Another worthwhile mnemonic is "EFG": (E) for **e**levated, (F) for **f**irm, and (G) for **g**rowing. A mole that begins to bleed or itch is another warning sign. Skin lesions meeting any of these criteria should be inspected by a physician or health professional. *Photo from the National Cancer Institute.*

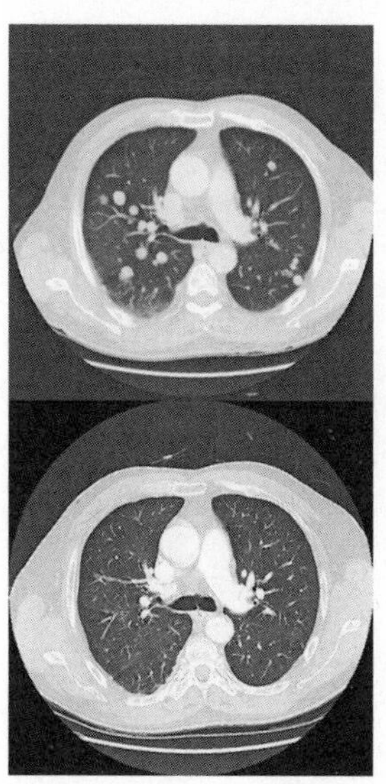

Figure 2: These CT (computed tomography) images are horizontal slices through the lungs at the level of the carina where the windpipe or trachea splits into right and left mainstem bronchi. The image above shows several large, round tumors consistent with metastatic hepatocellular carcinoma. The image below is the same patient four months after seventy gray in fifteen fractions of stereotactic body radiation therapy (SBRT) directed to the liver tumor alone. Apparently an abscopal effect (oncological action at a distance) occurred. The radiation therapy to the liver tumor somehow induced an immunological response which resulted in regression of the metastatic cancer at other (unirradiated) sites such as these lung lesions. *Photo from Michael Lock et al., "Abscopal Effects: Case Report and Emerging Opportunities,"* Cureus *7, no. 10 (October 7, 2015): e344, doi:10.7759/cureus.344.*

Figure 3: This handsome devil is *Sarcophilus harrisii*, the Tasmanian devil. It is currently the largest extant carnivorous marsupial. However, thanks to a bizarre epidemic of contagious cancer, the Tasmanian devil could soon join its cousin the Tasmanian tiger, or thylacine, in extinction. *Photo by Wayne McLean, from Wikimedia Commons.*

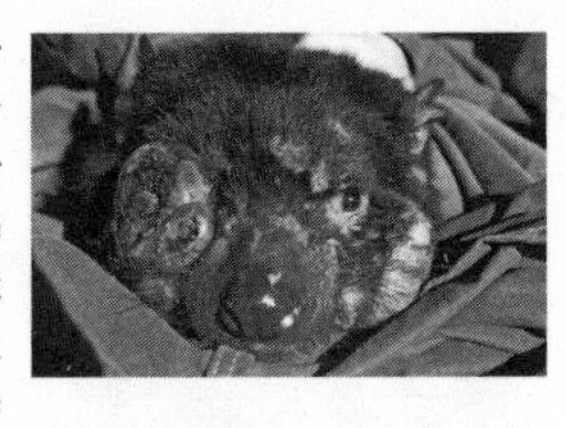

Figure 4: Devil facial tumor disease (DFTD) is an invariably fatal form of contagious cancer that appeared in the mid-1990s. Because the tumors affect the face and mouth—and because Tasmanian devils tend to fight and bite each other on the face and mouth (and tumors) with abandon—the cancerous cells often become implanted in the gums and the disease spreads relentlessly. Unless the Tasmanian devil evolves into a "Tasmanian angel" and ceases its aggressive behavior, it could very well face extinction in the wild in short order. Heroic conservation efforts are underway. The immunobiology of this disease is also under intense study. *Photo by Menna Jones, from Wikimedia Commons.*

Figure 5: The last known thylacine, Benjamin (shown here in 1933), died alone in captivity in the mid-twentieth century. *Photo from Wikimedia Commons.*

Figure 6: This jackal (*Canis aureus*) is closely related to but not identical to the domesticated dog (*Canis lupus familiaris* or *Canis familiaris*). Nevertheless, this species can be afflicted with the contagious malignant cancer known as canine transmissible venereal tumor, or CTVT. Unlike in Tasmanian devils and clams, the contagious cancer that affects canids is not uniformly fatal. Spontaneous regression can occur in animals with healthy immune systems through an intricate immunological phenomenon. Incredibly, the CTVT cancer-cell line has been passing itself along from animal to animal for an astonishing eleven thousand years. *Photo by Mariomassone, from Wikimedia Commons.*

Figure 7: Pictured here in 1892, Dr. William Coley (1862–1936), a surgeon and researcher at what would later become the Memorial Sloan-Kettering Cancer Center, was one of the true pioneers of cancer immunotherapy. Through his practice of injecting living or heat-killed bacteria (Coley's toxins) into cancer patients, raging fevers developed. The unrestrained immunological response took no prisoners—in fiercely fighting the bacteria, the immune system occasionally also vanquished the malignant tumors. His approach fell out of favor, but today, over a century later, investigators are reexamining his ideas and attempting to revive such strategies in a modern form. *Photo from Wikimedia Commons.*

Figure 8: People, dinosaurs, dogs, and other animals are not the only creatures that can get tumors. Even plants can develop neoplasms. This is Robin's pincushion, or rose bedeguar gall, a plant tumor found on certain rose plants. This particular type of plant tumor is induced by certain wasps capable of parthenogenesis (meaning the females can reproduce without males). *Photo copyright Anne Burgess, from Geograph.org.uk.*

Figure 9: When gall wasps pierce plant tissues with their long, sharp ovipositors and lay eggs, tumors (galls) are induced. These galls serve as tiny wasp nests where the eggs are kept safe and the larvae can grow by feeding on the neoplastic plant tissue. But the gall wasp's eggs and larvae are not entirely safe. . . . Other wasps, incapable of inducing protective tumors of their own, exploit the galls created by gall wasps and lay their eggs within. This cuckoo wasp (*Chrysura refulgens*) lays its eggs in the "nest" made by gall wasps in a manner similar to how cuckoo birds lay their eggs in the nests of other birds. *Photo by Alvesgaspar, from Wikimedia Commons.*

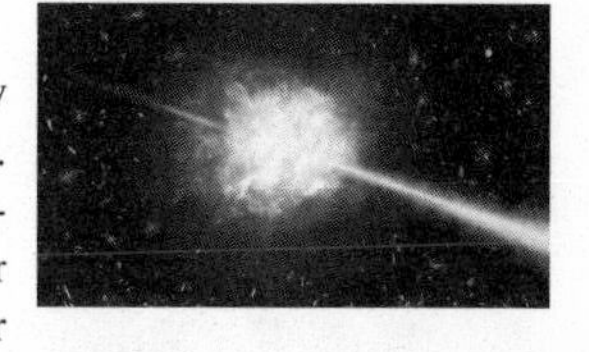

Figure 10: The birth of a black hole can lead to a gamma ray burst. Gamma ray bursts are brief but intense and highly focused. One is tempted to speculate what might happen if a nearby supermassive star were to create a gamma-ray burst headed in our direction. Fortunately, it seems that our galaxy does not appear to host many stars with such potential, and those few that could produce gamma ray bursts are not pointed directly our way. *Photo by ESO / A. Roquette.*

Figure 11: Soft-shell clams (*Mya arenaria*) are under attack! There is an outbreak of contagious cancer plaguing clams along the Eastern seaboard of the United States and Canada. How long the epidemic has been raging and just how the cancerous cells are getting from clam to clam remain uncertain. Also unknown is why the clams' immune systems do not promptly reject the foreign, malignant cells. It is possible that invertebrates such as clams and other mollusks that do not possess the major histocompatibility complex (MHC) system of immunological recognition are at a distinct disadvantage when it comes to contagious cancer. Some have speculated that the MHC found in man and other vertebrates might have evolved expressly for the prevention of transmission of malignant cells from animal to animal. *Photo by Kirsten Poulsen, from Wikimedia Commons.*

Figure 12: While sharks, skates, rays, and their relatives do not frequently get tumors, they are certainly not immune. And in contrast to what was once vigorously promulgated, eating their cartilage cannot prevent or cure cancer. This great white shark has what appears to be a large tumor growing on its mandible. *Photo used with permission from Sam Cahir / Predapix.*

Figure 13: While this naked mole rat might not win any beauty contests, it does hold several titles—including *Science* magazine's 2013 Vertebrate of the Year—thanks to its ability to ward off cancer. One way naked mole rats avoid malignancy is through their high-molecular-weight hyaluronan molecules that physically separate their cells and thereby facilitate a biological phenomenon called contact inhibition. Additionally, naked mole rats amplify their cancer-fighting contact inhibition through expression of another molecule called p16. *Photo by Roman Klementschitz, from Wikimedia Commons.*

Figure 14: This little ball of fur—a blind mole rat—has no eyes or ears. It also has no malignant cells. This is thanks to multiple copies of the gene for the cancer-fighting cytokine beta-interferon. In addition, it appears that a mutation in their gene for the p53 protein has a paradoxical effect of avoiding malignant transformation. Cells with excessive DNA damage are directed towards necrosis as a means of eliminating them through this combination of excess beta-interferon and mutant p53. *Photo by Bassem18, from* Wikipedia.

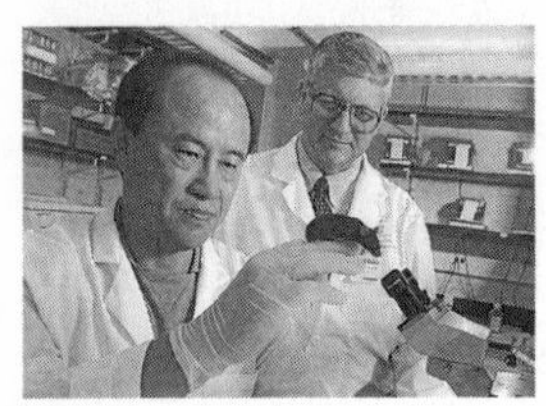

Figure 15: Dr. Zheng Cui and colleagues serendipitously came upon a mouse that just could not be given cancer through standard methods. Most intriguingly, white blood cells taken from this "mighty mouse" could actually cure the cancers of other mice when transfused into them. Could something similar be used in people? Might there be "mighty men" out there who could similarly save us from cancer via blood transfusions? More research into this idea is certainly justified. *Image from* Popular Mechanics.

Figure 16: This skull, dated at approximately eighteen thousand years old, was found on the island of Flores in Indonesia. Although highly controversial, the specimen has been labeled by some researchers as *Homo floresiensis*, a newly discovered dwarf species of man, a "hobbit." Others have argued that the specimen is nothing more than a person with microcephaly, Down syndrome, cretinism, or perhaps Laron syndrome. *Photo by Ryan Somma, from Wikimedia Commons.*

Figure 17: Drs. Arlan Rosenbloom and Guevara-Aguirre with some of "the little women of Loja." The Ecuadorean individuals with Laron syndrome exhibit a short stature but also an extremely low incidence of cancer. *Photo courtesy of Dr. Arlan Rosenbloom.*

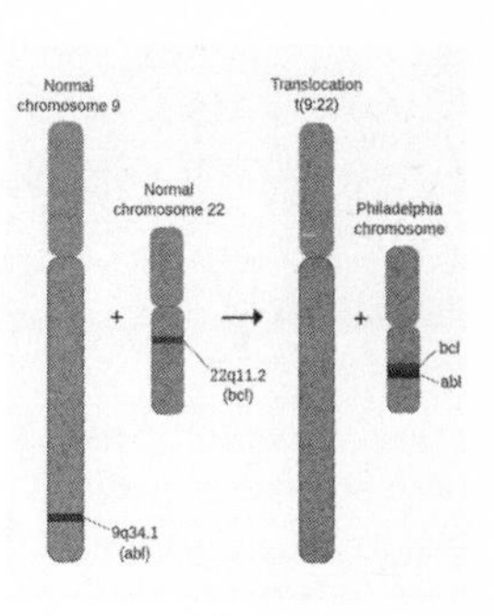

Figure 18: Chromosomal translocations, in which a part of one chromosome detaches and joins another chromosome, can lead to certain types of cancer. The Philadelphia chromosome, in which a part of chromosome 9 is swapped with a part of chromosome 22, can cause chronic myelogenous leukemia. Basketball legend Kareem Abdul-Jabbar announced in 2009 that he was diagnosed with this form of cancer. Thanks to modern "molecular targeted therapies" that specifically interfere with the abnormal tyrosine kinase enzyme that is created by the Philadelphia chromosome, individuals with this type of previously lethal leukemia are now often living long, cancer-free lives. *Image by Aryn89, from Wikimedia Commons.*

Figure 19: This little cutie is a Syrian hamster. Decades ago a unique but highly disconcerting outbreak of mosquito-transmitted leukemia occurred in a captive colony of an inbred strain of these rodents. *Photo by Peter Maas, from Wikimedia Commons.*

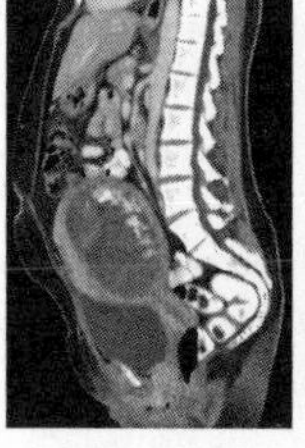

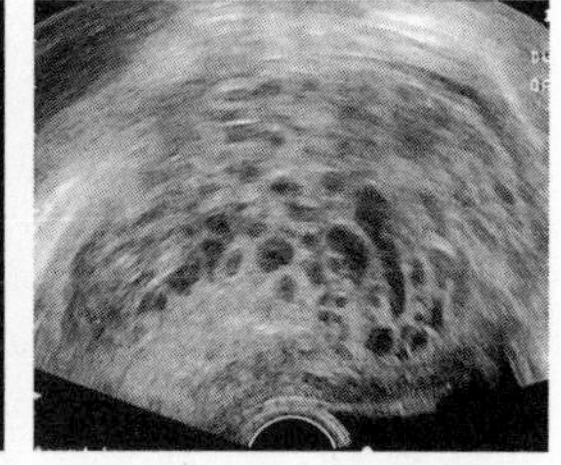

Figure 20, A & B: Hyperemesis gravidarum (severe morning sickness), an enlarging belly, and a positive pregnancy test (positive HCG—human chorionic gonadotropin) would all point to a baby on the way. However, they could also represent a *molar pregnancy* in which, rather than a fetus, the conceptus has turned into a type of tumor. Some molar pregnancies can evolve into a highly malignant (but fortunately highly curable) cancer called choriocarcinoma. How these tumors—which are largely composed of foreign cells—escape the immune system of the mother remains a mystery. *CT image (A) by Hellerhoff, from Wikimedia Commons. Ultrasound image (B) by Mikael Häggström, from Wikimedia Commons.*

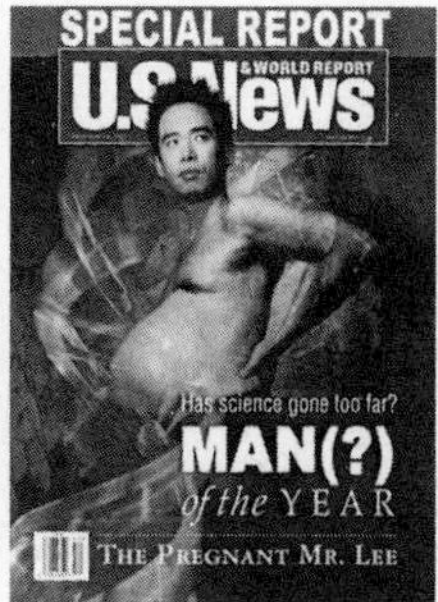

Figure 21, A & B: Could a man get pregnant? Although "the pregnant Mr. Lee" was nothing more than a well-crafted Internet hoax, some argue that an ectopic pregnancy intentionally implanted into the abdominal cavity of a male remains hypothetically possible if the requisite hormonal milieu of pregnancy could be faithfully duplicated. But why would the man's immune system not reject the embryo just as it would reject any foreign tissue? How the immune system "tolerates" embryos remains a subject of intense study. Clues to understanding how cancers evade the immune system might come from such studies. *Images from http://www.malepregnancy.com.*

Figure 22: According to the most popular model of cancer development, we should all get cancer if we live long enough. Simple probability dictates that if one has enough cells dividing enough times, eventually one cell with a dangerous mutation will arise and propagate. That cell's progeny will accumulate additional mutations until eventually an unequivocally malignant cell arises by chance alone. Therefore, very large and very long-lived animals should all eventually succumb to cancer. Dinosaurs, however, did not exhibit the high rates of cancer as predicted. One might dismiss this observation because of the different physiology between humans and dinosaurs. Yet large, rapidly growing mammals such as elephants do have fairly similar basic physiology to humans, and given their enormous number of cells, they should all die of cancer. The fact is they do not. This is in part thanks to multiple copies of the gene encoding the cancer-fighting p53 protein. *Photo by Godot13, from Wikimedia Commons.*

Figure 23: Some might argue that simply being huge is not enough to inevitably lead to cancer; more *time* is required for the requisite number of cell divisions to assuredly lead to malignant neoplasms. According to this concept, very large animals should not possess long life spans. Nevertheless, with a lifespan of over two hundred years, the bowhead whale is perhaps the longest-lived mammal. In order for a human being to have the same number of cells and cell divisions as a two-hundred-year-old bowhead whale, we might have to live to about thirty thousand years. The finding of archaic harpoon tips that haven't been made or used for over two centuries in the skin and blubber of surviving whales is one means of estimating the venerable ages of the leviathans. The fact that dinosaurs, elephants, and whales do not all succumb to cancer as theory would predict is a classic example of Peto's Paradox. *"Harpooning the Greenland Whale," 1876, courtesy of the Freshwater and Marine Image Bank of the University of Washington.*

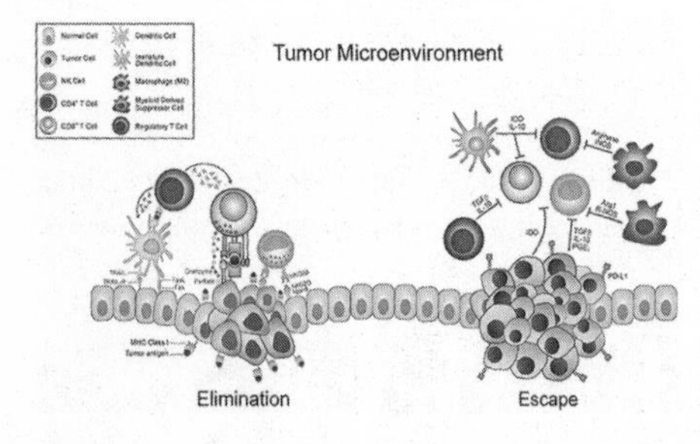

Figure 24: The tumor microenvironment plays a large and previously unappreciated role in how the immune system interacts with malignant tumors. On the left, the immune system has the upper hand: Light-blue mature dendritic cells interact with dark-blue CD4+ helper T cells, which in turn activate green cytotoxic CD8+ T cells. The CD8+ cells, along with orange natural killer cells cooperate in the elimination of malignant cells. On the right side of the panel, the tumor has gotten the upper hand: Pink immature dendritic cells and dark-brown regulatory T cells antagonize CD4+ and CD8+ T cells. Blue myeloid-derived suppressor cells inhibit the function of natural killer cells. Maroon M2 macrophages also get into the fray and inhibit T cell activity. Cancer cells might also express ligands such as PD-L1 to facilitate their ultimate escape. *Image from A. M. Monjazeb et al., "Immunoediting and Antigen Loss: Overcoming the Achilles Heel of Immunotherapy with Antigen Non-Specific Therapies,"* Frontiers in Oncology *3, no. 197 (2013, doi: 10.3389/fonc.2013.00197.*

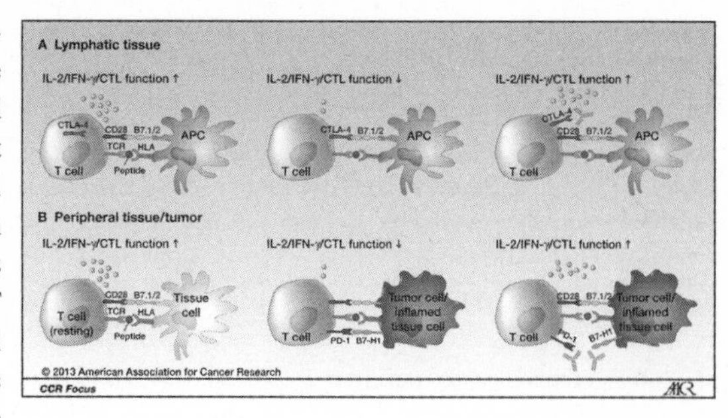

Figure 25: The top panel demonstrates what happens in the lymph nodes during the so-called priming phase. In the upper-left frame, an antigen presenting cell (e.g., a dendritic cell) with an MHC molecule (which in humans is called HLA) interacts with a T cell expressing a specific T cell receptor (TCR) with a specific tumor peptide. This part of the activation step is augmented by interactions between CD28 on the T cell and B7.1 or B7.2 on the dendritic cell. But as seen in the upper-middle frame, competing with the activating CD28 molecule is the repressive CTLA-4 molecule. When CTLA-4 on the T cell binds B7.1 or B7.2 on the dendritic cell, anticancer function is diminished. The upper-right frame shows that antibodies (such as ipilimumab, represented by the yellow Y-shaped structure) directed against CTLA-4 can abrogate the inhibition and once again allow anticancer activity.

The lower panel demonstrates what happens in the normal tissues and tumor during the so-called effector phase. In the lower-left frame, the normal cell with its MHC molecule interacts with a T cell that is expressing a specific T cell receptor and a specific peptide. CD28 on the T cell interacts with B7.1 or B7.2 on the tissue cell. This leads to increased production of inflammatory cytokines and increased cytotoxic T cell activity. While this might be useful during acute inflammation, prolongation of this pattern (i.e., chronic inflammation) is deleterious, and the body strives to reverse the inflammation. In the middle frame on the lower panel, the inflamed cell begins to express B7-H1, which interacts with PD-1 on the T cell. The interaction of PD-1 with B7-H1 (or PD-L1 or PD-L2) quells the inflammation. Malignant cells emulate inflamed cells and express various ligands for PD-1. In this manner, the T cells call off the attack. Antibodies directed against PD-1 on T cells (such as nivolumab or pembrolizumab) or antibodies directed against B7-H1 or PD-L1 or PD-L2 might reestablish the anticancer activities of the previously suppressed T cells. *Photo reprinted by permission from the American Association for Cancer Research: Patrick A. Ott, F. Stephen Hodi, and Caroline Robert, "CTLA-4 and PD-1/PD-L1 Blockade: New Immunotherapeutic Modalities with Durable Clinical Benefit in Melanoma Patients,"* Clinical Cancer Research *19 (October 1, 2013): 5300.*

NOTES

CHAPTER 1: WHAT JUST HAPPENED?

1. R. H. Mole, "Whole Body Irradiation; Radiobiology or Medicine?," *British Journal of Radiology* 26, no. 305 (May 1953): 234–41. This was the first time the term "abscopal" was mentioned in the radiotherapy literature.

2. James Welsh, "Aggressive Cancer Performs Strange Disappearing Act," *Discover*, March 2014, pp. 24–26.

CHAPTER 2: ACTION AT A DISTANCE

1. M. A. Postow et al., "Immunologic Correlates of the Abscopal Effect in a Patient with Melanoma," *New England Journal of Medicine* 366, no. no. 10 (March 2012): 925–31.

This is a fascinating and encouraging case study in which a patient on ipilimumab experienced an impressive abscopal effect following stereotactic body radiation therapy. This suggested that perhaps combining radiotherapy with checkpoint inhibitors could be synergistic rather than just additive in effect.

2. Susan M. Hiniker, Daniel S. Chen, and Susan J. Knox, "Abscopal Effect in a Patient with Melanoma," *New England Journal of Medicine* 366 (May 2012): 2035–36.

3. Michael A. Postow, Margaret K. Callahan, and Jedd D. Wolchok, "Abscopal Effect in a Patient with Melanoma—Author's Reply," *New England Journal of Medicine* 366 (May 2012): 2035–36.

4. Jacob Aron, "Quantum Weirdness Is Reality. A Groundbreaking Experiment Proves Einstein Wrong," *New Scientist*, September 5–11, 2015, pp. 8–9.

This is a recent article on the confirmation of entanglement. An online version is also available, https://www.newscientist.com/article/dn28112-quantum-weirdness-proved-real-in-first-loophole-free-experiment/ (accessed November 17, 2015).

5. Brian Greene, "Spooky Action at a Distance," NOVA, September 22, 2011, http://www.pbs.org/wgbh/nova/physics/spooky-action-distance.html (accessed September 27, 2015).

CHAPTER 3: DISAPPEARING DEVILS

1. Julie Rehmeyer, "The Immortal Devil," *Discover*, May 2014, pp. 43–49.

2. Ibid., pp. 44–45; David Quammen, "Contagious Cancer: The Evolution of a Killer," *Harper's Magazine*, April 2008, pp. 33–43.

CHAPTER 4: THE DEVIL HIMSELF: SOME DIABOLICAL BIOLOGY

1. "Thylacine or Tasmanian Tiger, Thylacinus Cynocephalus," Parks and Wildlife Service, Tasmania, *2006,* http://www.parks.tas.gov.au/index.aspx?base=4765 (accessed October 20, 2015); Amy Ziniak, "Sad Tale of the Tasmanian Tiger: How Benjamin, The Last of His Kind, Died of Exposure at Hobart Zoo After Being Left Out in the Cold," *Daily Mail Australia*, October 2014), http://www.dailymail.co.uk/news/article-2746016/Left-cold-die-The-Tasmania-Tiger-extinct-result-human-neglect.html#ixzz3n8S7XW7t (accessed October 20, 2015).

2. Loveleena Rajeev, "Tasmanian Devil Facts," Buzzle, http://www.buzzle.com/articles/tasmanian-devil-facts.html (accessed October 20, 2015).

3. Katrina Morris, Jeremy J. Austin, and Katherine Belov, "Low Major Histocompatibility Complex Diversity in the Tasmanian Devil Predates European Settlement and May Explain Susceptibility to Disease Epidemics," *Biology Letters*, 9 (December 5, 2012): 20120900, http://dx.doi.org/10.1098/rsbl.2012.0900.

4. Alexandre Kreiss et al., "Allorecognition in the Tasmanian Devil (Sarcophilus harrisii), an Endangered Marsupial Species with Limited Genetic Diversity," *PLoS One* 6, no. 7 (2011): e22402. This study explored the functional significance of the limited genetic diversity in Tasmanian devils by seeing whether or not skin grafts would take. As a rule, they do not. This illustrates that although there is limited genetic diversity, the population of animals are not all like identical twins, which can indiscriminately accept skin grafts (and therefore other organ "donations" such as tumors).

CHAPTER 5: DEVIL OF A DISEASE

1. "National Cancer Institute Fact Sheets. Alcohol and Cancer Risk," National Cancer Institute, http://www.cancer.gov/about-cancer/causes-prevention/risk/alcohol/alcohol-fact-sheet (accessed October 20, 2015); Cancer.Net Editorial Board, "Head and Neck Cancer: Risk Factors and Prevention," Cancer.net, May 2015, http://www.cancer.net/cancer-types/head-and-neck-cancer/risk-factors-and-prevention (accessed October 20, 2015).

2. A. M. Pearse and K. Swift, "Allograft Theory: Transmission of Devil Facial-

Tumor Disease," *Nature* 439, no. no.7076 (February 2006): 549. This study, via detailed cytogenetic analysis, proved that all the DFTD cases were clonal. Thus, they were transmitted directly from animal to animal and not virally induced.

3. David Quammen, "Contagious Cancer: The Evolution of a Killer," *Harper's Magazine*, April 2008, pp. 33–43. This is an extremely well-researched early account aimed for the general reader written in nontechnical terms. It reviews much of the history that led to the conclusion that this frightening new cancer epidemic was actually a contagious disease.

4. Julie Rehmeyer, "The Immortal Devil," *Discover*, May 2014, pp. 43–49. Another well-researched but more up-to-date overview published in a popular science magazine for the general reader. This article also details much of the relevant history of DFTD.

CHAPTER 6: THE PERFECT PARASITE

1. George Johnson, *The Cancer Chronicles: Unlocking Medicine's Deepest Mystery* (New York: Alfred A. Knopf, 2013). The first chapter of this book goes into a good deal of detail about ancient cancers, including tumors in dinosaurs. This is an excellent popular book written for the educated reader.

2. "Surveillance, Epidemiology and End Results (SEER)," National Cancer Institute, http://seer.cancer.gov/statistics/summaries.html (accessed October 20, 2015).

3. Ibid.; Alexandra Sifferlin, "4 Diseases Making a Comeback Thanks to Anti-Vaxxers," *TIME* Health, March 17, 2014, http://time.com/27308/4-diseases-making-a-comeback-thanks-to-anti-vaxxers (accessed October 20, 2015).

4. Erin Conway-Smith, "Ebola 'Has Killed a Third of the World's Chimpanzees and Gorillas,'" *Telegraph*, January 22, 2015, http://www.telegraph.co.uk/news/worldnews/ebola/11363433/Ebola-has-killed-a-third-of-the-worlds-chimpanzees-and-gorillas.html (accessed October 20, 2015).

5. "Famous Food Fetishes, Cockroaches and Butterflies?," CBS News, http://www.cbsnews.com/pictures/famous-food-fetishes-cockroaches-and-butterflies/11/ (accessed October 20, 2015). This is one of the many accounts in the popular media regarding the still undocumented rumor that Maria Callas used a tapeworm to drop some pounds.

6. J. A. Turton, "IgE, Parasites, and Allergy," *Lancet* 308, no. 7987 (September 1976): 686. This was the first article I could find on the concept of an inverse relationship between parasitic infections and allergies and asthma. See also: "References and Clinical Studies – Helmintherapy," BioTherapeutics, Education & Research (BTER) Foundation, http://www.bterfoundation.org/helminthrefs (accessed October 20, 2015). This website provides a long list of references on the concept of the potential value of helminths for helping in the management of asthma and other autoimmune disorders.

7. Robert L. Dorit, "Breached Ecological Barriers and the Ebola Outbreak: The Epidemic May Be Waning, but the Social and Ecological Context That Brought It About Remains," *American Scientist* 103, no. 4 (July–August 2015): 256.

8. Eliot Barford, "Parasite Makes Mice Lose Fear of Cats Permanently: Behavioral Changes Persist after Toxoplasma Infection Is Cleared," *Nature*, September 18, 2013, http://www.nature.com/news/parasite-makes-mice-lose-fear-of-cats-permanently-1.13777 (accessed October 20, 2015).

9. R. H. Yolken, F. B. Dickerson, and Torrey E. Fuller, "Toxoplasma and Schizophrenia," *Parasite Immunology* 31, no. 11 (2009): 706–15.

10. Mayo Clinic Staff, "Diseases and conditions: Toxoplasmosis," Mayo Clinic, http://www.mayoclinic.org/diseases-conditions/toxoplasmosis/basics/definition/con-20025859 (accessed Dec 17, 2015).

11. Ali Nawaz Khan, "Imaging in CNS Toxoplasmosis," Medscape, http://emedicine.medscape.com/article/344706-overview#a4 (accessed Dec 17, 2015).

CHAPTER 7: A MALIGNANT MALADY IN MAN'S BEST FRIEND

1. S. Mukaratirwa and E. Gruys, "Canine Transmissible Venereal Tumour: Cytogenetic Origin, Immunophenotype, and Immunobiology. A Review," *Veterinary Quarterly* 25, no. 3 (September 2003): 101–11. Although written over a decade ago, this remains a good general scientific-review article, especially for the veterinary readership. See also: J. E. Prier and R. S. Brodey, "Canine Neoplasia: A Prototype for Human Cancer Study," *Bulletin World Health Organization* 29, no. 3 (1963): 331–44. This is one of the older papers on the topic. It hints at ways this contagious cancer might be of relevance for the study of human cancer.

2. V. K. Gandotra, F. S. Chauhan, and R. D. Sharma, "Occurrence of Canine Transmissible Venereal Tumor and Evaluation of Two Treatments," *Indian Veterinary Journal* 70, no. 9 (1993): 854–57.

3. V. Wunderlich, "Anton Sticker (1861–1944) and the Sticker's Sarcoma of Dogs: A Current Model for Infectious Cancer Cells," *Tierarztl Prax Ausg K Kleintiere Heimtiere* 40, no. 6 (2012): 432–37. This is a historical account of Anton Sticker and the discovery of this contagious cancer (written in German except for the abstract).

4. M. A. Novinski, "Zur Frage Uber die Impfung der Krebsigen Geschwulste," *Zentralbl Med Wissensch* 14 (1876): 790–91.

5. T. J. Yang and J. B. Jones, "Canine Transmissible Venereal Sarcoma: Transplantation Studies in Neonatal and Adult Dogs," *Journal of the National Cancer Institute* 51, no. 6 (December 1973): 1915–18.

6. Mukaratirwa and Gruys, "Canine Transmissible Venereal Tumour."

7. S. Kurbel, S. Plestina, and D. Vrbanec, "Occurrence of the Acquired Immunity in Early Vertebrates Due to Danger of Transmissible Cancers Similar to Canine Venereal Tumors," *Medical Hypotheses* 68 (2007): 1185–86.

8. C. Murgia et al., "Clonal Origin and Evolution of a Transmissible Cancer," *Cell* 126, no. 3 (August 11, 2006): 477–87. This is one of the important papers that definitively

proved through cytogenetic analysis (i.e., detailed study of the chromosomes) that CTVT is actually a transmissible tumor.

9. B. M. VonHoldt and E. A. Ostrander, "The Singular History of a Canine Transmissible Tumor," *Cell* 126, no. 3 (August 11, 2006): 445–47. This short article describes the importance of the LINE-1 element and its proximity to c-myc in the oncogenesis process.

10. E. P. Murchison et al., "Transmissible [corrected] Dog Cancer Genome Reveals the Origin and History of an Ancient Cell Lineage," *Science* 343, no. 6169 (January 24, 2014): 437–40. This is the original article that details the methodology used to conclude the astonishing eleven thousand-year-old age of this tumor line. See also: I. Lokody, "Cancer Genetics: The Origin and Evolution of an Ancient Cancer," *Nature Reviews Genetics* 15, no. 3 (March 2014): 144; and H. G. Parker and E. A. Ostrander, "Cancer. Hiding in Plain View—An Ancient Dog in the Modern World," *Science* 343, no. 6169 (January 24, 2014): 376–78.

11. Katrina Morris, Jeremy J. Austin, and Katherine Belov, "Low Major Histocompatibility Complex Diversity in the Tasmanian Devil Predates European Settlement and May Explain Susceptibility to Disease Epidemics," *Biology Letters*, 9 (2013): 20120900. This is one of many scientific studies that have confirmed the inbreeding and relative loss of genetic diversity in the island population of *Sarcophilus harrisii*. This limited genetic diversity was once held as the primary reason why Tasmanian devils were susceptible to contagious cancers whereas humans and most other animals were not. More recent data refutes this hypothesis.

12. Alexandre Kreiss et al., "Allorecognition in the Tasmanian Devil (*Sarcophilus harrisii*): An Endangered Marsupial Species with Limited Genetic Diversity," *PLoS One* 6, no. 7 (2011): e22402. This study explored the functional significance of the limited genetic diversity in Tasmanian devils by seeing whether or not skin grafts would take. As a rule they do not. This illustrates that although there is limited genetic diversity, the population of animals are not all like identical twins, which can indiscriminately accept skin grafts (and therefore other organ "donations" such as tumors).

13. U. Das and A. K. Das, "Review of Canine Transmissible Venereal Sarcoma," *Veterinary Research Communications* 24, no. 8 (December 2000): 545–56. This is a good review on the topic that includes some discussion on historically interesting treatments.

CHAPTER 8: THE CURIOUS CASE OF COLEY'S TOXINS (OR SOMETIMES THE TREATMENT WORKED)

1. Michael Martinez, "After Ravages of Flesh-Eating Bacteria, Aimee Copeland Uses New Bionic Hands," CNN, May 20, 2013, http://www.cnn.com/2013/05/17/us/georgia-aimee-copeland/ (accessed October 20, 2015).

2. Matthew Tontonoz, "The Legacy of Bessie Dashiell," Cancer Research Institute News&Publications, December 13, 2013, http://www.cancerresearch.org/news-publications/our-blog/december-2013/the-legacy-of-bessie-dashiell (accessed October 22, 2015).

3. "Saint Peregrine, The Cancer Saint—Biography," National Shrine of Saint Peregrine, http://stperegrine.org/biography/ (accessed October 20, 2015).

4. R. Jackson, "Saint Peregrine, OSM—The Patron Saint of Cancer Patients," *Canadian Medical Association Journal* 111 (1974): 824–27. See also: S. A. Hoption Cann, J. P. van Netten, and C. van Netten, "Dr. William Coley and Tumour Regression: A Place in History or in the Future," *Postgraduate Medical Journal* 79 (2003): 672–80.

5. Uwe Hobohm, "Killing Cancer with Fever: An Old Therapy Revisited," *New Scientist*, December 31, 2013, https://www.newscientist.com/article/mg22129500-400-killing-cancer-with-fever-an-old-therapy-revisited/ (accessed October 22, 2015). This is an "opinion piece" describing Uwe Hobohm's hopes to modernize Coley's toxins for twenty-first century cancer treatment. On the first page it describes how Wilhelm Busch was a clinical predecessor of William Coley.

6. Charlie O. Starnes, "Coley's Toxins in Perspective," *Nature* 357 (1992): 11–12. This is a brief recounting of Coley's toxin's in a prestigious journal during the midst of the relatively ineffective first generation of modern cancer immunotherapy. See also: E. F. McCarthy, "The Toxins of William B. Coley and the Treatment of Bone and Soft-Tissue Sarcomas," *Iowa Orthopaedic Journal* 6 (2006): 154–58; available online at: http://www.ncbi.nlm.nih.gov/pmc/articles/PMC1888599/ (accessed October 22, 2015). This is a brief but excellent historical overview that includes some discussion on Mr. Stein's peculiar case.

7. Ibid., 155–58. McCarthy also went into the battles between Ewing and Coley.

8. Ibid.

9. Hobohm, "Killing Cancer with Fever." See also: Uwe Hobohm, "Fever Therapy Revisited," *British Journal of Cancer* 92, no. 30 (February 14, 2005): 421–25; available online at: http://www.nature.com/bjc/journal/v92/n3/full/6602386a.html (accessed October 22, 2015). This is one of the several articles by Hobohm describing the concepts of "fever therapy."

CHAPTER 9: THE DOG KNOWS

1. James S. Welsh, "Olfactory Detection of Human Bladder Cancer by Dogs: Another Cancer Detected by 'Pet Scan,'" *British Medical Journal* 329, no. 7477 (November 25, 2004): 1286–87. This very brief account is also available at: http://www.bmj.com/content/329/7477/1286.3 (accessed October 22, 2015). This is my own encounter with a cancer-sniffing-canine case. See also: J. S. Welsh, D. Barton, and H. Ahuja, "A Case of Breast Cancer Detected by a Pet Dog," *Community Oncology* 2, no. 4 (2005): 324–25.

2. Amanda Cable, "Daisy, the Dog Who's Sniffed Out over 500 Cases of Cancer: She Even Saved the Woman Whose Research Revealed Her Uncanny Skill," *Daily Mail*, July 2014, http://www.dailymail.co.uk/health/article-2700561/Daisy-dog-whos-sniffed-500-cases-cancer-She-saved-woman-research-revealed-uncanny-skill.html#ixzz3bq0aGcLy (ac-

cessed October 22, 2015). This was the incredible story about a researcher in the field whose own dog detected her breast cancer.

3. Ibid.

CHAPTER 11: COULD BROWN FAT BE THE SECRET TO WEIGHT LOSS?

1. Jung-whan Kim and Chi V. Dang, "Cancer's Molecular Sweet Tooth and the Warburg Effect," *Cancer Research* 66, no. 18 (September 15, 2006): 8927–30.

Also available online at: http://cancerres.aacrjournals.org/content/66/18/8927.full.html (accessed October 22, 2015). This is an excellent summary of the Warburg effect and its potential relevance to modern oncology.

2. Ibid., 8927.

3. Q. S. Chu et al., "A Phase I Open-Labeled, Single-Arm, Dose-Escalation, Study of Dichloroacetate (DCA) in Patients with Advanced Solid Tumors," *Investigational New Drugs* 33, no. 3 (June 2015): 603–10. This is one of several publications form the University of Alberta. This phase I study established a maximal tolerable safe dose. Unfortunately, the clinical impact on the advanced cancer patients was only fair. The abstract mentioned: "No responses were observed and eight patients had stable disease."

4. T. F. Hany et al., "Brown Adipose Tissue: A Factor to Consider in Symmetrical Tracer Uptake in the Neck and Upper Chest Region," *European Journal of Nuclear Medicine and Molecular Imaging* 29, no. 10 (October 2002): 1393–98. This is one of the several early papers documenting brown fat as a potential false positive error on PET scans.

5. O. Hashimoto et al., "Brown Adipose Tissue in Cetacean Blubber," *PLoS One* 10, no. 2 (February 26, 2015): e0116734. This is an interesting article on the role of brown fat in cetaceans. From the abstract: "We believe delphinoid BAT might also function like an electric blanket, enabling animals to frequent waters cooler than blubber as an insulator alone might otherwise allow an animal to withstand."

6. Chloe Lambert, "Growing Fat to Stay Slim," *New Scientist*, April 18, 2015, pp. 32–35. Although beige fat has yet to be definitively proven to exist in humans, if it does indeed exist, it could hold promise as a novel means of weight loss (assuming it could somehow be activated).

7. Ibid. This is a very interesting personal experiment on the author's own activation of brown fat.

8. Michael J. Tisdale, "Cachexia in Cancer Patients," *Nature Reviews Cancer*, 2 (November 2002): 862–71. Also available at: http://www.nature.com/nrc/journal/v2/n11/full/nrc927.html (accessed October 22, 2015). This is an excellent review on cachexia that even addresses the roles of uncoupling proteins in brown fat—but this article is a bit advanced for those without a background in biology.

9. John M. Pawelek and Ashok K. Chakraborty, "Fusion of Tumour Cells with Bone

Marrow-Derived Cells: A Unifying Explanation for Metastasis," *Nature Reviews Cancer* 8 (May 2008): 377–86. Also available at: http://www.nature.com/nrc/journal/v8/n5/full/nrc2371.html (accessed October 22, 2015). This is one of several provocative articles by John Pawelek and colleagues on the idea that fusion of tumor cells with macrophages or other marrow-derived cells could confer them with the capacity for metastasizing. The abstract mentions: "BMDC–tumour hybrids have been detected in numerous animal models and recently in human cancer. Molecular studies indicate that gene expression in such hybrids reflects a metastatic phenotype. Should BMDC–tumour fusion be found to underlie invasion and metastasis in human cancer, new approaches for therapy would surely follow."

CHAPTER 12: GAMMA RAYS AND DINOSAUR CANCER

1. Robert M. Hazen, *The Story of Earth: The First 4.5 Billion Years from Stardust to Living Planet* (New York: Penguin, 2012), p. 271.

2. R. W. Klebesadel, I. B. Strong, and R. A. Olson, "Observations of Gamma-Ray Bursts of Cosmic Origin," *Astrophysical Journal* 182 (1973): L85–88. Upon declassification of the information, scientists were able to disclose the amazing truth about these previously unknown gamma-ray-emitting extraterrestrial explosions.

3. "Compton Gamma Ray Observatory," NASA Science Missions, National Aeronautics and Space Administration (NASA), May 28, 2015, http://science.nasa.gov/missions/cgro/ (accessed October 22, 2015).

4. "Welcome to the BeppoSax Home Page," BeppoSAX Mission, http://www.asdc.asi.it/bepposax/ (accessed October 22, 2015).

5. P. Andrew Karam, "Gamma and Neutrino Radiation Dose from Gamma Ray Bursts and Nearby Supernovae," *Health Physics* 82 (2002): 491.

6. Charles Q. Choi, "Did Deadly Gamma-Ray Burst Cause a Mass Extinction on Earth?" Live Science, http://www.livescience.com/49040-gamma-ray-burst-mass-extinction.html (accessed December 22, 2015).

7. Bruce M. Rothschild, Brian J. Witzke, and Israel Hershkovitz, "Metastatic Cancer in the Jurassic," *Lancet* 354 (1999): 398.

8. Ibid.

9. George Johnson, *The Cancer Chronicles: Unlocking Medicine's Deepest Mystery* (New York: Alfred A. Knopf, 2013). This is one of the finest popular reads in the field of oncology for the educated general science aficionado.

10. B. M. Rothschild et al., "Epidemiologic Study of Tumors in Dinosaurs," *Naturwessenschaften* 90 (2003): 459–500.

11. Ibid.

12. L. C. Natarajan et al., "Bone Cancer Rates in Dinosaurs Compared with Modern Vertebrates," *Transactions Kansas Academy of Science* 110, no. 3/4 (Fall 2007): 155–58.

A PDF version of this paper is available at: arxiv.org/vc/arxiv/papers/0704/0704.1912v2 .pdf (accessed October 22, 2015).

13. C. Tomasetti and B. Vogelstein, "Variation in Cancer Risk among Tissues Can Be Explained by the Number of Stem Cell Divisions," *Science* 347, no. 6217: 78–81.

14. Carl Zimmer, "Elephants: Large, Long-Living and Less Prone to Cancer," *NY Times*, October 8, 2015, http://www.nytimes.com/2015/10/13/science/why-elephants-get -less-cancer.html?_r=0 (accessed December 17, 2015).

15. L. M. Abegglen, A. F. Caulin, A. Chan et al. "Potential Mechanisms for Cancer Resistance in Elephants and Comparative Cellular Response to DNA Damage in Humans," *JAMA* 314, no. 17 (2015):1850–60.

16. R. J. Wiese and K. Willis, "Calculation of Longevity and Life Expectancy in Captive Elephants," *Zoo Biology* 23 (2004): 365–73.

17. Michael Sulak et al., "TP53 Copy Number Expansion Correlates with the Evolution of Increased Body Size and an Enhanced DNA Damage Response in Elephants," bioRxiv, http://biorxiv.org/content/early/2015/10/06/028522 (accessed December 17, 2015).

18. William A. Lindsay et al., "Immune Responses of Asian Elephants (Elephas maximus) to Commercial Tetanus Toxoid Vaccine," *Veterinary Immunology and Immunopathology* 133 (2010): 287–89.

CHAPTER 13: CANCER OF THE CLAM!

1. E. P. Metzger et al., "Horizontal Transmission of Clonal Cancer Cells Causes Leukemia in Soft-Shell Clams," *Cell* 161 (2015): 255–63. See also: Robin A Weiss and Ariberto Fassati, "The Clammy Grip of Parasitic Tumors," *Cell* 161 (2015): 191–92.

2. Ibid.

3. K. S. Sfanos et al., "XMRV and Prostate Cancer—A 'Final' Perspective," *Nature Reviews Urology* 9, no. 2 (January 10, 2012): 111–18. This article succinctly states: "There is no reason to believe that it has any role in the etiology of prostate cancer or other diseases."

4. L. J. Kleinsmith, *Principles of Cancer Biology* (San Francisco, CA: Pearson Benjamin Cummings, 2006).

5. Metzger et al., "Horizontal Transmission."

6. Erick Stokstad, "Infectious Cancer Found in Clams," *Science* 348 (2015): 170.

7. Ibid.

8. Ibid.

9. Ibid.

10. Ibid.

CHAPTER 14: SHARKS DO GET CANCER (OR HOW SHARK CARTILAGE CAN KILL YOU)

1. Gary K. Ostrander et al., "Shark Cartilage, Cancer and the Growing Threat of Pseudoscience," *Cancer Research* 64 (December 1, 2004): 8485–91.

2. Christie Wilcox. "Mythbusting 101: Sharks Will Cure Cancer," *Scientific American* blogs, September 1, 2011, http://blogs.scientificamerican.com/science-sushi/mythbusting-101-sharks-will-cure-cancer/ (accessed October 22, 2015).

3. H. Brem and J. Folkman, "Inhibition of Tumor Angiogenesis Mediated by Cartilage," *Journal of Experimental Medicine* 141 (1975): 427–39.

4. Wilcox, "Mythbusting 101."

5. Ibid. See also: R. Langer et al., "Isolations of a Cartilage Factor That Inhibits Tumor Neovascularization," *Science* 193 (1976): 70–72.

6. C. A. Luer and W. H. Luer, "Acute and Chronic Exposure of Nurse Sharks to Aflatoxin B_1," *Federation American Society Experimental Biology, Federation Proceedings* 41 (1982): 925.

7. L. B. Ellwein and S. M. Cohen, "The Health Risks of Saccharin Revisited," *Critical Reviews Toxicology* 20, no.5 (1990): 311–26.

8. Douglas Main, "Sharks Do Get Cancer: Tumor Found in Great White," *Scientific American*, December 5, 2013, http://www.scientificamerican.com/article/sharks-do-get-cancer-tumor-found-great-white/ (accessed October 22, 2015).

9. Ibid.

10. "Guess What? Sharks Do Get Cancer," WebMD Health News, April 5, 2000, http://www.webmd.com/cancer/news/20000405/sharks-cancer-cartilage (accessed October 22, 2015).

11. J. B. Finkelstein, "Sharks Do Get Cancer: Few Surprises in Cartilage Research," *Journal of the National Cancer Institute* 97, no. 21 (November 2, 2005): 1562–63. See also: C. L. Loprinzi et al., "Evaluation of Shark Cartilage in Patients with Advanced Cancer: A North Central Cancer Treatment Group Trial," *Cancer* 104, no. 1 (July 1, 2005): 176–82; National Center for Complementary and Alternative Medicine (NCCAM), "Powdered Shark Cartilage for Advanced Breast and Colorectal Cancer," Department Health and Human Services, National Institutes Health, July 1, 2005, https://nccih.nih.gov/research/results/spotlight/sharkcartilage_rr.htm (accessed October 22, 2015).; C. Lu et al., "Chemoradiotherapy With or Without AE-941 in Stage III Non-Small Cell Lung Cancer: A Randomized Phase III Trial," *Journal of the National Cancer Institute* 102, no. 12 (June 16, 2010): 859–65; and D. R. Miller et al., "Phase I/II Trial of the Safety and Efficacy of Shark Cartilage in the Treatment of Advanced Cancer," *Journal of Clinical Oncology* 16, no. 11 (November 1998): 3649–55.

12. Lu et al., "Chemoradiotherapy."

13. Federal Trade Commission, "'Operation Cure.all' Nets Shark Cartilage Promoters: Two Companies Charged with Making False and Unsubstantiated Claims for Their

Shark Cartilage and Skin Cream as Cancer Treatments," Federal Trade Commission, June 29, 2000, https://www.ftc.gov/news-events/press-releases/2000/06/operation-cureall-nets-shark-cartilage-promoters-two-companies (accessed October 22, 2015).

14. Main, "Sharks Do Get Cancer."

CHAPTER 15: WHO TRULY DOESN'T GET CANCER?— MEET THE MOLE RATS

1. Xiao Tian et al., "High-Molecular-Mass Hyaluronan Mediates the Cancer Resistance of the Naked Mole Rat," *Nature* 499 (July 18, 2013): 346–49.

2. "Vertebrate of the Year: The Rat That Ages Beautifully," *Science* 342, no. 6165 (December 20, 2013): 1444.

3. "Naked Mole Rat Named Vertebrate of the Year," Science Daily, December 23, 2013, http://www.sciencedaily.com/releases/2013/12/131223181145.htm (accessed October 22, 2015).

4. "Blind Mole Rat," Encyclopedia Britannica, http://www.britannica.com/animal/blind-mole-rat (accessed October 22, 2015).

5. L. J. Kleinsmith, *Principles Cancer Biology* (San Francisco, CA: Pearson Benjamin Cummings, 2006), p. 179. Well written and beautifully illustrated, this is one of the finest undergraduate-level texts on cancer biology. Although now somewhat dated, it can serve as a useful introductory platform for those interested in further learning.

6. V. Gorbunova et al., "Cancer Resistance in the Blind Mole Rat is Mediated by Concerted Necrotic Cell Death Mechanism," *Proceedings of the National Academy of Sciences* 109, no. 47 (November 20, 2012): 19392–96. See also: Jorge Azpurua and Andrei Seluanov, "Long-Lived Cancer-Resistant Rodents as New Model Species for Cancer Research," *Frontiers in Genetics* 3 (2012): 319. This is a brief review of the anti-cancer mechanisms in naked and blind mole rats (NMR and BMR respectively).

CHAPTER 16: PAR FOR THE COURSE

1. Y. Zhao et al., "Cancer Resistance in Transgenic Mice Expressing the SAC Module of Par-4," *Cancer Research* 67, no. 19 (October 1, 2007): 9276–85. See also: Y. Zhao and V. M. Rangnekar, "Apoptosis and Tumor Resistance Conferred by Par-4," *Cancer Biology and Therapy* 12 (December 7, 2008): 1867–74; and N. Hebbar, C. Wang and V. M. Rangnekar, "Mechanisms of Apoptosis by the Tumor Suppressor Par-4," *Journal of Cellular Physiology* 227, no. 12 (December 2012): 3715–21.

2. Roland Piquepaille, "A Cancer-Resistant Mouse?," ZDNet, November 27, 2007, http://www.zdnet.com/article/a-cancer-resistant-mouse/ (accessed October 22, 2015).

3. J. Butler and V. M. Rangnekar, "Par-4 for Molecular Therapy of Prostate Cancer," *Current Drug Targets* 4, no. 3 (April 2003): 223–30.

4. M. S. Frasinyuk et al., "Development of *6H*-Chromeno[3,4-*c*]pyrido[3',2':4,5] thieno[2,3-e]pyridazin-6-ones as Par-4 Secretagogues," *Tetrahedron Letters* 56, no. 23 (June 2015): 3382–84.

CHAPTER 17: MIGHTY MOUSE TO THE RESCUE!

1. Z. Cui, "The Winding Road to the Discovery of the SR/CR Mice," *Cancer Immunity* 3, no. 14 (October 16, 2003). This is an interesting and personal account of Dr. Zheng Cui's discovery of the "mighty" (SR/CR—spontaneous remission/complete regression) mouse.

2. A. M. Hicks et al., "Transferable Anticancer Innate Immunity in Spontaneous Regression/Complete Resistance Mice," *Proceedings of the National Academy of Sciences* 103, no. 20 (May 16, 2006): 7753–58. This study demonstrated that cancer resistance from mighty mouse could be transferred to other mice.

3. A. M. Soto and C. Sonnenschein, "The Somatic Mutation Theory of Cancer: Growing Problems with the Paradigm?," *Bioessays* 26, no. 10 (October 2004): 1097–107.

4. A. M. Sanders et al., "Cancer Resistance of SR/CR Mice in the Genetic Knockout Backgrounds of Leukocyte Effector Mechanisms: Determinations for Functional Requirements," *BMC Cancer* 10 (March 31, 2010): 121. See also: Brian Wang, "Update on New Cancer Treatment Reported at SENS3," Next Big Future, September 21, 2007, http://nextbig future.com/2007/09/update-on-new-cancer-treatment-reported.html (accessed October 22, 2015).

5. M. J. Blanks et al., "Novel Innate Cancer Killing Activity in Humans," *Cancer Cell International* 11 (August 3, 2011): 26. Some humans might also be "mighty."

CHAPTER 18: FRODO OF FLORES

1. M. J. Morwood et al., "Archaeology and Age of a New Hominin from Flores in Eastern Indonesia," *Nature* 431, no. 7012 (October 28, 2004): 1087–91. See also: P. Brown et al., "A New Small-Bodied Hominin from the Late Pleistocene of Flores, Indonesia," *Nature* 431, no. 7012 (October 28, 2004): 1055–61.

2. G. D. van den Bergh et al., "The Liang Bua Faunal Remains: A 95k.yr. Sequence from Flores, East Indonesia," *Journal of Human Evolution* 57, no. 5 (November 2009): 527–37.

3. For quick examples of the controversy see: P. J. Obendorph et al., "Are the Small Human-Like Fossils on Flores Human Endemic Cretins?," *Proceedings Royal Society B: Biological Sciences* 275 (2008): 1287–96; and P. Brown, "LB1 and LB6 Homo floresiensis are not modern human (Homo sapiens) cretins." *Journal of Human Evolution,* 62 (2009):

201-24. For an excellent summary of the controversies see: Joseph Castro, "Homo Floresiensis: Facts about the 'Hobbit,'" *Live Science*, July 28, 2015, http://www.livescience.com/29100-homo-floresiensis-hobbit-facts.html (accessed October 22, 2015).

4. Cleveland Clinic, "Microcephaly in Children," Cleveland Clinic-Children's, January 4, 2011, http://my.clevelandclinic.org/childrens-hospital/health-info/diseases-conditions/hic-Microcephaly (accessed October 22, 2015).

5. C. Oxnard, P. J. Obendorf, and B. J. Kefford, "Post-Cranial Skeletons of Hypothyroid Cretins Show a Similar Anatomical Mosaic as Homo floresiensis," *PLoS One* 5, no. 9 (2010): e13018.

6. For articles refuting the concept of endemic cretinism as the explanation for the hobbit, see: Jungers et al., "The Hobbits (Homo floresiensis) Were Not Cretins," *Journal of Physical Anthropology*, Supplement 48 (2009): 244; and Brown, "LB1 and LB6."

7. D. Falk et al., "Brain Shape in Human Microcephalics and Homo floresiensis," *Proceedings of the National Academy of Sciences* 104, no. 7 (2007): 2513–18.

8. R. C. Vannucci, T. F. Barron, and R. L. Holloway, "Craniometric Ratios of Microcephaly and LB1, Homo floresiensis, using MRI and Endocasts," *Proceedings of the National Academy of Sciences* 108, no. 34 (August 23, 2011): 14043–48.

9. M. Dembo et al., "Bayesian Analysis of a Morphological Supermatrix Sheds Light on Controversial Fossil Hominin Relationships," *Proceedings of the Royal Society B: Biological Sciences* 282, no. 1812 (July 22, 2015): 20150943.

10. Michael Carrington et al., "Mandibular Evidence Supports Homo floresiensis as a Distinct Species," *Proceedings of the National Academy of Sciences* 112, no. 7 (February 17, 2015): E604-E605.

11. Robin McKie, "Homo floresiensis: Scientists Clash over Claims 'Hobbit Man' Was Modern Human with Down's Syndrome," *Guardian*, August 16, 2014, http://www.theguardian.com/science/2014/aug/16/flores-hobbit-human-downs-syndrome-claim-homo-floresiensis (accessed October 22, 2015).

12. Castro, "Homo Floresiensis." See also: Maciej Henneberg et al., "Evolved Developmental Homeostasis Distributed in LB1 from Flores Indonesia Denoted Down Syndrome and Not Diagnostic Traits of the Invalid Species Homo floresiensis," *Proceedings of the National Academy of Sciences* 111, no. 33 (August 19, 2014): 11961–66; and McKie, "Homo floresiensis."

13. McKie, "Homo floresiensis."

14. "'Hobbit' Remains Kept out of Reach," *New Zealand Herald*, February 11, 2005, http://www.nzherald.co.nz/technology/news/article.cfm?c_id=5&objectid=10010714 (accessed October 22, 2015).

15. Tabitha Powledge, "No Microcephaly for Hobbit," *Scientist*, March 4, 2005, http://www.the-scientist.com/?articles.view/articleNo/23288/title/No-microcephaly-for-Hobbit/ (accessed October 22, 2015).

16. "Hobbit Cave Digs Set to Restart," BBC News, January 25, 2007, http://news.bbc.co.uk/go/pr/fr/-/2/hi/science/nature/6294101.stm (accessed October 22, 2015).

17. "Hobbits Triumph Tempered by Tragedy," *Sydney Morning Herald*, March 5, 2005, http://www.smh.com.au/news/Science/Hobbits-triumph-tempered-by-tragedy/2005/03/04/1109700677461.html (accessed October 22, 2015).

18. "Hobbit Cave Digs Set to Restart."

19. McKie, "Homo floresiensis."

20. McKie, "Homo floresiensis." See also: Dawn Papple, "Homo floresiensis: Modern Humans and Real-Life Hobbits Lived Simultaneously on Earth, Says Anthropologists," *Inquisitr*, January 11, 2015, http://www.inquisitr.com/1740996/homo-floresiensis-hobbits/ (accessed October 22, 2015). See also: McKie, "Homo floresiensis."

CHAPTER 19: A CANCER-FREE CLAN?

1. Z. Laron, A. Pertzelan, and S. Mannheimer, "Genetic Pituitary Dwarfism with High Serum Concentration of Growth Hormone—A New Inborn Error of Metabolism," *Israel Journal of Medical Sciences* 2, no. 2 (March–April 1966): 152–55.

2. "Dwarfism," WebMD, October 26, 2014, http://www.webmd.com/children/dwarfism-causes-treatments (accessed October 22, 2015).

3. Mayo Clinic Staff, "Diseases and Conditions, Dwarfism: Causes," Mayo Clinic, September 11, 2014, http://www.mayoclinic.org/diseases-conditions/dwarfism/basics/causes/con-20032297 (accessed October 22, 2015).

4. W. H. Daughaday and B. Trivedi, "Absence of Serum Growth Hormone Binding Protein in Patients with Growth Hormone Receptor Deficiency (Laron Dwarfism)," *Proceedings of the National Academy of Sciences* 84, no. 13 (July 1987): 4636–40. See also: R. Eshet et al., "Defect of Human Growth Hormone Receptors in the Liver of Two Patients with Laron-Type Dwarfism," *Israel Journal of Medical Sciences* 20, no. 1 (January 1984): 8–11.

5. Gary Taubes, "Rare Form of Dwarfism Protects against Cancer," *Discover*, March 27, 2013, http://discovermagazine.com/2013/april/19-double-edged-genes (accessed October 22, 2015).

6. Ibid.

7. A. L. Rosenbloom et al., "The Little Women of Loja—Growth Hormone-Receptor Deficiency in an Inbred Population of Southern Ecuador," *New England Journal of Medicine* 323, no. 20 (November 15, 1990): 1367–74.

8. Nicholas Wade, "Ecuadorean Villagers May Hold Secret to Longevity," *New York Times*, February 16, 2011, http://www.nytimes.com/2011/02/17/science/17longevity.html (accessed October 22, 2015).

9. Personal communication from Dr. Arlan Rosenbloom.

10. J. Guevara-Aguirre et al., "Growth Hormone Receptor Deficiency Is Associated with a Major Reduction in Pro-Aging Signaling, Cancer, and Diabetes in Humans," *Science Translational Medicine* 3, no. 70 (February 16, 2011): 70ra13.

11. Y. Zhou et al., "A Mammalian Model for Laron Syndrome Produced by Targeted

Disruption of the Mouse Growth Hormone Receptor/Binding Protein Gene (The Laron Mouse)," *Proceedings of the National Academy of Sciences* 94, no. 24 (November 25, 1997): 13215–20.

12. J. J. Kopchick and Zvi Laron, "Is the Laron Mouse an Accurate Model of Laron Syndrome?," *Molecular Genetics and Metabolism* 68 (1999): 232–36.

13. D. Gobel, "The Methuselah Mouse Prize," Sens Research Foundation, http://www.sens.org/outreach/conferences/methuselah-mouse-prize (accessed October 22, 2015).

14. D. B. Friedman and T. E. Johnson, "A Mutation in the Age-1 Gene in Caenorhabditis Elegans Lengthens life and reduces hermaphrodite fertility," *Genetics* 118, no. 1 (January 1988): 75–86. See also: G. J. Lithgow et al., "Thermotolerance and Extended Life-Span Conferred by Single-Gene Mutations and Induced by Thermal Stress," *Proceedings of the National Academy of Sciences* 92, no. 16 (August 1, 1995): 7540–44.

15. Guevara-Aguirr, "Growth Hormone Receptor Deficiency."

16. J. Green et al., "Height and Cancer Incidence in the Million Women Study: Prospective Cohort, and Meta-Analysis of Prospective Studies of Height and Total Cancer Risk," *Lancet Oncology* 12, no. 8 (August 2011): 785–94.

17. O. Shevah and Z. Laron, "Patients with Congenital Deficiency of IGF-I Seem Protected from the Development of Malignancies: a Preliminary Report," *Growth Hormone IGF Research Journal* 17, no. 1 (February 2007): 54–57.

18. Rachel Steuerman, Orit Shevah, and Zvi Laron, "Congenital IGF1 Deficiency Tends to Confer Protection against Post-Natal Development of Malignancies," *European Journal of Endocrinology* 164 (April 2011): 485–89.

19. E. Braverman et al., "Low and Normal IGF-1 Levels in Patients with Chronic Medical Disorders (CMD) Is Independent of Anterior Pituitary Hormone Deficiencies: Implications for Treating IGF-1 Abnormal Deficiencies with CMD," *Journal of Genetic Syndromes and Gene Therapy* 4, no. 123 (February 2013).

CHAPTER 20: CANCER: A DISEASE OF IMMUNE FAILURE?

1. "Cervical Cancer Health Center, Cervical Cancer," WebMD, May 26, 2014, http://www.webmd.com/cancer/cervical-cancer/cervical-cancer?print=true (accessed October 22, 2015).

2. Biography.com Editors, "Henrietta Lacks Biography," Biography.com, http://www.biography.com/people/henrietta-lacks-21366671 (accessed October 22, 2015).

3. Rebecca Skloot, *The Immortal Life of Henrietta Lacks* (New York: Crown, 2010).

4. Maryland Commission for Women, "Henrietta Lacks, 1920–1951," Maryland State Archives, Maryland Women's Hall of Fame, 2014, http://msa.maryland.gov/msa/educ/exhibits/womenshall/html/lacks.html (accessed October 22, 2015).

5. L. Hayflick and P. S. Moorhead, "The Serial Cultivation of Human Diploid Cell Strains," *Experimental Cell Research* 25, no. 3 (December 1961): 585–621.

6. Lijing Jiang, "Alexis Carrel's Immortal Chick Heart Tissue Cultures (1912–1946)," Embryo Project Encyclopedia, July 3, 2012, http://embryo.asu.edu/pages/alexis-carrels-immortal-chick-heart-tissue-cultures-1912-1946 (accessed October 22, 2015).

7. L. Hayflick, "Living Forever and Dying in the Attempt," *Experimental Gerontology* 38, no. 11–12 (November–December, 2003): 1231–41.

8. E. A. Engels et al., "Spectrum of Cancer Risk among US Solid Organ Transplant Recipients," *JAMA* 306, no. 17 (November 2, 2011): 1891–901.

9. S. A. Birkeland et al., "Cancer Risk after Renal Transplantation in the Nordic Countries, 1964–1986," *International Journal of Cancer* 60, no. 2 (January 17, 1995): 183–89.

10. E. Ducroux et al., "Skin Cancers after Liver Transplantation: Retrospective Single-Center Study on 371 Recipients," *Transplantation* 98, no. 3 (August 15, 2014): 335–40.

11. M. L. Gillison et al., "Epidemiology of Human Papillomavirus-Positive Head and Neck Squamous Cell Carcinoma," *Journal of Clinical Oncology* 33, no. 29 (2015): 3235–42.

12. A. Psyrri, T. Rampias, and J. B. Vermorken, "The Current and Future Impact of Human Papillomavirus on Treatment of Squamous Cell Carcinoma of the Head and Neck," *Annals of Oncology* 25, no. 11 (November 2014): 2101–15.

13. Indranil Mallick, "Non-Hodgkin Lymphoma (NHL) after Organ Transplantation," About Health, June 18, 2014, http://lymphoma.about.com/od/nonhodgkinlymphoma/p/transplantlymph.htm (accessed October 22, 2015).

14. "X-Linked Lymphoproliferative Disease," Genetics Home Reference, November 23, 2015, http://ghr.nlm.nih.gov/condition/x-linked-lymphoproliferative-disease (accessed November 24, 2015).

15. Karen Ross, "For Organ Transplant Recipients, Cancer Threatens Long-term Survival," *Journal of the National Cancer Institute* 99, no. 6 (2007): 421–22.

CHAPTER 21: MALIGNANT CARGO

1. J. F. Buell et al., "Donor Transmitted Malignancies," *Annals of Transplantation* 9 (2004): 53–56.

2. I. Penn, "Transmission of Cancer from Organ Donors," *Nefrologia* 15 (1995): 205–13. See also: I. Penn, "Transmission of Cancer from Organ Donors," *Annals Transplant* 2, no. 4 (1997): 7–12.

3. S. Pandanaboyana et al., "Transplantation of Liver and Kidney from Donors with Malignancy at the Time of Donation: An Experience from a Single Centre," *Transplant International* (September 24, 2015), Epub ahead of print.

4. S. A. Birkeland and H. H. Storm, "Risk for Tumor and Other Disease Transmission by Transplantation: A Population-Based Study of Unrecognized Malignancies

and Other Diseases in Organ Donors," *Transplantation* 74, no. 10 (November 27, 2002): 1409–13.

5. D. C. Strauss and J. M. Thomas, "Transmission of Donor Melanoma by Organ Transplantation," *Lancet Oncology* 11, no. 8 (August 2010): 790–96.

CHAPTER 22: THE POWER OF THE IMMUNE SYSTEM

1. D. C. Strauss and J. M. Thomas, "Transmission of Donor Melanoma by Organ Transplantation," *Lancet Oncology* 11, no. 8 (August 2010): 790–96.

2. M. J. Gandhi and D. M. Strong, "Donor Derived Malignancy Following Transplantation: A Review," *Cell Tissue Bank* 8, no. 4 (2007): 267–86.

3. G. B. Forbes et al., "Accidental Transplantation of Bronchial Carcinoma from a Cadaver Donor to Two Recipients of Renal Allografts," *Journal of Clinical Pathology* 34, no. 2 (February 1981): 109–15.

4. R. Wieczorek-Godlewska et al., "Dramatic Recurrence of Cancer in a Patient Who Underwent Kidney Transplantation—Case Report," *Transplant Proceedings* 46, no. 8 (October 2014): 2897–902.

5. G. P. Dunn et al., "Cancer Immuno-Editing: From Immunosurveillance to Tumor Escape," *Nature Immunology* 3, no. 11 (November 2002): 991–98.

6. Paul Ehrlich, *Chemotherapy* (London: Pergamon, 1960). See also: Stefan H. E. Kaufmann, "Paul Ehrlich: Founder of Chemotherapy," *Nature Reviews Drug Discovery* 7, no. 373 (May 7, 2008).

7. Dunn et al., "Cancer Immuno-Editing."

8. F. M. Burnet, "Cancer: A Biological Approach," *British Medical Journal* 1, no. 5023 (April 13, 1957): 841–47.

CHAPTER 23: MAN DIES OF OVARIAN CANCER

1. G. S. Lipshutz et al., "Death from Metastatic Donor-Derived Ovarian Cancer in a Male Kidney Transplant Recipient," *American Journal of Transplantation* 9, no. 2 (February 2009): 428–32.

2. Jennifer Peltz, "Man Dies of Uterine Cancer Linked to Transplant," Cancer on NBC News, May 27, 2010, http://www.nbcnews.com/id/37381460/ns/health-cancer/t/man-dies-uterine-cancer-linked-transplant/#.Viw2kiuvFSM (accessed October 22, 2015).

3. Basile Pasquier, "Accidental Transplantation of Tumour Cells," *Journal of Clinical Pathology* 34, no. 9 (September 1981): 1065–66. See also: S. Jonas et al., "Liver Graft-Transmitted Glioblastoma Multiforme. A Case Report and Experience with 13 Multiorgan Donors Suffering from Primary Cerebral Neoplasia," *Transplant International* 9, no.

4 (1996): 426–29; F. Val-Bernal et al., "Glioblastoma Multiforme of Donor Origin after Renal Transplantation: Report of a Case," *Human Pathology* 24, no. 11 (November 1993): 1256–59; and M. Y. Armanios et al., "Transmission of Glioblastoma Multiforme Following Bilateral Lung Transplantation from an Affected Donor: Case Study and Review of the Literature," *Journal of Neuro-Oncology* 6, no. 3 (July 2004): 259–63.

4. Pasquier, "Accidental Transplantation."

5. Y. Lazebnik and G. E. Parris, "Comment on: 'Guidelines for the Use of Cell Lines in Biomedical Research': Human-to-Human Cancer Transmission as a Laboratory Safety Concern," *British Journal of Cancer* 112, no. 12 (June 9, 2015): 1976–77.

6. E. F. Scanlon et al., "Fatal Homotransplanted Melanoma: A Case Report," *Cancer* 18, no. 6 (June 1965): 782–89.

CHAPTER 24: MAN'S LIFE SAVED BY MOSQUITO BITE

1. C. M. Southam, "Homotransplantation of Human Cell Lines," *Bulletin of the New York Academy of Medicine* 34, no. 6 (June 1958): 416–23.

2. Alice E. Moore, Cornelius P. Rhoads, and Chester M. Southam, "Homotransplantation of Human Cell Lines," *Science* 125 (January 25, 1957): 158–60.

3. Kent Sepkowitz, "A Virus's Debut in a Doctor's Syringe," *New York Times*, August 24, 2009, http://www.nytimes.com/2009/08/25/health/25nile.html?mwrsm=Email&_r=0 (accessed October 22, 2015).

4. Ibid.

5. Ibid.

6. Daniel J. DeNoon, "Kareem Abdul-Jabbar Has CML Form of Leukemia," WebMD Leukemia & Lymphoma, http://www.webmd.com/cancer/lymphoma/news/20091110/kareem-abdul-jabbar-leukemia-is-cml (accessed October 22, 2015).

7. E. A. Gugel and M. E. Sanders, "Needle-Stick Transmission of Human Colonic Adenocarcinoma," *New England Journal of Medicine* 315, no. 23 (December 4, 1986): 1487.

8. H. V. Gartner et al., "Genetic Analysis of a Sarcoma Accidentally Transplanted from a Patient to a Surgeon," *New England Journal of Medicine* 335, no. 20 (November 14, 1996): 1494 –97.

9. J. M. Pawelek, "Fusion of Bone Marrow-Derived Cells with Cancer Cells: Metastasis as a Secondary Disease in Cancer," *Chinese Journal of Cancer* 33, no. 3 (March 2014): 133–39.

10. F. Annunziato and S. Romagnani, "Heterogeneity of Human Effector CD4+ T Cells," *Arthritis Research and Therapy* 11 (December 9, 2009): 257.

11. James S. Welsh et al., "Association between Thymoma and Second Neoplasms," *JAMA* 283, no. 9 (March 1, 2000): 1142–43.

12. P. Christopoulos et al., "A Novel Thymoma-Associated Immunodeficiency with

Increased Naive T cells and Reduced CD247 Expression," *Journal of Immunology* 194, no. 7 (April 1, 2015): 3045–53.

13. J. S. Welsh and S. P. Howard, "Comment on 'A Novel Thymoma-Associated Immunodeficiency with Increased Naive T Cells and Reduced CD247 Expression,'" *Journal of Immunology* 195, no. 8 (October 15, 2015): 3505.

14. James. S. Welsh, "Contagious Cancer," *Oncologist* 16, no. 1 (January 2011): 1–4.

15. Y. Lazebnik, and G. E. Parris, "Comment on: 'Guidelines for the Use of Cell Lines in Biomedical Research': Human-to-Human Cancer Transmission as a Laboratory Safety Concern," *British Journal of Cancer* 112, no. 12 (June 9, 2015): 1976–77.

16. W. G. Banfield et al., "Mosquito Transmission of a Reticulum Cell Sarcoma of Hamsters," *Science* 148, no. 3674 (May 28, 1965): 1239–40.

CHAPTER 25: MOLES, MOLES, AND MORE MOLES

1. S. A. Khanlian and L. A. Cole, "Management of Gestational Trophoblastic Disease and Other Cases with Low Serum Levels of Human Chorionic Gonadotropin," *Journal of Reproductive Medicine* 51, no. 10 (October 2006): 812–18. This is an elementary review of the role of "the pregnancy test" (hCG) in gestational trophoblastic disease and related conditions such as quiescent GTD, placental site trophoblastic tumor, and nontrophoblastic neoplasms (e.g., germ cell tumors).

2. Lisa E Moore and Enrique Hernandez, "Hydatidiform Mole," Medscape Drugs and Diseases, September 22, 2014, http://emedicine.medscape.com/article/254657-overview#a6 (accessed October 22, 2015). This article has some basic information for the interested reader on invasive moles including some material on epidemiology.

3. L. Braun-Parvez et al., "Gestational Choriocarcinoma Transmission following Multiorgan Donation," *American Journal of Transplantation* 10, no. 11 (November 2010): 2541–46.

CHAPTER 26: TUMORS THROUGH THE WORMHOLE

1. "Cystic Echinococcosis (CE) FAQ," CDC, http://www.cdc.gov/parasites/echinococcosis/gen_info/ce-faqs.html (accessed December 15, 2015).

2. N. Turhan et al., "Co-existence of *Echinococcus granulosus* Infection and Cancer Metastasis in the Liver Correlates with Reduced Th1 Immune Responses" *Parasite Immunology* 37 (2015): 16–22.

3. S. Tez and M. Tez, "Echinococcus and Cancer: Unsolved Mystery," *Parasite Immunology* 37, no. 8 (August 2015): 426.

4. H. Akgül et al., "Echinococcus against Cancer: Why Not?," *Cancer* 98 (2003): 1999–2000.

5. Atis Muehlenbachs et al., "Malignant Transformation of *Hymenolepis nana* in a Human Host," *New England Journal of Medicine* 373 (November 2015):1845–52, http://www.nejm.org/doi/full/10.1056/NEJMoa1505892?af=R&rss=currentIssue&#t=articleTop (accessed January 3, 2016).

6. Ibid.

7. Ibid.

CHAPTER 27: THE IMPOSTER

1. For some basic information on AFP and birth defects see: "Testing for Birth Defects, Editorial Review," MedicineNet.com, October 3, 2005, http://www.medicinenet.com/script/main/art.asp?articlekey=10106 (accessed October 22, 2015).

2. For a nice introduction for those curious about the new immunopharmacology that we hear about every day via television advertisements see: D. Boraschi and G. Penton-Rol, "Perspectives in Immunopharmacology: The Future of Immunosuppression," *Immunology Letters* 161, no. 2 (October 2014): 211–15.

3. "Pop! The First Male Pregnancy," RYT Hospital, http://malepregnancy.com/ (accessed October 22, 2015). This very cleverly designed hoax drew a lot of attention and provoked a lot of debate (among my personal colleagues that is).

4. Danilda Hufana-Duran et al., "Full-Term Delivery of River Buffalo Calves (2n = 50) from In Vitro-Derived Vitrified Embryos by Swamp Buffalo Recipients (2n = 48)," *Livestock Science* 107, no. 2–3 (April 2007): 213–19.

5. A. Fernandez-Arias et al., "Interspecies Pregnancy of Spanish Ibex Fetus in Domestic Goat Recipients Induces Abnormally High Plasmatic Levels of Pregnancy-Associated Glycoprotein," *Theriogenology* 51, no. 8 (June 1999): 1419–30.

6. D. Y. Chen et al., "Interspecies Implantation and Mitochondria Fate of Panda-Rabbit Clone Embryos," *Biology of Reproduction* 6, no. 2 (August 2002): 637–42.

7. I didn't realize it, but this has already been done! Here is the case involving the bucardo, an extinct wild goat: J. Folch et al., "First Birth of an Animal from an Extinct Subspecies (Capra pyrenaica pyrenaica) by Cloning," *Theriogenology* 71, no. 6 (April 2009): 1026–34.

8. Domenico Ribatti, "Peter Brian Medawar and the Discovery of Acquired Immunological Tolerance," *Immunology Letters* 167, no. 2 (October 2015): 63–66. This is an excellent brief biography I have read on the scientific achievements of the father of transplantation and the father of reproductive immunology.

9. G. Androutsos et al., "Joseph-Claude-Anthelme Récamier (1774–1852): Forerúnner in Surgical Oncology," *Journal of BUON* 16, no. 3 (July–September 2011): 572–76.

In addition to excelling in the field of oncology and coining the term "metastasis," Récamier was the undisputed founder of modern gynecologic surgery. He performed the first successful vaginal hysterectomy for cancer.

10. Mariusz Z. Ratajczak et al., "The Embryonic Rest Hypothesis of Cancer Development—An Old XIX Century Theory Revisited," *Journal of Cancer Stem Cell Research*, 2 (April 2014): e1001.

11. Peer reviewed information with a (relatively) modern biological interpretation on Beard's trophoblastic theory can be found here: C. Gurchot, "The Trophoblast Theory of Cancer (John Beard, 1857–1924) Revisited," *Oncology* 31, no. 5–6 (1975): 310–33.

12. W. David Billington, "The Immunological Problem of Pregnancy: 50 Years with the Hope of Progress. A Tribute to Peter Medawar," *Journal of Reproductive Immunology* 60, no. 1 (October 2003): 1–11. Medawar's concepts and follow-up ideas by other luminaries can be found in the subsection, "The Conceptus as an Allograft."

13. For a more modern interpretation of the immunological paradox of pregnancy see: J. C. Warning, S. A. McCracken and J. M. Morris, "A Balancing Act: Mechanisms by which the Fetus Avoids Rejection by the Maternal Immune System," *Reproduction* 141, no. 6 (June 2011): 715–24; S. G. Holtan and D. J. Creedon, "Mother Knows Best: Lessons from Fetomaternal Tolerance Applied to Cancer Immunity," *Frontiers in Bioscience (Schol Ed)* 3 (June 1, 2011): 1533–40.

14. S. G. Holtan et al., "Cancer and Pregnancy: Parallels in Growth, Invasion, and Immune Modulation and Implications for Cancer Therapeutic Agents," *Mayo Clinic Proceedings* 84, no. 11 (November 2009): 985–1000.

15. Ibid., 988.

16. Ibid., 989.

17. Ibid.

18. O. B. Christiansen, "Reproductive Immunology," *Molecular Immunology* 55, no. 1 (August 2013): 8–15.

19. Holtan et al., "Cancer and Pregnancy," 990.

20. Ibid.

21. B. Fu, Z. Tian, and H. Wei, "TH17 Cells in Human Recurrent Pregnancy Loss and Pre-Eclampsia," *Cellular & Molecular Immunology* 11, no. 6 (November 2014): 564–70. This is a brief review article on the idea that immunological abnormalities are the root cause of pre-eclampsia and recurrent spontaneous abortion with a focus on TH17 cells.

22. Ibid., 568.

23. Y. Hirohashi et al., "Cytotoxic T Lymphocytes: Sniping Cancer Stem Cells," *Oncoimmunology* 1, no. 1 (January 1, 2012): 123–25. This is an introduction to cytotoxic T lymphocytes for readers who would like a bit more information.

24. L. T. Medeiros et al., "Monocytes from Pregnant Women with Pre-Eclampsia Are Polarized to a M1 Phenotype," *American Journal of Reproductive Immunology* 72, no. 1 (July 2014): 5–13.

25. S. A. Almatroodi et al., "Characterization of M1/M2 Tumour-Associated Macrophages (TAMs) and Th1/Th2 Cytokine Profiles in Patients with NSCLC," *Cancer Microenvironment* (August 30, 2015), Epub ahead of print.

26. S. G. Holtan and D. J. Creedon, "Mother Knows Best: Lessons from Fetoma-

ternal Tolerance Applied to Cancer Immunity," *Frontiers in Bioscience (Schol Ed)* 3 (June 1, 2011): 1533–40.

27. G. Chaouat, "The Th1/Th2 Paradigm: Still Important in Pregnancy?," *Seminars in Immunopathology* 29, no. 2 (June 2007): 95–113. The old Th1/Th2 paradigm is being challenged.

28. B. Saifi et al., "T Regulatory Markers Expression in Unexplained Recurrent Spontaneous Abortion," *Journal of Maternal-Fetal and Neonatal Medicine* (June 5, 2015): 1–6, Epub ahead of print. Tregs and CTLA-4 (which are very familiar to cancer specialists) have recently been found to have a possible role in recurrent spontaneous abortions.

CHAPTER 28: COMPETITION: THE CAUSE OF THE CELLULAR DISEASE

1. Bruce Alberts et al., "Cancer as a Microevolutionary Process," chap. 23 in *Molecular Biology of the Cell*, 4th ed. (New York: Garland Science, 2002). Although not the latest edition, I found this description of the cellular evolutionary process in carcinogenesis to be the clearest I have ever read. Despite its age, I would recommend this short subchapter as an excellent introduction for anyone interested in learning more about the aberrant cellular processes in cancer. Also available online: *NCBI*, http://www.ncbi.nlm.nih.gov/books/NBK26891/ (accessed October 22, 2015).

2. Ibid.

3. C. C. Harris, "Molecular Basis of Multistage Carcinogenesis," *Princess Takamatsu Symposium* 22 (1991): 3–19. This is an early paper on the role of mutations in carcinogenesis with an emphasis on p. 53.

CHAPTER 29: A STANDARD MODEL OF MOLECULAR ONCOLOGY

1. Klaus Hentschel, "Atomic Models, Nagaoka's Saturnian Model," in *Compendium of Quantum Physics*, ed. Daniel Greenberger, Klaus Hentschel, Friedel Weinert (Berlin Heidelberg: Springer, 2009).

2. Brian Cathcart, The Fly in the Cathedral: How a Small Group of Cambridge Scientists Won the Race to Split the Atom (New York: Farrar, Straus and Giroux, 2005).

3. James S. Welsh, "Quarks, Leptons, Fermions, Bosons: The Subatomic Pharmacology of Radiation Therapy," *Science & Medicine* 10, no. 2 (April 2005): 124–36, https://www.sciandmed.com/sm/journalviewer.aspx?issue=-1&year=2005#issue1164 (Accessed November 23, 2015).

CHAPTER 30: RUNAWAY TRAIN!

1. P. Rous, "A Transmissible Avian Neoplasm (Sarcoma of the Common Fowl)," *Journal of Experimental Medicine* 12, no. 5 (September 1, 1910): 696–705.

2. Ibid.

3. Ibid.

4. Jessica Wapner, "The Story of Peyton Rous and Chicken Cancer," *PLOS* blogs, February 9, 2012, http://blogs.plos.org/workinprogress/2012/02/09/the-story-of-peyton-rous-and-chicken-cancer/ (accessed October 22, 2015).

5. For specific information on v-src and cancer see: L. J. Kleinsmith, *Principles of Cancer Biology* (San Francisco, CA: Pearson Benjamin Cummings, 2006), pp. 130–33. This is one of the finest undergraduate-level texts on cancer biology. This is very well written and beautifully illustrated. Although somewhat dated, it can serve as a useful introductory platform for those interested in further learning.

6. Ibid., 133.

7. Ibid.

8. Ibid., 120–23.

9. Ibid., 142–44. See also: Cancer.net Editorial Board, "Retinoblastoma—Childhood: Statistics," Cancer.net, August 2014, http://www.cancer.net/cancer-types/retinoblastoma-childhood/statistics (accessed October 22, 2015).

10. David Sadava, "What Science Knows about Cancer," *Teaching Company* (2013), Course number 1956. The finest discussion on Knudsen and his two-hit hypothesis might be in these lectures by David Sadava.

11. "Retinoblastoma Treatment–for health professionals (PDQ), General Information about Retinoblastoma" National Cancer Institute, Cancer Types, August 14, 2015, http://www.cancer.gov/types/retinoblastoma/hp/retinoblastoma-treatment-pdq (accessed October 22, 2015).

12. Kleinsmith, *Principles of Cancer Biology*, p. 143–44.

13. Ibid., 144.

14. "Li-Fraumeni Syndrome," Genetics Home Reference, November 23, 2015, http://ghr.nlm.nih.gov/condition/li-fraumeni-syndrome (accessed November 24, 2015).

15. D. P. Lane, "Cancer. p53, Guardian of the Genome," *Nature* 358, no. 6381 (July 2, 1992): 15–16.

16. G. P. Pfeifer et al., "Tobacco Smoke Carcinogens, DNA Damage and p53 Mutations in Smoking-Associated Cancers," *Oncogene* 21, no. 48 (October 21, 2002): 7435–51.

17. Ed Payne, "Angelina Jolie Undergoes Double Mastectomy," CNN, May 16, 2013, http://www.cnn.com/2013/05/14/showbiz/angelina-jolie-double-mastectomy/index.html (accessed October 22, 2015).

18. Annabel Fenwick Elliott, "'You Can Get Better Boobs Than You Had Before!' Christina Applegate on the Perks (and the Pitfalls) of Having a Double Mastectomy," MailOnline, October 8, 2014, http://www.dailymail.co.uk/femail/article-2785613/You-better-boobs-Christina-Applegate-perks-pitfalls-having-double-mastectomy.html (accessed October 22, 2015).

19. M. C. King et al., "Breast and Ovarian Cancer Risks Due to Inherited Mutations in BRCA1 and BRCA2," *Science* 302, no. 5645 (October 24, 2003): 643–46.

20. J. Pollock and J. S. Welsh, "Clinical Cancer Genetics: Part I: Gastrointestinal," *American Journal of Clinical Oncology* 34, no. 3 (June 2011): 332–36.

21. A. Lugli et al., "The Medical Mystery of Napoleon Bonaparte: An Interdisciplinary Exposé," *Advances Anatomic Pathology* 18, no. 2 (March 2011): 152–58.

22. Pollock and Welsh, "Clinical Cancer Genetics."

23. Ibid.

24. Kleinsmith, *Principles of Cancer Biology*, p. 166.

25. Bruce Alberts et al., "Cancer as a Microevolutionary Process," chap. 23 in *Molecular Biology of the Cell*, 4th ed. (New York: Garland Science, 2002).

26. "Definition of Differentiation Therapy," MedicineNet.com, September 20, 2012, http://www.medicinenet.com/script/main/art.asp?articlekey=19760 (accessed October 22, 2015).

27. A. Weston and C. C. Harris, "The Mutator Phenotype Concept," in *Holland-Frei Cancer Medicine*, 6th ed., ed. D. W. Kufe et al. (Hamilton, ON: BC Decker; 2003). Online version: "Mutator Phenotype," NCBI, http://www.ncbi.nlm.nih.gov/books/NBK13432/ (accessed October 22, 2015).

28. Miriam Falco, "Basketball Great Abdul-Jabbar Has Cancer," CNN, November 10, 2009, http://www.cnn.com/2009/US/11/10/abdul.jabbar.cancer/index.html?_s=PM:US (accessed October 22, 2015).

29. "Philadelphia Chromosome 50th Anniversary Symposium," Fox Chase Cancer Center, September 28, 2010, http://pubweb.fccc.edu/philadelphiachromosome/history.html (accessed October 22, 2015).

30. Jessica Wapner, The Philadelphia Chromosome: A Genetic Mystery, a Lethal Cancer, and the Improbable Invention of a Lifesaving Treatment (New York: Experiment, 2014).

CHAPTER 31: ORDER OUT OF CHAOS

1. D. Hanahan and R. A. Weinberg, "The Hallmarks of Cancer," *Cell* 100, no. 1 (January 2000): 57–70. Among the most significant papers in the history of cancer biology, this article put all the cellular and molecular features of cancer (known at that time) together into a coherent framework.

2. Ibid.

3. "Definition of Acute Promyelocytic Leukemia," MedicineNet.com, June 14, 2012, http://www.medicinenet.com/script/main/art.asp?articlekey=19758 (accessed October 22, 2015).

This is a very brief (one page) online medical article about acute promyelocytic leukemia for interested readers.

4. Alison Palkhivala, "Birth Defects Still Happening with Accutane," WebMD Health News, August 17, 2001, http://www.webmd.com/news/20010817/birth-defects-still-happening-with-accutane (accessed October 22, 2015).

5. "What Is Accutane? Its Uses and Interactions," Drugwatch, October 5, 2015, http://www.drugwatch.com/accutane/ (accessed October 22, 2015).

6. D. Hanahan and R. A. Weinberg, "Hallmarks of Cancer: The Next Generation," *Cell* 144, no. 5 (March 4, 2011): 646–74.

7. Jung-whan Kim and Chi V. Dang, "Cancer's Molecular Sweet Tooth and the Warburg Effect," *Cancer Research* 66, no. 18 (September 15, 2006): 8927–30. Also available online at: http://cancerres.aacrjournals.org/content/66/18/8927.full.html (accessed October 22, 2015).

8. "Goldie-Coldman Hypothesis," Drugs.com, http://www.drugs.com/dict/goldie-coldman-hypothesis.html (accessed October 22, 2015).

9. E. Shacter and S. A. Weitzman, "Chronic Inflammation and Cancer," *Oncology* 16, no. 2 (2002): 217–26, 229.

CHAPTER 32: IMMUNE THEORY OF CANCER

1. Bruce Alberts et al., "Cancer as a Microevolutionary Process," chap. 23 in *Molecular Biology of the Cell*, 4th ed. (New York: Garland Science, 2002).

2. A. I. Rozhok and J. DeGregori, "Toward an Evolutionary Model of Cancer: Considering the Mechanisms that Govern the Fate of Somatic Mutations," *Proceedings of the National Academy of Sciences* 112, no. 29 (July 21, 2015): 8914–21.

3. R. Peto et al., "Cancer and Ageing in Mice and Men," *British Journal of Cancer* 32, no. 4 (October 1975): 411–26.

4. Author-based hypotheses.

5. D. F. Merlo et al., "Cancer Incidence in Pet Dogs: Findings of the Animal Tumor Registry of Genoa, Italy," *Journal of Veterinary Internal Medicine* 22, no. 4 (July–August 2008): 976–84.

6. Carl Zimmer, "Elephants: Large, Long-Living and Less Prone to Cancer," *New York Times*, October 8, 2015, http://www.nytimes.com/2015/10/13/science/why-elephants-get-less-cancer.html?mabReward=CTM&action=click&pgtype=Homepage®ion=CColumn&module=Recommendation&src=rechp&WT.nav=RecEngine&_r=1 (accessed October 22, 2015).

7. Ibid. See also: L. M. Abegglen et al., "Potential Mechanisms for Cancer Resistance in Elephants and Comparative Cellular Response to DNA Damage in Humans," *JAMA* 314, no. 17 (2015): 1850–60.

8. Danielle Elliot, "Ming the Clam, World's Oldest Animal, Was Actually 507 Years Old," CBS News, November 14, 2013, http://www.cbsnews.com/news/ming-the-clam-worlds-oldest-animal-was-actually-507-years-old/ (accessed October 22, 2015).

9. Abegglen et al., "Potential Mechanisms for Cancer Resistance in Elephants."

10. G. E. Goodman et al., "The Beta-Carotene and Retinol Efficacy Trial: Incidence of Lung Cancer and Cardiovascular Disease Mortality during 6-Year Follow-Up after Stopping Beta-Carotene and Retinol Supplements," *Journal of the National Cancer Institute* 96, no. 23 (December 1, 2004): 1743–50.

11. K. Smigel, "Beta Carotene Fails to Prevent Cancer in Two Major Studies; CARET Intervention Stopped," *Journal of the National Cancer Institute* 88, no. 3–4 (February 21, 1996): 145.

12. E. A. Klein et al., "Vitamin E and the Risk of Prostate Cancer: Results of the Selenium and Vitamin E Cancer Prevention Trial (SELECT)," *JAMA* 306, no. 14 (2011): 1549–56.

13. Author-based hypotheses.

14. J. B. Swann and M. J. Smyth, "Immune Surveillance of Tumors," *Journal of Clinical Investigation* 117, no. 5 (2007): 1137–46.

15. For two studies that have demonstrated elevated cancer rates in immunosuppressed populations, specifically those with HIV/AIDS and organ transplant recipients, see: A. E. Grulich et al., "Incidence of Cancers in People with HIV/AIDS Compared with Immunosuppressed Transplant Recipients: A Meta-Analysis," *Lancet* 370, no. 9581 (2007): 59–67; and Oliveira Cobucci et al., "Comparative Incidence of Cancer in HIV-AIDS Patients and Transplant Recipients," *Cancer Epidemiology* 36, no. 2 (April 2012): e69–73.

16. D. J. Waters and K. Wildasin, "Cancer Clues from Pet Dogs," *Scientific American* 295 (November 2006): 94–101. According to this article, very long-lived dogs and people (e.g., those over one hundred years of age) have a reduced likelihood of ultimately succumbing to cancer.

17. F. Boccardo et al., "Interleukin-2, Interferon-Alpha and Interleukin-2 plus Interferon-Alpha in Renal Cell Carcinoma. A Randomized Phase II Trial," *Tumori* 84, no. 5 (September–October 1998): 534–39.

CHAPTER 33: WHAT CAN COWS TEACH US ABOUT CONQUERING CANCER?

1. Heather Brannon, "The History of Smallpox: The Rise and Fall of a Disease," About Health, December 16, 2014, http://dermatology.about.com/cs/smallpox/a/smallpoxhx.htm (accessed October 22, 2015).

2. Ibid. See also: Donald R. Hopkins, *The Greatest Killer: Smallpox in History, with a New Introduction* (Chicago, IL: University of Chicago Press, 2002). More details can be found in this authoritative history of smallpox.

3. "Rinderpest Eradicated—What Next? Eradication of the Deadly Virus Is a Model for Other Diseases," Food and Agriculture Organization of the United Nations, June 28, 2011, http://www.fao.org/news/story/en/item/80894/icode/ (accessed October 22, 2015).

4. Brannon, "History of Smallpox."

5. For two references for two cancer-related virus vaccines see: "Recombivax," RxList, September 2, 2014, http://www.rxlist.com/recombivax-drug.htm (accessed October 22, 2015), and "Gardasil," RxList, October 21, 2015, http://www.rxlist.com/gardasil-drug.htm (accessed October 22, 2015).

6. Robert Jesty and Gareth Williams, "Who Invented Vaccination?," *Malta Medical Journal* 23, no. 2 (2011).

7. Harry W. Herr and Alvaro Morales, "History of Bacillus Calmette-Gurin and Bladder Cancer: An Immunotherapy Success Story," *Journal of Urology* 179, no. 1 (2008): 53–56. This is a historical account written by researchers who were actually involved with the original discovery decades ago.

8. B. J Hawgood, "Albert Calmette (1863–1933) and Camille Guérin (1872–1961): The C and G of BCG Vaccine," *Journal of Medical Biography* 15, no. 3 (August 2007): 139–46.

9. R. Pearl, "Cancer and Tuberculosis," *American Journal of Hygiene* 9 (1929): 97.

10. Lloyd J. Old, "Father of Modern Tumor Immunology," Cancer Research Institute, September 23, 1933–November 28, 2011, http://www.cancerresearch.org/about/lloyd-j-old (accessed October 22, 2015).

11. Herr and Morales, "History of Bacillus Calmette-Guérin."

12. Robert C. Wittes, "Immunology of Bacille Calmette-Guérin and Related Topics," *Clinical Infectious Diseases* 31, supplement 3 (2000): S59–S63.

13. M. J. Silverstein, J. DeKernion, and D. L. Morton, "Malignant Melanoma Metastatic to the Bladder. Regression Following Intratumor Injection of BCG Vaccine," *JAMA* 229, no. 6 (August 1974): 688.

14. Herr and Morales, "History of Bacillus Calmette-Guérin."

15. "Cancer Immunotherapy: Bladder Cancer," Cancer Research Institute, December 2014, http://www.cancerresearch.org/cancer-immunotherapy/impacting-all-cancers/bladder-cancer#sthash.eYnxAUfM.dpuf (accessed October 22, 2015).

CHAPTER 34: TOUGH MOTHERS AND JUVENILE DELINQUENTS

1. Antigen presenting cells: See Glossary.

2. K. Murphy, *Janeway's Immunobiology*, 8th ed. (New York: Garland Science, 2012).

3. Kohrt Holbrookt, "Concepts in Immuno-Oncology: Understanding the Key Players," Medscape Oncology, April 30, 2014, http://www.medscape.org/viewarticle/823638_3 (accessed October 22, 2015).

4. For two of the several thousand scientific papers on the tumor microenvironment see: S. L. Shiao et al., "Immune Microenvironments in Solid Tumors: New Targets for Therapy" *Genes and Development* 25 (2011): 2559–72; and F. Balkwill and A. Mantovani, "Inflammation and Cancer: Back to Virchow?," *Lancet* 357 (2001): 539–45.

CHAPTER 35: AS CRAZY AS THE QUANTUM CAFÉ

1. Jim Vadeboncoeur Jr., "Walt Kelly," JVJ Publishing: Illustrators, 2011, http://www.bpib.com/kelly.htm (accessed October 22, 2015).

2. For a couple of papers on the role of macrophages in the tumor microenvironment see: M. R. Galdiero et al., "Tumor Associated Macrophages and Neutrophils in Cancer," *Immunobiology* 218, no. 11 (November 2013): 1402–10; and J. Condeelis and J. W. Pollard, "Macrophages: Obligate Partners for Tumor Cell Migration, Invasion, and Metastasis," *Cell* 124, no. 2 (January 27, 2006): 263–66.

3. IL-10 and TGFβ do a lot more than just affect macrophages. For more information see: R. Sabat et al., "Biology of Interleukin-10," *Cytokine Growth Factor Review* 21, no. 5 (2010): 331–44; and L. Yang, "TGFbeta, a Potent Regulator of Tumor Microenvironment and Host Immune Response, Implication for Therapy," *Current Molecular Medicine* 10 (2010): 374–80.

4. T. J. Curiel, "Regulatory T Cells and Treatment of Cancer," *Current Opinion Immunology* 20, no. 2 (April 2008): 241–46. Although slightly dated, this article serves as a nice introduction to the concept of regulatory T cells and their various roles in cancer.

5. T. J. Waldron et al., "Myeloid Derived Suppressor Cells: Targets for Therapy," *Oncoimmunology* 2, no. 4 (April 1, 2013): e24117.

6. M. R. Galdiero et al., "Tumor Associated Macrophages and Neutrophils in Cancer," *Immunobiology* 218, no. 11 (November 2013): 1402–10.

7. L. L. Lanier, "Missing Self, NK Cells, and the White Album," *Journal of Immunology* 174, no. 11 (June 1, 2005): 6565.

8. P. Cirri and P. Chiarugi, "Cancer-Associated-Fibroblasts and Tumour Cells: A Diabolic Liaison Driving Cancer Progression," *Cancer Metastasis Review* 31, no. 1–2 (June 2012): 195–208.

9. B. Bierie and H. L. Moses, "Tumour Microenvironment: TGFbeta: The Molecular Jekyll and Hyde of Cancer," *Nature Reviews Cancer* 6 (2006): 506–20.

10. Francisco A Bonilla, "The Humoral Immune Response," *UpToDate*, October 2015, http://www.uptodate.com/contents/the-humoral-immune-response (accessed October 22, 2015).

11. Y He et al., "The Roles of Regulatory B Cells in Cancer," *Journal of Immunology Research* (2014): 215471.

CHAPTER 36: CONNECTED DOTS: LOOKING BACK AND GLIMPSING THE FUTURE

1. Ira Mellman, George Coukos, and Glenn Dranoff, "Cancer Immunotherapy Comes of Age," *Nature* 480 (December 22, 2011): 480–89. This is not just my own opinion.

2. Drew M. Pardol, "The Blockade of Immune Checkpoints in Cancer Immunotherapy," *Nature Reviews Cancer* 12 (April 2012): 252–64. This is an excellent introduction and review on immunological checkpoints in cancer.

3. Ibid., 253. CTLA4, the first immune checkpoint receptor to be clinically targeted, is expressed exclusively on T cells where it primarily regulates the amplitude of the early stages of T cell activation.

4. Ibid.

5. D. Mavilio D and E. Lugli, "Inhibiting the Inhibitors: Checkpoints Blockade in Solid Tumors," *Oncoimmunology* 2, no. 9 (September 1, 2013): e26535.

6. Pardol, "The Blockade of Immune Checkpoints," 253.

7. Ibid., 256.

8. "Nivolumab," US Food and Drug Administration, December 23, 2014, http://www.fda.gov/Drugs/InformationOnDrugs/ApprovedDrugs/ucm427807.htm (accessed October 22, 2015). See also: "Pembrolizumab," US Food and Drug Administration, September 1, 2015, http://www.fda.gov/drugs/InformationOnDrugs/approveddrugs/ucm412861.htm (accessed October 22, 2015).

9. "FDA Expands Approved Use of Opdivo in Advanced Lung Cancer," US Food and Drug Administration, October 9, 2015, http://www.fda.gov/newsevents/newsroom/pressannouncements/ucm466413.htm (accessed October 22, 2015).

10. "A Phase I Study of an Anti-KIR Antibody in Combination with an Anti PD1 Antibody in Patients with Advanced Solid Tumors," CancerTrials.gov US National Institutes of Health, July 2015, https://clinicaltrials.gov/ct2/show/NCT01714739 (accessed October 22, 2015).

11. J. Larkin et al., "Combined Nivolumab and Ipilimumab or Monotherapy in Untreated Melanoma," *New England Journal of Medicine* 373, no. 1 (July 2, 2015): 23–34.

12. C. Robert et al., "Pembrolizumab versus Ipilimumab in Advanced Melanoma," *New England Journal of Medicine* 372, no. 26 (June 25, 2015): 2521–32.

13. P. W. Kantoff et al., "Sipuleucel-T immunotherapy for Castration-Resistant Prostate Cancer," *New England Journal of Medicine* 363, no. 5 (July 29, 2010): 411–22.

14. Todd Ackerman. "Jim Allison Confronts Cancer, Critics with Immunotherapy," *SFGate*, April 16, 2014, http://www.sfgate.com/health/article/Jim-Allison-confronts-cancer-critics-with-5405290.php (accessed October 22, 2015).

GLOSSARY

A: (See adenine)

Adaptive immunity: a form of immunity characterized by immunological memory and mediated by specific B cells and T cells that have been modified to recognize and attack certain following prior exposure.

Adenine: a purine nucleoside base present in DNA and RNA. A forms complementary base pairs with thymine (T) in DNA and uracil (U) in RNA.

Adenocarcinoma: malignant tumor originating in a gland or glandular tissue.

Acquired immunodeficiency syndrome: a disease caused by the human immunodeficiency virus (HIV) which weakens the immune system. Patients with AIDS have an increased risk of several types of cancers including many that are virus-induced and some that are not.

Aflatoxin: a carcinogenic compound produced by the fungus *Aspergillus flavus* that contaminates certain foods, including peanuts and grains.

AIDS: (See acquired immunodeficiency syndrome).

Alkylating agent: a chemical that introduces an alkyl group on to DNA. Alkylating agents can act as carcinogens but can also be used as chemotherapeutic agents an alternative form of a gene at the same locus or position in a chromosomal pair. One allele may be dominant over another.

Allele: any of the alternative forms of a given gene.

Allograft: a transplant of an organ or tissue from one individual to another in the same species.

Alpha particle: the type of radiation composed of two neutrons and two protons (i.e. a helium nucleus).

Anaplastic: tissue that is poorly differentiated and abnormal in cellular appearance.

Anchorage dependence: a characteristic of normal cells and tissues in which the cells must be attached to a solid surface (such as the extracellular matrix or the surfaces of a Petri dish) before they can propagate.

Anchorage independence: a characteristic of cancer cells in which they can grow well not just when they are attached to a solid surface but also when they are freely suspended in a liquid (or semi-solid) medium.

Aneuploid: an abnormal number of chromosomes.

Angiogenesis: the process of forming new blood vessels. The induction of angiogenesis is a hallmark of cancer. (Also See neovasculature).

Anoikis: apoptosis triggered in a response to a lack of contact with a solid surface or the extracellular matrix.

Antibody: also known as immunoglobulins, antibodies are a class of proteins produced by B lymphocytes or plasma cells that bind to substances with high specificity. The substance to which they bind is called an antigen.

Antigen: any foreign or abnormal molecule capable of triggering an immune response.

Antigen presenting cell: Any cell that dismantles and processes antigens and displays the fragments on its surface to trigger an immune response. Examples include dendritic cells and macrophages.

Antioxidant: a molecule that inhibits oxidation reactions, often working by reacting with free radicals or directly reducing other molecules.

APC: (See antigen presenting cell).

Apoptosis: a mechanism of cellular death that involves an organized series of molecular events that leads to the dismantling of the internal contents of the cells. Also known as programmed cell death. Apoptosis plays a role in tumor suppression, and inhibition of apoptosis is a hallmark of cancer.

Ataxia telangiectasia: an inherited syndrome characterized by loss of coordination, dilation of small blood vessels, immune system deficiencies, & a 40% risk of developing cancer. Abbreviated AT, ataxia telangiectasia is caused by mutant alleles of the ATM tumor suppressor gene and is inherited in an autosomal recessive fashion.

ATM gene: a tumor suppressor gene coding for the ATM protein kinase which plays a central role in DNA damage response pathways. Inheritance of two mutant alleles of the ATM gene causes the disease ataxia telangiectasia.

Autoimmunity: a condition in which an individual's immune system reacts against his or her own tissues, often resulting in disease.

Autophagy: a process whereby proteins and organelle components that are no longer required are targeted to the lysosomes for degradation. Excessive autophagy leads to a specific type of non-apoptotic cell death program.

Autosomal: alluding to a chromosome that is not one of the gender-determining chromosomes (X or Y in humans).

Autosomal dominant: a pattern of inheritance in which a given gene or allele will determine how the trait in question will appear in an organism irrespective of whether it is present in the heterozygous or homozygous state. In other words, the allele dominates over the recessive allele.

Autosomal recessive: a pattern of inheritance in which a given gene or allele will determine how the trait in question will appear in an organism only when present in the homozygous state. In other words, both versions of a recessive allele must be present (or no dominant allele must be present) in order for expression.

B cell: a type of lymphocyte responsible for producing antibodies. Also called B lymphocytes, B cells can mature into plasma cells.

BCG: (See Bacillus Calmette-Guerin).

Bacillus Calmette-Guerin (BCG): A bacterial strain related to the tuberculosis bacterium that does not cause disease but elicits a strong immune response and is sometimes used in cancer immunotherapy (especially in superficial bladder cancer).

Basal cell carcinoma: a type of nonmelanoma skin cancer. This accounts for approximately 75% of all skin cancers but results in very few deaths since it rarely metastasizes.

BCR- ABL: a novel oncogene created by reciprocal translocation between chromosomes 9 and chromosomes 22, (the Philadelphia chromosome). The oncogene codes for an abnormal tyrosine kinase that contributes to the development of chronic myelogenous leukemia.

Benign: referring to a tumor that does not invade surrounding tissues or metastasize. A benign tumor grows only locally.

Beta particle: a type of radiation composed of electrons or positrons.

Biomarker: a biochemical or genetic feature that can be used to measure the progression of disease or the effects of treatment.

Blood fluke: parasitic flatworms (of Phylum Platyhelminthes, Class Trematoda) that can cause chronic inflammation of the blood vessels of the organ they are infesting. Schistosomiasis is a blood fluke infestation which can occasionally lead to cancer.

Bloom syndrome: an inherited syndrome characterized by short stature, sun-induced facial rashes, immunodeficiency, decreased fertility, and an increased risk of developing cancer before the age of twenty. This is caused by mutations in both alleles of the BLM tumor suppressor gene which codes for a DNA helicase involved in DNA repair. It is inherited in an autosomal recessive fashion.

BRCA1 and BRCA2 2 genes: tumor suppressor genes which when mutated or inactivated may confer a high risk for breast and/or ovarian cancer. These genes code for proteins involved in repairing double strand DNA breaks and mutations are inherited in autosomal dominant fashion.

Burkitt lymphoma: a type of lymphoma associated with infection by Epstein-Barr virus along with a chromosomal translocation in which the myc gene is activated by moving it from chromosome 8 to chromosome 14 or another chromosome involved in antibody production.

C: (See cytosine).

Cachexia: a metabolic disorder often associated with cancer that is characterized by progressive weight loss despite adequate caloric intake.

Cancer: an uncontrolled proliferation of cells capable of invading adjacent tissues and spreading to distant tissues (metastasizing). A localized cancer is called a malignant tumor.

Cancer stem cells: cells within a tumor that have the ability to self-renew and to give rise to phenotypically diverse cancer cells.

Carcinogen: any agent capable of causing cancer.

Carcinogenesis: the process of inducing cancer.

Carcinoma: a malignant tumor (cancer) originating from the epithelial cells that cover external surfaces and/or line internal body surfaces.

Carcinoma in situ: an epithelial cancer that has not yet evolved the capability of invading or metastasizing.

Caretaker: the class of tumor suppressor genes involved in repairing chromosomes or DNA. Loss of function mutations in caretaker genes contributes to genetic instability.

Caspase: any of a family of proteases that degrade other cellular proteins during the process of apoptosis.

Cell cycle: A sequence of steps that a cell passes through between one cell division and the next. The phases of the cell cycle are called mitosis (M), G1, S, and G2.

Cellular differentiation: (See differentiation).

Chromosomal translocation: a process in which a section of one chromosome is broken off and moved to another chromosome.

Chromosome: a structure composed of a DNA molecule and associated

RNA and proteins. Humans have 46 chromosomes (23 pairs) in the nucleus of their somatic cells.

Chronic: referring to a long lasting condition. The opposite of acute.

Clinical trial: a scientific procedure involving human or veterinary patients which test drugs or other treatments for safety, dose and effectiveness. Clinical trials can consist of Phase 1, Phase 2, and Phase 3 studies.

Clonal: originating from a single cell.

Clone (noun): a population of genetically identical cells produced from the reproduction of a single cell.

Codon: the sequence of three nucleotides in DNA or RNA molecules that serve as a coding unit for amino acids or a start or stop signal.

Combination chemotherapy: treatment of cancer with several drugs in combination rather than a single agent alone.

Complementary base pairing: in DNA or RNA, the pairing of the base G (guanine) with the base C (cytosine) and the base A (adenine) with the base T (thymine in DNA) or U (uracil in RNA). This is also known as Watson-Crick base pairing.

Complete carcinogen: an agent that can trigger both the initiation stage and the promotion stage of carcinogenesis. A complete carcinogen can cause cancer all by itself.

Computed tomography (CT): an imaging technique in which x-ray pictures taken from multiple different angles are combined buy a computer into a series of anatomically detailed cross-sectional images.

Cytokine: any protein produced by cells of the immune system to stimulate an immune response to foreign infectious organisms or abnormal cells.

Cytosine (C): a pyrimidine base found in DNA and RNA that forms complementary base pairs with guanine (G).

Cytotoxic T lymphocyte (CTL): a class of lymphocyte specialized in attacking foreign cells, cells that have been infected with a virus or bacterium, or abnormal cells (including cancer cells). Cytotoxic T lymphocytes typically have the marker CD8 on their surfaces.

Delaney amendment: a law in existence from 1958 to 1996 that required the FDA to ban any food additive or contaminant if it was found capable of causing cancer in any animal studies at any dose.

Deletion: a chromosomal abnormality involving the loss of nucleotides from the DNA. A deletion can range in size from a single nucleotide to large segments of DNA containing multiple genes.

Dendritic cell: a type of antigen presenting cell (APC) that engulfs antigens and break them down into smaller fragments and then presents those fragments on their cell surfaces. Once on the surfaces of the dendritic cells these fragments can then potentially incite lymphocytes to react against those antigens.

Density dependent inhibition of growth: a tendency of (normal) cells to stop dividing when they reach a certain high population density in a cell culture.

Differentiating agent: any substance that promotes the process of differentiation or maturation in cells.

Differentiation: the process in which cells acquire the specialized appearances and functions that distinguish them from immature cells and other cells. The process of cellular maturation.

Diploid: containing two sets of chromosomes and therefore two copies or alleles of each gene.

Dysplasia: an abnormal condition in which the organization of cells and tissues are disrupted. Dysplasia may be an early stage in the development of cancer.

DNA methylation: the addition of methyl groups to nucleotides in DNA which can lead to epigenetic silencing of genes.

Dominant: (See autosomal dominant).

Dominant negative: a mutation that produces a protein that interacts with and/or interferes with the function of a wild-type protein.

Double helix: two intertwined helical chains of a DNA molecule held together by complementary (Watson Crick) base pairing between A & T and G & C.

E6: a protein produced by an oncogene found in some strains of HPV that binds to a cell's p53 protein thereby targeting it for destruction. This is one of the ways that certain HPV viruses can cause cancer.

E7: a protein produced by an oncogene found in some strains of HPV that binds to a cell's Rb protein and interfering with its ability of the RV protein to restrain cell proliferation. This is one of the ways that certain HPV viruses can cause cancer.

Electromagnetic radiation: energy in the form of waves and photons that propagate through a vacuum at the speed of light. The characteristics of the radiation depend on its wavelength. Electromagnetic radiation if high enough in energy can be ionizing.

Endostatin: a protein that inhibits the growth of blood vessels.

Endothelial cell: a type of cell that lines the internal surfaces of blood vessels or lymphatic vessels.

Epidemiology: the branch of science that investigates the frequency and distribution of conditions or diseases in populations.

Epidermal growth factor (EGF): a protein that stimulates the growth and division of a wide variety of epithelial cell types.

Epigenetic: referring to heritable information that is encoded by modifications (such as methylation) to the DNA which can affect gene expression. It does not include gross chemical changes (mutations) in the base sequence of DNA.

Epigenetic changes: alterations in cellular properties induced by changes in gene expression through modifications of DNA (such as methylation) rather than by genetic mutation.

Epithelial cell: a type of cell that forms the covering layers of external and internal body surfaces.

Epithelial-mesenchymal transition (EMT): the process in which cells leave an epithelial layer and become a loose mass of mesenchymal cells which can migrate individually. EMT is crucial early embryonic development and appears to be involved in tumor formation.

Epstein-Barr virus (EBV): a virus associated with certain lymphomas such as Burkitt lymphoma and a subtype of nasopharyngeal carcinoma (especially in China) as well as mononucleosis.

Familial adenomatous polyposis: (also called FAP), an inherited condition in which numerous polyps develop in the colon eventually culminating in cancer. FAP is caused by the autosomal dominant inheritance of a single defective or missing copy of the APC gene.

Familial cancer: hereditary cancer. Cancers that arise as a result of an inherited mutation that leads to a predisposition towards developing cancer.

Free radical: an atom or molecule with an unpaired electron. The unpaired electron makes them highly reactive chemically and potentially hazardous to biological molecules.

Fusion gene: a gene derived from two different genes spliced together.

Fusion protein: a protein containing amino acid sequences and coded by portions of two different teams that have been fused together.

G: (See guanine).

G protein: a class of protein molecules whose activity is regulated by binding to GTP and GDP.

Gamma rays: the most energetic form of electromagnetic radiation. Gamma rays are capable of ionizing atoms and molecules and are thus a form of ionizing radiation.

Gatekeeper: the class of tumor suppressor genes involved in restraining cellular replication. Loss of function mutations in gatekeepers may lead to excessive cellular proliferation and tumor formation.

Gene: a nucleotide base sequence in DNA that codes for a functional product.

Gene amplification: replication of a section of DNA that results in the production of multiple copies of individual genes. If a potentially cancer-inducing gene is amplified it increases the odds of that cell eventually turning malignant.

Gene expression: the process by which the information encoded by a gene is converted into a protein. In terms of molecular biology this usually refers to transcription.

Genetic instability: a trait of malignant cells in which unusually high mutation rates are observed. This is often caused by defects in DNA repair mechanisms such as caretakers.

Germline mutation: a mutation in either egg or sperm cell DNA. Only germline mutations can be passed on to the next generation.

Gray (Gy): the unit of dose for ionizing radiation. It corresponds to 1 Joule of energy deposited in one kilogram of matter.

Growth factor: a class of extracellular signaling proteins that stimulate the proliferation of cells by binding to specific receptor proteins on the outer surfaces of certain cells.

Guanine (G): a purine nucleoside base present in DNA and RNA. G forms complementary base pairs with cytosine (C).

Helicobacter pylori: a type of bacteria that causes stomach inflammation (gastritis), peptic ulcers and potentially gastric adenocarcinoma or gastric MALT lymphoma.

Hepatitis: inflammation or infection of the liver.

Hepatitis B virus (HBV): a DNA virus that can cause hepatitis and liver cancer.

Hepatitis C virus (HCV): an RNA virus that can cause hepatitis and liver cancer.

Herceptin® (Trastuzumab): a monoclonal antibody used in the treatment of certain breast cancers and other cancers that over expresses the HER2 gene.

Hereditary non-polyposis colon cancer (HNPCC): also known as Lynch syndrome. An inherited cancer predisposition syndrome caused by defects in any of several DNA mismatch repair genes. In addition to colon cancer several other types of cancers develop in people with HNPCC.

Heterozygous: having two different alleles for a given gene on homologous chromosomes.

HIV: (See human immunodeficiency virus).

Hodgkin disease: a type of lymphoma characterized by the presence of Reed Sternberg cells having a distinct clinical behavior and management strategy.

Homologous chromosomes: the two copies of a specific chromosome derived from each parent. Homologous chromosomes contain the same sets of genes arranged in the same order.

Homozygous: having two identical alleles for a given gene on homologous chromosomes.

HPV: (See human papilloma virus).

Human immunodeficiency virus (HIV): the retrovirus(es) that causes AIDS or acquired immunodeficiency syndrome.

Human papillomavirus (HPV): the family of non-enveloped, DNA viruses that causes warts (papillomas). Some strains of HPV are linked to certain cancers. HPV viruses possess oncogenes that code for proteins which interfere with the function of the Rb and p53 proteins.

Human T lymphotropic virus 1 (HTLV1): the retrovirus associated with adult T cell leukemia/lymphoma.

Hybrid cell: a cell whose nucleus contains chromosomes derived from two different cells that were joined together.

Hyperplasia: tissue growth due to an increase in the number of cells but in which the cellular organization is still normal.

Hypertrophy: tissue growth due to an increase in cellular size.

Immune surveillance hypothesis: the hypothesis that under ordinary circumstances, immunological rejection of malignant cells protects people against the development of cancer.

Immunosuppressive drug: any drug that inhibits the immune system.

Immunosuppressive drugs may be given to organ transplant recipients in order to prevent immune rejection of transplanted organs.

Immunotherapy: treatment of a disease (such as cancer) by stimulating the immune system or the administration of agents such as antibodies that are made by the immune system.

Incomplete carcinogen: an agent that can function in the initiation or promotion stage of carcinogenesis but not both. Incomplete carcinogens are unable to induce cancer by themselves.

Inflammation: the response of tissues to injury, irritation or infection characterized by swelling, redness, pain, and heat. Swelling, redness, pain and heat are sometimes referred to by the Greek words tumor, rurbor, dolor, and calor respectively.

Informed consent: the process in which individuals who wish to participate in a clinical trial are given information regarding risks, benefits alternative options and the purpose of the clinical trial. The participant then signs a document indicating their understanding of these conditions and their voluntary consent.

Initiation: the stage of carcinogenesis in which a cell is converted into a precancerous state by an agent that causes DNA damage.

Innate immune cells: white blood cells that mediate innate immunity such as basophils, dendritic cells, eosinophils, Langerhans cells, mast cells, macrophages, monocytes, natural killer cells and neutrophils.

Innate immunity: non-specific defense mechanisms that includes physical and chemical barriers (such as the skin, lysozyme in the tears, and complement in the blood) as well as immune cells that do not have to undergo prior exposure to foreign cells or substances before gaining the ability to recognize them as enemies and attack them.

Insertion: a chromosomal abnormality involving the addition of nucleotides to the DNA. An insertion can range in size from a single nucleotide to large segments of DNA containing multiple genes.

Insertional mutagenesis: a change in gene structure or activity caused by the integration of DNA derived from another source (such as a virus) into the host's own chromosomal DNA.

Invasion: the spread of malignant cells into surrounding normal tissues.

Ionizing radiation: radiation energetic enough to cause ionization (removing electrons from atoms and molecules to create ions).

Kaposi sarcoma: a cancer arising from blood vessels in the skin caused by Kaposi sarcoma-associated herpesvirus (a virus in the herpes virus family). The incidence of Kaposi's sarcoma is very low but is greatly increased in people who are severely immunosuppressed.

Kinase: an enzyme that phosphorylates (transfers phosphate groups) to other molecules.

Laetrile: a substance extracted from apricot pits that was once purported to be a cure for cancer. Clinical trials of laetrile have not proven any beneficial effects.

Latent virus: the condition in which a virus remains alive but concealed within a cell and not producing or releasing any new virus particles.

Leukemia: cancer of blood cells. In the early phases of the disease the malignant cells typically reside within the bloodstream rather than growing as tumors.

Li-Fraumeni syndrome: an inherited cancer predisposition syndrome that confers a high risk of developing a broad range of cancers. It is caused by the inheritance of a defective copy of the *TP53* gene and is passed along in an autosomal dominant inheritance pattern.

Liver fluke: parasitic flatworms (of Phylum Platyhelminthes, Class Trematoda) that cause chronic inflammation of the bile ducts, which may lead to cholangiocarcinoma (cancer of the bile ducts).

Long term repeats (LTRs): repetitive nucleotide sequences located at the ends of retrovirus genomes. Long terminal repeats are involved in the integration of the virus into the host chromosomal DNA and in activating the transcription of viral genes. Long terminal repeats can also inappropriately activate nearby host proto-oncogenes thereby leading to cancer.

Loss of heterozygosity: the condition in which one of the two alleles initially present in a heterozygous state is lost, silenced or mutated and the cell is thereby converted to the homozygous state (i.e. only one form of the allele is present).

Lymphocyte: a class of white blood cells involved in immune responses. Lymphocytes can be B cells, T cells or natural killer cells.

Lymphoma: a cancer of lymphocytes in which the cells grow mostly as a solid tumors rather than as cells in the bloodstream (leukemia).

Lysis: a form of cell death caused by rupturing of the plasma membrane.

MHC: (See Major histocompatibility complex).
MRI: (See magnetic resonance imaging).
mRNA: (See messenger RNA).
Macrophage: a type of white blood cell that recognizes, engulfs and digests foreign or abnormal cells through phagocytosis. Macrophages tend to remain in one place whereas their precursors, monocytes, roam freely in the bloodstream.
Magnetic resonance imaging (MRI): a medical imaging technique in which radio waves and magnetic field are used instead of x-rays to create a series of anatomically detailed cross-sectional images of the body.
Major histocompatibility complex (MHC): a set of proteins expressed on the surfaces of cells that play a key role in distinguishing one individual's cells from that of another and are also recognized by the immune system during the process of inducing an immune response. Also refers to the set of genes that encodes these proteins.
Malignant tumor: a tumor capable of invading neighboring normal tissues and spreading throughout the body (metastasizing).
Melanin: the dark brown pigments synthesized by melanocytes responsible for skin color and the tanning process that occurs in response to sunlight.
Melanoma: a type of cancer arising from melanocytes, the cells which produce melanin. Melanomas usually arise on the skin but can appear nearly anywhere in the body including internal sites.
Mesothelioma: a form of cancer derived from mesothelial cells that cover the interior surfaces of the chest (pleura) and abdominal cavities (peritoneum). Mesothelioma can be caused by exposure to asbestos.
Messenger RNA (mRNA): an RNA molecule whose base sequence codes for the amino acid sequence of a protein chain. mRNA molecules are transcribed from genes in the DNA.
Metastases: the spread of malignant cells from one part of the body to another. Metastases can occur via the bloodstream or the lymphatic system.
Mismatch repair: the DNA repair mechanism that detects and corrects incorrectly matched base pairs. If A is not bound to T, or if G is not bound to C, this is a DNA mismatch.
Mitosis: division of the nucleus during M phase of the cell cycle. Mitosis is colloquially synonymous with cellular division.

Mitotic cell death: cellular death caused by high dose radiation which produces chromosomal damage so severe that cells cannot progressed through mitosis.

Mitotic index: the percentage of cells in a population of cells that are undergoing mitosis at any given time.

Monoclonal antibody: a purified and highly specific antibody directed against a single antigen.

Mononucleosis: a usually mild "flu-like" disease caused by the Epstein-Barr virus and characterized by the non-malignant proliferation of lymphocytes.

Multidrug resistance transport proteins: a family of plasma membrane proteins that actively pump a variety of drugs out of cells. P-glycoprotein is the classic example.

Mutagen: a chemical or physical agent that is capable of inducing mutations.

Mutation: a change in the base sequence of a DNA molecule.

***MYC* gene:** a normal human proto-oncogene that codes for the myc protein, a transcription factor that initiates cellular proliferation. Upon mutation, *MYC* can act as an oncogene and may induce cancer.

Neoplasia: the abnormal growth process in which cells proliferate in an uncontrolled, autonomous fashion.

Neoplasm: a tumor. Neoplasms and tumors are abnormally growing masses in which cells proliferate in an uncontrolled and autonomous fashion, leading to continual increase in the number of dividing cells. Neoplasms may be benign or malignant.

NF-kappa B: a transcription factor activated in tissues where inflammation is ongoing. NF-kappa B can initiate the transcription of certain genes that produce proteins which stimulate cell division and make cells resistant to apoptosis.

Nucleoside: a composite molecule consisting of a nitrogen-containing base (adenine, guanine, thymine, cytosine, uracil) linked to a five-carbon sugar (ribose or deoxyribose) that is NOT attached to a phosphate group. Nucleosides are nucleotides without a phosphate group.

Nucleotide: a composite molecule consisting of a nitrogen-containing base (adenine, guanine, thymine, cytosine, uracil) linked to a five-carbon sugar (ribose or deoxyribose) that is attached to a phosphate group. Nucleotides are the building blocks of DNA and RNA.

Oncogene: any gene whose presence can potentially lead to cancer. Some oncogenes are introduced by viruses but in most human cancers oncogenes arise by activating (rather than silencing) mutations of normal cellular genes called proto-oncogenes.

Oncogenic virus: a virus that can cause cancer.

Oncology: the scientific study of cancer and the medical field of managing patients with cancer. Oncology is broken down into several subdisciplines including medical oncology, pediatric oncology, surgical oncology and radiation oncology among others.

Organic compound: a carbon-containing chemical compound.

Ozone: the molecule composed of three oxygen atoms (the molecule composed of two oxygen atoms is called "molecular oxygen," diatomic oxygen," or just "oxygen"). Atmospheric ozone in the stratosphere filters out much of the harmful ultraviolet radiation from the Sun.

PAH: (see polycyclic aromatic hydrocarbon)

p53 protein: a molecule coded for by the *TP53* gene that accumulates in response to DNA damage and in turn activates genes that halt the cell cycle, trigger apoptosis and stimulate DNA repair among other functions.

Pap smear: a cancer screening technique for the detection of cervical cancer. Cells are obtained from a cervix sample and are examined under the microscope.

Philadelphia chromosome: an abnormal version of chromosome 22 found in the malignant cells of most patients with chronic myelogenous leukemia. It is generated by a reciprocal translocation of DNA between chromosomes 9 and 22 and leads to the formation of the BCR-ABL oncogene.

Phosphorylation: in chemistry or biochemistry, the addition of a phosphate group to another molecule.

Point mutation: a mutation involving a single nucleotide. Point mutations can be either transitions (pyrimidine to pyrimidine or purine to purine) or transversions (pyrimidine to or vice versa).

Polycyclic aromatic hydrocarbon (PAH): a group of hundreds of organic compounds containing only carbon and hydrogen and constructed from multiple fused benzene rings. Many are carcinogenic. Some PAHs are found in space leading some scientists to speculate a role for them in the origin of life.

Polypeptide: a short chain of amino acids (historically smaller than an insulin molecule).

Protein: a large chain of amino acids (historically larger than an insulin molecule)

Promoter: regulatory region of a gene that initiates transcriptions usually DNA sequences located 5 prime to the coding sequences but which made me look up located in other regions such as in Franz and 3 pine sequences.

Protease: an enzyme that degrades proteins.

Proto-oncogene: a normal cellular gene that can be converted into an oncogene and cause tumors through various gain of function mutations.

Purine: a double ring nitrogen-containing organic molecule found in nucleic acids (DNA and RNA) that pairs up with a complementary pyrimidine. Purines in biological nucleic acids, the purines include adenine and guanine.

Pyrimidine: a single ring nitrogen-containing organic molecule found in nucleic acids (DNA and RNA) that pairs up with a complementary purine. Pyrimidines in biological nucleic acids include cytosine, thymine and uracil.

Radiation oncologist: a physician who specializes in the diagnosis and treatment of cancer particularly focusing on the application of ionizing radiation in a therapeutic fashion. In the distant past radiation oncologists we're called therapeutic radiologists and the field of radiation oncology was called therapeutic radiology.

Receptor: a transmembrane cytoplasmic, or nuclear molecule that binds to a specific molecule such as growth factor or hormone.

Recessive: (See autosomal recessive).

Relapse: reappearance of a disease, especially cancer.

Remission: reduction in the extent or severity of cancer, usually as a result of treatment. Remissions can be partial or complete.

Retinoblastoma: A malignant tumor that arises from the retinal cells of the eye and is related to a loss of function in the *RB* tumor suppressor gene.

Retrovirus: an RNA virus that uses reverse transcriptase to make a DNA copy of its RNA.

Reverse transcriptase: an enzyme found in retroviruses that uses an RNA template to synthesize a complementary molecule of double-stranded DNA.

Rous sarcoma virus (RSV): a virus that induces sarcomas in chickens. It was the first cancer virus discovered.

S phase: the phase of the cell cycle during which DNA synthesis occurs.

Sarcoma: a malignant tumor of mesenchymal tissue such as muscle, bone, cartilage or fat.

Senescence: irreversible cell cycle arrest (i.e. when a cell can no longer replicate)

Somatic cells: all cells other than egg or sperm cells. Mutations in somatic cells are not passed on to the next generation.

Sporadic cancer: a non-hereditary cancer.

Squamous cell carcinoma: a subtype of cancer arising from epithelial cells and exhibiting a squamous (a thin, flattened square) shape.

***SRC* gene:** a proto-oncogene that is related to the src oncogene of the Rous sarcoma virus. *SRC* codes for a tyrosine kinase involved in growth signaling pathways.

Stem cell: a cell that can self-renew and give rise to more differentiated cell types.

Synergistic: the condition where two or more systems act together in a such way that the two agents or systems in combination produce an effect that is greater than the sum of the effects of the individual agents acting alone.

T: (See thymine)

Telomerase: an enzyme that extends telomere length. Elevated levels of telomerase are observed in many cancer cells.

Telomere: repeated sequences of DNA that are located at the end of chromosomes. Telomeres shorten up on each round of replication and act as cellular clocks.

Thymine (T): a pyrimidine base found in DNA (not RNA) that forms complementary base pairs with adenine (A).

Transcription: the process of transferring the information encoded by DNA into RNA (i.e. the process by which RNA is synthesized using DNA as a template).

Transcription factor: a protein that binds to DNA and stimulates the activation of a specific gene or set of genes.

Transformation: the process of converting a normal cell into a malignant cell.

Transforming growth factor-beta (TGF-β): A growth factor relevant for tumor development because it is an inhibitor of epithelial cell proliferation and 90% of human cancers are of epithelial origin.

Translation: the process of transferring the information encoded by RNA into protein (i.e. the process by which a protein or polypeptide chain is synthesized using RNA as a template). The process of translation is mediated by ribosomes.

Transition: a point mutation in which a pyrimidine is replaced by another pyrimidine or a purine is replaced by another purine.

Transversion: a point mutation in which a pyrimidine is replaced by a purine or a purine is replaced by a pyrimidine.

Translocation: (See chromosomal translocation).

Tumor: an that can be either benign or malignant.

Tumor: also known as a neoplasm, an abnormal growth of cells created by an aberrant process in which cells divide in an uncontrolled, autonomous fashion resulting in a continuous increase in the number of proliferating cells. Tumors may be benign or malignant.

Tumor angiogenesis: process by which cancer cells stimulate the development of a blood supply.

Tumor dormancy: the condition in which tumor masses remain dormant (i.e. clinically inactive) for prolonged periods of time.

Tumor progression: a change in tumor properties observed over time as cancer cells acquire more aberrant traits and become increasingly aggressive. Also, when a tumor that was previously in remission again becomes clinically active.

Tumor suppressor gene: a gene whose loss or inactivation by deletion, mutation or epigenetic silencing can potentially lead to cancer. Tumor suppressor genes can be gatekeepers or caretakers.

Two-hit model: a model for tumor formation in which both alleles of the same gene must undergo deletion or inactivation before a cancer will arise. This model applies to tumor suppressor genes.

Tyrosine kinase: a protein kinase that catalyzes the phosphorylation of the amino acid tyrosine in target proteins. Tyrosine kinases play a prominent role in signaling pathways, including those important in cancer formation.

Ultrasound imaging: a technique in which sound waves are there echoes are used to produce a picture of internal bottle e structures.

Ultraviolet radiation (UV): Electromagnetic radiation with wavelengths between that of visible light and x-rays. Ultraviolet radiation can cause sunburn or potentially lead to skin cancer.

Uracil: a pyrimidine base found in RNA (not DNA) that forms complementary base pairs with adenine (A).

Vaccine: a preparation containing antigens that stimulates an immune response towards those antigens.

Vascular endothelial growth factor (VEGF): the signaling protein that plays a central role in stimulating the growth of new blood vessels.

Viral oncogene: a cancer causing gene found in the genome of a virus.

Virus: a microscopic particle composed of DNA or RNA along with a protein coat that is capable of reproducing independently only when inside a cell.

Warburg effect: the observation originally made by Otto Warburg that tumor cells utilize fermentation for glucose metabolism even in the presence of ample oxygen levels. Also known as aerobic glycolysis.

Xenograft: a transplant of an organ or tissue from one individual to another in a different species. A common xenograft model used in cancer research is the transfer of human tumor cells into immunodeficient mice.

Xeroderma pigmentosum: an inherited disease caused by defects in thymine dimer DNA excision repair which confers susceptibility to cancer (mainly skin cancers).

X-rays: a type of energetic, ionizing electromagnetic radiation with a wavelength shorter than that of ultraviolet radiation. X-rays are commonly used in diagnostic radiology and in radiation therapy.

BIBLIOGRAPHY

GENERAL REFERENCES FOR THE INTERESTED READER:

DeVita Jr., Vincent T., Theodore S. Lawrence and Steven A. Rosenberg, eds. *DeVita, Hellman, and Rosenberg's Cancer: Principles & Practice of Oncology*, 10th edition. Philadelphia, Pennsylvania: Wolters Kluwer, Lippincott Williams & Wilkins, 2014. *Note: THE definitive oncology textbook. 2280 pages of very up to date material on almost all cancer-related subjects. Must reading for all students of oncology.*

Johnson, George. *The Cancer Chronicles. Unlocking Medicine's Deepest Mystery.* New York, NY: Alfred A. Knopf, a division of Random House, Inc, 2013. (In Canada by Random House of Canada Limited, Toronto, 2013). *Note: One of the finest reads in the field. A popular book written for the educated reader.*

Kleinsmith, Lewis J. *Principles of Cancer Biology*. San Francisco, California: Pearson Benjamin Cummings, 2006. *Note: One of the finest undergraduate level texts on cancer biology. This is very well written and beautifully illustrated. Although somewhat dated, it can still serve as a useful introductory platform for those interested in further learning.*

Mukherjee, Siddhartha. *The Emperor of All Maladies. A Biography of Cancer*. New York, New York: Scribner, 2010. *Note: Definitely the finest and most thorough history of cancer ever written.*

Halperin, Edward C., et al., eds. *Principles and Practice of Radiation Oncology, 6th edition*. Philadelphia, Pennsylvania: Wolters Kluwer, Lippincott Williams & Wilkins, 2013. *Note: The nearly 2,000 page standard textbook (Perez and Brady's) for those learning the field of radiation oncology.*

Weinberg, R. A. *The Biology of Cancer, 2nd edition*. New York: Garland Science, 2013. *Note: Among the finest of the more advanced scientific texts on this topic.*

Sadava, David. *"What Science Knows About Cancer,"* The Teaching Company, Great Courses. Course #1956. 2013. *Note: An excellent undergraduate level introduction to cancer biology for those who prefer lectures to standard textbooks.*

CHAPTER 1: WHAT JUST HAPPENED?

Golden, E. B., et al. "Local Radiotherapy and Granulocyte-Macrophage Colony-Stimulating Factor to Generate Abscopal Responses in Patients with Metastatic Solid Tumours: a Proof-of-Principle Trial." *Lancet Oncolology* 16 (2015). *Note: An exciting new study that suggests that GM-CFS can work synergistically with radiotherapy to induce abscopal effects. Patients experiencing abscopal effects tended to survive longer than those who did not.*

Mole, R. H. "Whole Body Irradiation; Radiobiology or Medicine?" *British Journal of Radiology* 26 (1953). *Note: The first time the name "abscopal" had been mentioned in the radiotherapy literature.*

Welsh, James. "Aggressive Cancer Performs Strange Disappearing Act," *Discover Magazine* 35, 2 (March 2014). *Note: Discover Magazine was where I first described the mystifying case of my melanoma patient with the abscopal effect.*

A couple of recent medical articles addressing abscopal effects:

Fernandez-Palomo, C. et al. "Use of Synchrotron Medical Microbeam Irradiation to Investigate Radiation-Induced Bystander and Abscopal Effects in Vivo." *Phys Med* 31, 6 (September 2015). *Note: A very interesting description of the use of synchrotron radiation to investigate bystander and abscopal effects. Their preliminary work suggests that bystander effects and the abscopal phenomenon are NOT one and the same.*

Grimaldi A. M., et al. "Abscopal Effects of Radiotherapy on Advanced Melanoma Patients Who Progressed after Ipilimumab Immunotherapy." *OncoImmunology* 14, 3 (May 2014).

Hutchinson, L. "Radiotherapy: Abscopal Responses: Pro-Immunogenic Effects of Radiotherapy." *Nature Reviews Clinical Oncology* 12, 9 (September 2015).

Lock, Michael, et al. "Abscopal Effects: Case Report and Emerging Opportunities." *Cureu,* 7, 10 (October 07, 2015). *Note: An extraordinary case of regression of metastatic liver cancer (hepatocellular carcinoma) that had spread to the lungs. After radiation therapy only to the liver tumor, the patient had an abscopal effect manifesting as a complete response in the lung.*

Postow, M. A., et al. "Immunologic Correlates of the Abscopal Effect in a Patient with Melanoma." *New England Journal of Medicine* 366, 10 (March 2012). *Note: An interesting case study in which a patient on ipilimumab experienced an impressive abscopal effect following stereotactic body radiation therapy.*

Stamell E.F., J.D. Wolchok, S. Gnjatic, N.Y. Lee and I. Brownell "The Abscopal Effect Associated with a Systemic Anti-Melanoma Immune Response." *International Journal Radiation Oncology Biology Physics,* 85, 2 (February 2013).

CHAPTER 2: ACTION AT A DISTANCE

Aron, Jacob. "Quantum Weirdness is Reality. A Groundbreaking Proved Einstein Wrong." *New Scientist* 227, 3037 (September 6, 2015). *Note: A succinct summary of a recent experiment that shows quantum entanglement to be a real phenomenon despite Einstein and others insistence that such common-sense violating predictions of quantum mechanics was evidence that the theory must be incomplete or inaccurate.*

Also available online as Aron, Jacob. "Quantum Weirdness Proved Real in First Loophole-Free Experiment." https://www.newscientist.com/article/dn28112-quantum-weirdness-proved-real-in-first-loophole-free-experiment/ (accessed November 17, 2015).

Aspect, Alain, Philippe Grangier and Gérard Roger. "Experimental Tests of Realistic Local Theories via Bell's Theorem." *Physical Review Letters* 47 (1981). *Note: One of the several early articles challenging the EPR paradox and testing Bell's theorem.*

Bell, John S. "On the Einstein-Podolsky-Rosen Paradox." *Physics* 1 (1964). *Note: The original paper outlining "Bell's theorem" which holds that no theory of local hidden variables can account for all of the predictions of quantum mechanics.*

Einstein, A., B. Podolsky, and N. Rosen. "Can Quantum-Mechanical Description of Physical Reality Be Considered Complete?" *Physical Review* 47 (1935). *Note: The first account of the famous "EPR paradox" and why these authors felt that quantum mechanics could not be a complete theory.*

Fickler, Robert, et al. "Real-Time Imaging of Quantum Entanglement." *Scientific Reports* 3 (2013). *Note: One of the weirdest predictions of quantum mechanics, the concept of entanglement, has new corroborating scientific evidence.*

Greene, Brian. "Spooky Action at a Distance." *NOVA*, September 22, 2011. http://www.pbs.org/wgbh/nova/physics/spooky-action-distance.html (accessed November 17, 2015) *Note: An excellent and brief (two page) summary of entanglement by the author of* The Fabric of the Cosmos.

Horodecki, Ryszard, Pawel Horodecki, Michal Horodecki and Karol Horodecki. "Quantum Entanglement." *Reviews of Modern Physics* 81 (2009).

A series of encouraging articles in the *New England Journal of Medicine* focusing on the abscopal phenomenon and future possibilities of combining radiation therapy with checkpoint inhibitors to induce abscopal effects:

Hiniker, Susan M., Daniel S. Chen, and Susan J. Knox. "Abscopal Effect in a Patient with Melanoma." *New England Journal of Medicine* 366 (2012).

Postow, M. A., et al. "Immunologic Correlates of the Abscopal Effect in a Patient with Melanoma." *New England Journal of Medicine* 366 (2012).

Postow, Michael A, Margaret K. Callahan, and Jedd D. Wolchok. "Abscopal Effect in a Patient with Melanoma—Author's Reply." *New England Journal of Medicine* 366 (2012).

CHAPTER 3: DISAPPEARING DEVILS

Welsh, James S. "Contagious Cancer." *The Oncologist* 16, 1 (2011). *Note: My own initial foray into this mesmerizing topic.*

A few well written reviews:

Belov, K. "Contagious Cancer: Lessons from the Devil and the Dog." *Bioessays* 34, 4 (April 2012).

O'Neill, I. D. "Concise Review: Transmissible Animal Tumors as Models of the Cancer Stem-Cell Process." *Stem Cells* 12 (December 29, 2011).

O'Neill, I. D. "Tasmanian Devil Facial Tumor Disease: Insights into Reduced Tumor Surveillance from an Unusual Malignancy." *International Journal of Cancer* 127, 7 (October 1, 2010).

Quammen, David. "Contagious Cancer: The Evolution of a Killer." *Harper's Magazine* (April, 2008).

Rehmeyer, Julie. "The Immortal Devil." *Discover* 35, 4 (May 2014). *Note: This is an extremely well-researched early account aimed for the general reader written in non-technical terms.*

Here are some more scientifically detailed original references:

McCallum, H. "Tasmanian Devil Facial Tumour Disease: Implications for Conservation Biology." *Trends in Ecology and Evolution* 23 (2008).

McCallum, H. D., et al. "Distribution and Impacts of Tasmanian Devil Facial Tumour Disease." *EcoHealth* 4 (2007).

Morris, K, and K. Belov. "Does the Devil Facial Tumour Produce Immunosuppressive Cytokines as an Immune Evasion Strategy?" *Veterinary Immunology and Immunopathology* 153, 1-2 (May 2013).

Murchison, E. P. "Clonally Transmissible Cancers in Dogs and Tasmanian Devils." *Oncogene* 27, 2 (2008).

Murgia, C., et al. "Clonal Origin and Evolution of a Transmissible Cancer." *Cell* 126 (2006).

Pearse, A.M., et al. "Evolution in a Transmissible Cancer: a Study of the Chromosomal Changes in Devil Facial Tumor (DFT) as it Spreads through the Wild Tasmanian Devil Population." *Cancer Genetics* 205, 3 (March 2012).

Siddle, Hannah V. and Jim Kaufman. "Immunology of Naturally Transmissible Tumours." *Immunology* 144, 1 (January 2015).

Siddle, H. V. and J. Kaufman. "A tale of Two Tumours: Comparison of the Immune Escape Strategies of Contagious Cancers." *Molecular Immunology* 55, 2 (September 2013).

Woods, Gregory M., et al. "Immunology of a Transmissible Cancer Spreading among Tasmanian Devils." *Journal of Immunology* 195, 1 (July 2015).

CHAPTER 4: THE DEVIL HIMSELF: SOME DIABOLICAL BIOLOGY

Kreiss, A., et al. "Allorecognition in the Tasmanian Devil (*Sarcophilus harrisii*), an Endangered Marsupial Species with Limited Genetic Diversity." *PLoS One* 6, 7 (2011). *Note: This study explored the functional significance of the limited genetic diversity in Tasmanian devils by seeing whether or not skin grafts would take. As a rule they do not. This illustrates that although there is limited genetic diversity, the population of animals are not all like identical twins which can indiscriminately accept skin grafts (and therefore other organ "donations" such as tumors).*

Morris, Katrina, Jeremy J. Austin, and Katherine Belov. "Low Major Histocompatibility Complex Diversity in the Tasmanian Devil Predates European Settlement and May Explain Susceptibility to Disease Epidemics." *Biology Letters* 9, 1 (2013). *Note: One of many scientific studies that have confirmed the inbreeding and relative loss of genetic diversity in the island population of Sarcophilus harrisii.*

Siddle, Hannah V. and Jim Kaufman. "Immunology of Naturally Transmissible Tumours." *Immunology* 144, 1 (January 2015).

Ziniak, Amy. "Sad Tale of the Tasmanian Tiger: How Benjamin, the Last of His Kind, Died of Exposure at Hobart Zoo after Being Left Out in the Cold." *Daily Mail Australia*, September 2014. Updated: October 2014. http://www.dailymail.co.uk/news/article-2746016/Left-cold-die-The-Tasmania-Tiger-extinct-result-human-neglect.html#ixzz3n8S7XW7t (accessed November 17, 2015) *Note: A sad account of the last thylacine.*

CHAPTER 5: DEVIL OF A DISEASE

Quammen, David. "Contagious Cancer: The Evolution of a Killer." *Harper's Magazine* (April 2008). *Note: Despite its age, this remains a fantastic introduction to the subject.*

Rehmeyer, Julie. "The Immortal Devil." *Discover* 35, 4 (May 2014).

CHAPTER 6: THE PERFECT PARASITE

Alberts, Bruce, Alexander Johnson, Julian Lewis, Martin Raff, Keith Roberts, and Peter Walter. *Molecular Biology of the Cell" 4th edition.* New York, NY: Garland Science, 2002. In chapter 23: Cancer—"Cancer as a Microevolutionary Process." *Note: Although not the latest edition, I found this description of the cellular evolutionary process in carcinogenesis to be quite thought-provoking. Despite its age, I would strongly recommend this subchapter as an excellent introduction for anyone seriously interested in learning more about the aberrant cellular processes in cancer.*

Dorit, Robert. "Breached Ecological Barriers and the Ebola Outbreak." *American Scientist* 3, 4 (Jul-August 2015). *Note: The epidemic may be waning, but the social and ecological context that brought it about, remains. An interesting account of how Ebola might have arisen as a consequence of deforestation.*

Muehlenbachs A., et al. "Malignant Transformation of Hymenolepis nana in a Human Host." *New England Journal of Medicine* 373, 19 (November 2015). *Note: One of the strangest clinical cases I have ever read. This is the description of a tapeworm that got "cancer" and its transformed cells went haywire in the immunosuppressed human host. The disease in the patient looked quite like cancer. Only after biopsy was the distinction finally made.*

Rozhok, A. I. and J. DeGregori. "Toward an Evolutionary Model of Cancer: Considering the Mechanisms that Govern the Fate of Somatic Mutations." *Proceedings of the National Academy of Sciences of the United States of America* 112, 29 (July 2015).

CHAPTER 7: A MALIGNANT MALADY IN MAN'S BEST FRIEND

Chu, R. M., et al. "Proliferation Characteristics of Canine Transmissible Venereal Tumor." *Anticancer Research* 21 (2001).

Frank, U. "The Evolution of a Malignant Dog." *Evolution & Development* 9, 6 (2007). *Note: An interesting article that was among the first to clearly conceptualize transmissible tumors as parasites.*

Murgia C., et al. "Clonal Origin and Evolution of a Transmissible Cancer." *Cell* 126, 3 (August 2006). *Note: One of the important papers that definitively proved through cytogenetic analysis that CTVT is actually a transmissible tumor.*

Mukaratirwa, S. and E. Gruys. "Canine Transmissible Venereal Tumour: Cytogenetic Origin, Immunophenotype, and Immunobiology. A Review." *Veterinary Quarterly* 25, 3 (2003). *Note: Although written over a decade ago, this remains a good general scientific review article, especially for the veterinary readership.*

Prier, J. E. and R. S. Brodey. "Canine Neoplasia: A Prototype for Human Cancer Study." *Bulletin of the World Health Organization* 29, 3 (1963). *Note: One of the older papers on the topic. This one hints at ways this contagious cancer might be of relevance for the study of human cancer.*

Vermooten, M. I. "Canine Transmissible Venereal Tumor (TVT): A Review." *Journal of the South African Veterinary Association* 58, 3 (September 1987).

VonHoldt, B. M. and E. A. Ostrander. "The Singular History of a Canine Transmissible Tumor." *Cell* 126, 3 (August 2006). *Note: This short article describes the importance of the LINE-1 element and its proximity to c-myc in the oncogenesis process.*

Wunderlich, V. "Anton Sticker (1861-1944) and the Sticker's Sarcoma of Dogs. A Current Model for Infectious Cancer Cells." [Article in German] *Tierarztl Prax Ausg K Kleintiere Heimtiere* 40, 6 (2012). *Note: A historical account of Anton Sticker and the discovery of this contagious cancer but written in German except for the abstract.*

Yang, T. J., J. P. Chandler and S. Dunne-Anway. "Growth Stage Dependent Expression of MHC Antigens on the Canine Transmissible Venereal Sarcoma." *British Journal of Cancer* 55 (1987). *Note: One of the papers that describes how MHC antigen expression varies over time and how that might be the key to how dogs reject the tumor.*

Some references regarding the age of CTVT:

Lokody, I. "Cancer Genetics: The Origin and Evolution of an Ancient Cancer." *Nature Reviews Genetics*, 15, 3 (March 2014).

Murchison, E. P., et al. "Transmissible [corrected] Dog Cancer Genome Reveals the Origin and History of an Ancient Cell Lineage." *Science* 343, 6169 (2014). *Note: This is the article that details the methodology used to conclude the astonishing 11,000 year old age of this tumor line.*

Parker, H. G. and E. A. Ostrander. "Cancer. Hiding in Plain View--an Ancient Dog in the Modern World" *Science* 343, 6169 (January 2014).

CHAPTER 8: THE CURIOUS CASE OF COLEY'S TOXINS (OR SOMETIMES THE TREATMENT WORKED)

"Crocodile Blood Could Hold Key to HIV Cure." *Daily Mail.com,* August 17, 2005. http://www.dailymail.co.uk/health/article-359556/Crocodile-blood-hold-key-HIV-cure.html (accessed November 17, 2015). *Note: A brief internet description of some of the surprisingly powerful immunological capabilities of crocodile blood.*

Hobohm, Uwe. "Killing Cancer with Fever: An Old Therapy Revisited." *New Scientist,* 2950 (December 31, 2013). http://webcache.googleusercontent.com/search?q=cache:JIdpqubqWvwJ:https://www.newscientist.com/article/mg22129500-400-killing-cancer-with-fever-an-old-therapy-revisited/+&cd=2&hl=en&ct=clnk&gl=us (accessed November 30, 2015). *Note: A short introduction to the topic by the author of the book "Healing Heat: An essay on cancer immune defence" which details hopes to modernize Coley's toxins for21st century cancer treatment.*

Hoption Cann, S. A., J. P. van Netten and C. van Netten. "Dr William Coley and Tumour Regression: A Place in History or in the Future." *Postgraduate Medical Journal* 79 (2003).

Hoption Cann, S. A., et al. "Spontaneous Regression: A Hidden Treasure Buried in Time." *Medical Hypotheses* 58, 2 (2002).

Kienle, G. S. "Fever in Cancer Treatment: Coley's Therapy and Epidemiologic Observations." *Global Advances in Health and Medicine* 1 (2012). *Note: Another comprehensive and well-written article that includes the historical background behind fever and cancer remission.*

McCarthy, E. F. "The Toxins of William B. Coley and the Treatment of Bone and Soft-Tissue Sarcomas." *Iowa Orthopaedic* Journal 6 (2006). *Note: A brief but excellent historical overview.*

Stames, Charlie O. "Coley's Toxins in Perspective." *Nature* 357 (1992). *Note: A brief recounting of Coley's toxin's in a prestigious journal during the midst of the relatively ineffective first-generation of modern cancer immunotherapy.*

CHAPTER 9: THE DOG KNOWS

Welsh, James S. "Olfactory Detection of Human Bladder Cancer by Dogs: Another Cancer Detected by "Pet Scan"." *British Medical Journal* 329, 7477 (November 2004). *Note: My own cancer-sniffing canine case, which piqued my curiosity on the subject. Online publication in the British Medical Journal encouraged me and my colleagues to formally publish our case.*

Welsh, James S., Darryl Barton and Harish Ahuja. "A Case of Breast Cancer Detected by a Pet Dog." *Community Oncology* 2, 4 (Jul/August 2005). *Note: To the best of our knowledge, this represents the first time an internal cancer (i.e. a cancer other than skin cancer) was detected by a pet dog. Therefore, despite the unfortunate outcome, our team elected to write the case up in a medical journal.*

These next two papers were the first to describe cancer-sniffing dogs. Both cases were skin cancers:

Church, J., and H. Williams. "Another Sniffer Dog for the Clinic?" *Lancet* 358 (2001).

Williams, H., and A. Pembroke. "Sniffer Dogs in the Melanoma Clinic?" *Lancet* 1 (1989).

These next two papers transformed the previous anecdotal case studies into a genuine scientific discipline:

Willis, C. M., et al. "Olfactory Detection of Human Bladder Cancer by Dogs: Proof of Principle Study." *British Medical Journal* 329 (2004).

Pickel, D., et al. "Evidence for Canine Olfactory Detection of Melanoma." *Applied Animal Behavior Science* 89 (2004).

Some recent articles of interest:

Bestic, Liz. "The Nose Knows." *New Scientist* 227, 3028 (July 2015). *Note: A recent and readable update on the topic.*

Cable, Amanda. "Daisy, the Dog Who's Sniffed Out Over 500 Cases of Cancer: She Even Saved the Woman whose Research Revealed her Uncanny Skill." *The Daily Mail* (July 2014). http://www.dailymail.co.uk/health/article-2700561/Daisy-dog-whos-sniffed-500-cases-cancer-She-saved-woman-research-revealed-uncanny-skill.html#ixzz3bq0aGcLy (accessed November 17, 2015). *Note: The incredible story of a researcher in the field whose own dog detected her breast cancer.*

Phillips, M., et al. "Volatile Organic Compounds in Breath as Markers of Lung Cancer: A Cross-Sectional Study." *Lancet* 353 (1999). *Note: This study provided some evidence that there were indeed volatile organic compounds emitted from cancer patients. In this case it was a volatile compound in the breath of lung cancer patients.*

CHAPTER 10: MALES NEED NOT APPLY

Clairborne Ray, C. "Can Plants Get Cancer?" *New York Times,* July 2013. http://www.nytimes.com/2013/07/16/science/can-plants-get-cancer.html?_r=0 (accessed November 17, 2015). *Note: A very short (one page) NY Times science article on plant tumors.*

Doonan, J. H., and R. Sablowski. "Walls Around Tumours - Why Plants do not Develop Cancer." *Nature Reviews Cancer* 10, 11 (November 2010).

Engber, Daniela. "Ask Anything: Do Plants Get Cancer? Taking a Look at the Disease in the Greenery." *Popular Science,* January 23, 2014. http://www.popsci.com/article/science/ask-anything-do-plants-get-cancer (accessed November 17, 2015). *Note: Another very brief but interesting non-technical article on plant tumors.*

Gohlke, J. and R. Deeken. "Plant Responses to Agrobacterium Tumefaciens and Crown Gall Development." *Frontiers in Plant Science* 5 (2014). *Note: A more technical review article on plant tumors due to A. tumefasciens.*

Polidori, C., A. J. García and J. L. Nieves-Aldrey. "Breaking Up the Wall: Metal-Enrichment in Ovipositors, But Not in Mandibles, Co-Varies with Substrate Hardness in Gall-Wasps and Their Associates." *PLoS One* 8, 7 (July 2013). *Note: This paper describes how little gall making wasps can pierce woody plants and induce galls by concentrating metals in their ovipositors.*

Ronquist, F., et al. "Phylogeny, Evolution and Classification of Gall Wasps: the Plot Thickens." PLoS One 10, 5 (May 2015). *Note: A more scientific article on gall wasps for those interested in greater detail.*

Stone, G. N., et al. "The Population Biology of Oak Gall Wasps (Hymenoptera: Cynipidae)." *Annual Review of Entomology* 47 (2002). *Note: A good review article on the fascinating biology of gall wasps, focusing on oak gall wasps.*

CHAPTER 11: COULD BROWN FAT BE THE SECRET TO WEIGHT LOSS?

Feron, O. "Pyruvate into Lactate and Back: From the Warburg Effect to Symbiotic Energy Fuel Exchange in Cancer Cells." *Radiotherapy and Oncology* 92, 3 (September 2009). *Note: A worthwhile review of the Warburg effect with insights into potential clinical applications.*

Lambert, Chloe. "Growing Fat to Stay Slim." *New Scientist* 226, 3017 (April 2015). *Note: A very interesting, personal experiment on activation of brown fat!*

Some of the early papers documenting brown fat as a potential false positive error on PET scans:

Cohade C., M. Osman, H. K. Pannu and R. L. Wahl. "Uptake in Supraclavicular Area Fat ("USA-Fat"): Description on 18F-FDG PET/CT." *Journal Nuclear Medicine* 44, 2 (2003).

Hany, T. F., et al. "Brown Adipose Tissue: A Factor to Consider in Symmetrical Tracer Uptake in the Neck and Upper Chest Region." *European Journal of Nuclear Medicine and Molecular Imaging*, 29, 10 (October 2002).

Yeung, H. W., et al. "Patterns of (18) F-FDG Uptake in Adipose Tissue and Muscle: A Potential Source of False-Positives for PET." *Journal Nuclear Medicine* 44, 11 (November 2003).

CHAPTER 12: GAMMA RAYS AND DINOSAUR CANCER

Hazen, Robert M. "*The Story of Earth. The First 4.5 Billion Years from Stardust to Living Planet.*" New York, New York: Viking, published by The Penguin Group, 2012. *Note: "the greatest of these episodes all of which coincide with global mass extinctions, produced hundreds of thousands to millions of cubic miles of lava. The biggest known event, now revealed by more than a half million square miles of basalt flows occurred in Siberia during Earth's greatest mass extinction, the great dying 251 million years ago. The demise of the dinosaurs 65 million years ago, so often ascribed to an asteroid impact, is also coincident with immense floods basalts in India- the Deccan Traps, almost 200,000 square miles in extent, representing more than 120,000 cubic miles of new rock." (p. 271).*

Johnson, George. *The Cancer Chronicles. Unlocking Medicine's Deepest Mystery.* New York, New York: Alfred A. Knopf, a division of Random House, Inc, 2013. (In Canada by Random House of Canada Limited, Toronto, 2013). *Note: Fascinating material on tumors found on dinosaur bones!*

Klebesadel, R. W., I. B. Strong and R. A. Olson. "Observations of Gamma-Ray Bursts of Cosmic Origin." *Astrophysical Journal* 182 (1973).

Melott, A. L. et al. "Did a Gamma Ray Burst Initiate the Late Ordovician Mass Extinction?" *International Journal of Astrobiology* 3, 1 (2004). *Note: The highly controversial and interesting possible explanation for one of the five major mass extinction events*. atarajan, L.C., A. L. Melott, B. M. Rothschild and L.D. Martin. "Bone Cancer Rates in Dinosaurs Compared with Modern Vertebrates." *Transactions of the Kansas Academy of Science* 110 (2007).

Available at: http://arxiv.org/abs/0704.1912 (accessed November 17, 2015).

Piran, Tsvi and Raul Jimenez. "Possible Role of Gamma Ray Bursts on Life Extinction in the Universe." *Physical Review Letters* 113, 23 (December 2014). *Note: The authors arrive at a most fascinating conclusion: "Early life forms must have been much more resilient to radiation."*

Rothschild, Bruce M, Brian J. Witzke and Israel Hershkovitz. "Metastatic Cancer in the Jurassic." *The Lancet* 354 (1999).

Rothschild, B. M., D. H. Tanke, M. Helbling and L. D. Martin. "Epidemiologic Study of Tumors in Dinosaurs." *Naturwessenschaften* 90 (2003).

CHAPTER 13: CANCER OF THE CLAM!

Metzger, E. P., C. Reinisch, J. Sherry and S. P. Goff. "Horizontal Transmission of Clonal Cancer Cells Causes Leukemia in Soft-Shell Clams." *Cell* 161 (2015).

Murgia, C., et al. "Clonal Origins and Evolution of a Transmissible Cancer." *Cell* 126, 3 (2006).

Stokstad, Erick. "Infectious Cancer Found in Clams." *Science* 348, 6231 (2015).

Weiss, Robin A. and Alberto Fassati. "The Clammy Grip of Parasitic Tumors." *Cell* 161, 2 (2015).

CHAPTER 14: SHARKS DO GET CANCER (OR HOW SHARK CARTILAGE CAN KILL YOU)

Folkman, J. "The Vascularization of Tumors." *Scientific American* 234 (May 1976).

Main, Douglas. "Sharks Do Get Cancer: Tumor Found in Great White." *Live Science*, December 03, 2013. http://www.livescience.com/41655-great-white-shark-cancer.html (accessed November 23, 2015). *Note: An excellent online article with fantastic photographs of several sharks with tumors by Andrew Fox and Sam Cahir.*

"Putting the Bite on Cancer. Biology of Sharks and Rays." *ReefQuest Centre for Shark Research.* http://www.elasmo-research.org/education/topics/p_bite_on_cancer.htm (accessed November 23, 2015). *Note: A brief (one page) on-line article on elasmobranchs (i.e. sharks and relatives) and the mechanisms behind their relatively low rates of cancer.*

CHAPTER 15: WHO TRULY DOESN'T GET CANCER?—MEET THE MOLE RATS

General references:

Callaway, Ewen. "Simple Molecule Prevents Mole Rats from Getting Cancer." *Nature News,* June 19, 2013, http://www.nature.com/news/simple-molecule-prevents-mole-rats-from-getting-cancer-1.13236 (accessed November 24, 2015).

Gorbunova, V., et al. "Comparative Genetics of Longevity and Cancer: Insights from Long-Lived Rodents." *Nature Reviews Genetics* 15, 8 (August 2014).

Detailed references:

Faulkes, C. G., K. T. Davies, S. J. Rossiter and N. C. Bennett. "Molecular Evolution of the Hyaluronan Synthase 2 Gene in Mammals: Implications for Adaptations to the Subterranean Niche and Cancer Resistance." *Biology Letters* 11, 5 (May 2015).

Some general references for readers interested in more scientific details:

Azpurua J, and A Seluanov. "Long-Lived Cancer-Resistant Rodents as New Model Species for Cancer Research." *Frontiers in Genetics* 3 (January 2013).

"Basic Research: Understanding Why the Naked Mole Rat is Cancer Resistant." *Nature Reviews Clinical Oncology* 10 (August 2013).

Lewis, K.N., J. Mele, P. J. Hornsby, R. Buffenstein. "Stress Resistance in the Naked Mole-Rat: The Bare Essentials - A Mini-Review." *Gerontology* 58, 5 (2012). *Note: A brief overview of the mechanisms underlying longevity in naked mole rats.*

Detailed references on naked mole rats:

Faulkes, C. G., et al."Molecular Evolution of the Hyaluronan Synthase 2 Gene in Mammals: Implications for Adaptations to the Subterranean Niche and Cancer Resistance." *Biology Letters* 11, 5 (May 2015).

Seluanov, A., et al. "Hypersensitivity to Contact Inhibition Provides a Clue to Cancer Resistance of Naked Mole-Rat." *Proceedings of the National Academy of Sciences of the United States of America* 106, 46 (November 2009). *Note: Contact inhibition is greatly exaggerated in naked mole rat cells and could be the way they ward off cancer.*

Tian, X., et al. "High-Molecular-Mass Hyaluronan Mediates the Cancer Resistance of the Naked Mole Rat." *Nature* 499, 7458 (July 2013).

Detailed references on blind mole rats:

Gorbunova V., et al. "Cancer Resistance in the Blind Mole Rat is Mediated by Concerted Necrotic Cell Death Mechanism." *Proceedings of the National Academy of Sciences of the United States of America* 109, 47 (November 2012).

Manov, I., et al. "Pronounced Cancer Resistance in a Subterranean Rodent, the Blind Mole-Rat, Spalax: In Vivo and In Vitro Evidence." *BMC Biology* 11 (August 2013).

CHAPTER 16: PAR FOR THE COURSE

Discovery and characterization of Par-4:

Johnstone, R.W., et al. "A Novel Repressor, Par-4, Modulates Transcription and Growth Suppression Functions of the Wilms' Tumor Suppressor WT1." *Molecular and Cellular Biology* 16, 12 (December 1996).

Potential of Par-4 for prostate cancer:

Butler, J, and V. M. Rangnekar. "Par-4 for Molecular Therapy of Prostate Cancer." *Current Drug Targets* 4, 3 (April 2003).

Early papers on function of Par-4:

Sells, S. F., et al. "Expression and Function of the Leucine Zipper Protein Par-4 in Apoptosis." *Molecular and Cellular Biology* 17, 7 (July 1997).

Scientific papers on Par-4 mice:

Hebbar, N., C. Wang, and V. M. Rangnekar. "Mechanisms of Apoptosis by the Tumor Suppressor Par-4." *Journal of Cellular Physiology* 227, 12 (December 2012).

Zhao, Y., et al. "Cancer Resistance in Transgenic Mice Expressing the SAC Module of Par-4." *Cancer Research* 67, 19 (October 2007).

Zhao Y, and V. M. Rangnekar. "Apoptosis and Tumor Resistance Conferred by Par-4." *Cancer Biology and Therapy* 7, 12 (December 2008).

Secretagogues for Par-4:

Frasinyuk, M. S., et al. "Development of *6H*-Chromeno[3,4-*c*]pyrido[3',2':4,5] thieno[2,3-e]pyridazin-6-ones as Par-4 Secretagogues." *Tetrahedron Letters* 56, 23 (June 2015).

Web articles for general readers:

"Cancer-Resistant Mouse Discovered." *EurekAlert! The Global Source for Science News*, *AAAS,* http://www.eurekalert.org/pub_releases/2007-11/uok-cmd112607.php (accessed November 23, 2015).

Smith, Michael. "Transgenic Mice Developed to Resist Cancer." *Medpage Today,* http://www.medpagetoday.com/HematologyOncology/OtherCancers/7493 (accessed November 23, 2015).

Wang, Brian. "NextBigFuture: Cancer Resistant Mouse with Par-4 Gene." *NextBigFuture,* 2007. http://www.nextbigfuture.com/2007/11/cancer-resistant-mouse-with-par-4-gee.html (accessed November 23, 2015).

Also at: http://www.eurekalert.org/pub_releases/2007-11/uok-cmd112607.php (accessed November 23, 2015).

CHAPTER 17: MIGHTY MOUSE TO THE RESCUE!

Scientific publications on SR/CR (spontaneous remission/ complete regression) mice ("mighty mouse"):

Blanks, M. J., et al. "Novel Innate Cancer Killing Activity in Humans." *Cancer Cell International* 11 (August 2011). *Note: Some humans might also be "mighty".*

Cui, Z., and M. C. Willingham. "The Effect of Aging on Cellular Immunity Against Cancer in SR/CR Mice." *Cancer Immunology and Immunotherapy* 53, 6 (June 2004). *Note: Not all mice are equally mighty—age plays an important role in anti-cancer immunity.*

Hicks, A. M., et al. "Transferable Anticancer Innate Immunity in Spontaneous Regression/Complete Resistance Mice." *Proceedings of the National Academy of Sciences U S A* 103, 20 (May 2006). *Note: Cancer resistance from mighty mouse could be transferred to other mice.*

Riedlinger, G., et al. "The Spectrum of Resistance in SR/CR Mice: The Critical Role of Chemo Attraction in the Cancer/Leukocyte Interaction." *BMC Cancer* 10 (May 2010). *Note: These mice are not completely resistant to all cancer cell challenges.*

Review article:

Cui, Z. "The Winding Road to the Discovery of the SR/CR Mice." *Journal of the Academy of Cancer Immunology* 3 (October 2003). *Note: An interesting and personal account of Dr Zheng Cui's discovery of the mighty mouse.*

General information for the interested reader:

"Wake Forest Scientists Develop Colony of Mice That Fight Off Virulent Cancer." *Wake Forest Baptist Medical Center*, Updated July 26, 2013. http://www.wakehealth.edu/News-Releases/2003/Wake_Forest_Scientists_Develop_Colony_of_Mice_That_Fight_Off_Virulent_Cancer.htm (accessed November 23, 2015).

Wang, Brian. "Update on New Cancer Treatment Reported at SENS3." *Next Big Future*, 2007. http://nextbigfuture.com/2007/09/update-on-new-cancer-treatment-reported.html (accessed November 23, 2015).

CHAPTER 18: FRODO OF FLORES

Castro, Joseph. "Homo floresiensis: Facts about the "Hobbit." *Live Science,* July 2015. http://www.livescience.com/29100-homo-floresiensis-hobbit-facts.html (accessed November 24, 2015).

Falk, D., et al. "The Type Specimen LB1 of Homo floresiensis did not have Laron Syndrome." *American Journal Physical Anthropology* 140, 1 (2009).

Henneberg, Maciej, et al. "Evolved Developmental Homeostasis Distributed in LB1 from Flores Indonesia Denoted Down Syndrome and Not Diagnostic Traits of the Invalid Species Homo floresiensis." *Proceedings of the National Academy of Sciences* 111, 33 (2014).

Jungers, W. L., et al. "The Hobbits (Homo floresiensis) Were Not Cretins." *American Journal of Physical Anthropology* S48 (2009). *Note: The title says it all!*

Jungers, W. L., et al. "The Foot of Homo floresiensis." *Nature* 459, 7243 (May 2009).

McKie, Robin. "Homo floresiensis: Scientists Clash over Claims "Hobit Man" was Modern Human with Down's Syndrome." *The Observer, The Guardian,* August 2014. http://www.theguardian.com/science/2014/aug/16/flores-hobbit-human-downs-syndrome-claim-homo-floresiensis (accessed November 24, 2015).

Morwood, M. J., et al. "Archaeology and Age of a New Hominin from Flores in Eastern Indonesia." *Nature* 431 (October 2004).

Obendorph, P. J. et al. "Are the Small Human-Like Fossils on Flores Human Endemic Cretins?" *Proceedings of the Royal Society B: Biological Sciences* 275 (2008).

Orr, C. M., et al. "New Wrist Bones of Homo floresiensis from Liang Bua (Flores, Indonesia)." *Journal of Human Evolution* 64, 2 (February 2013).

Waterman, Hillary. "The Evolutionary Timeline, Retooled." *Discover Magazine* 36, 9 (November 2015). *Note: "The very name of the oldest member of our species, Homo habilis, translates to "handyman" in reference to tool making."*

CHAPTER 19: A CANCER-FREE CLAN?

Green J., et al. "Height and Cancer Incidence in the Million Women Study: Prospective Cohort, and Meta-Analysis of Prospective Studies of Height and Total Cancer Risk." *Lancet Oncology* 12, 8 (2011). *Note: This study is one of several that have suggested that women with greater height might have higher risks risk of cancer.*

Kopchick, J. J. and Zvi Laron. "Is the Laron Mouse an Accurate Model of Laron syndrome?" *Molecular Genetics and Metabolism* 68 (1999).

Rosenbloom, Arlan L. ed. "Growth Hormone Resistance." *Medscape*, Updated September 15, 2015. http://emedicine.medscape.com/article/922902-overview (accessed November 24, 2015). *Note: A very thorough review on the subject, aimed for medical professionals.*

Steuerman, Rachel, Orit Shevah and Zvi Laron. "Congenital IGF1 Deficiency Tends to Confer Protection against Post-Natal Development of Malignancies." *European Journal of Endocrinology* 164 (2011).

Taubes, Gary. "Rare Form of Dwarfism Protects against Cancer." *Discover,* March 27, 2013. http://discovermagazine.com/2013/april/19-double-edged-genes#.UVpm_qt369c (accessed November 24, 2015).

Wade, Nicholas. "Ecuadorean Villagers May Hold Secret to Longevity." *New York Times,* February 16, 2011. http://www.nytimes.com/2011/02/17/science/17longevity.html?_r=0 (accessed November 24, 2015).

CHAPTER 20: CANCER: A DISEASE OF IMMUNE FAILURE?

Masters J. R. "HeLa Cells 50 Years On: The Good the Bad and the Ugly." *Nature Reviews Cancer 2* (2002).

CHAPTER 21: MALIGNANT CARGO

Birkeland, S. A. and H. H. Storm. "Risk for Tumor and Other Disease Transmission by Transplantation: A population-Based Study of Unrecognized Malignancies and Other Diseases in Organ Donors." *Transplantation* 74 (2002).

Buell, J. F., et al. "Donor Transmitted Malignancies." *Annals of Transplantation* 9 (2004).
Lipshutz, G. S., et al. "Death from Metastatic Donor-Derived Ovarian Cancer in a Male Kidney Transplant Recipient." *American Journal of Transplantation* 9 (2009).
Penn, I. "Transmission of Cancer from Organ Donors." *Nefrologia* 15 (1995).

CHAPTER 22: THE POWER OF THE IMMUNE SYSTEM

Burnet, F. M. "Cancer-a Biological Approach." *British Medical Journal* 1, 5023 (1957).
MacKie, R. M., R. Reid and B. Junor. "Fatal Melanoma Transferred in a Donated Kidney 16 Years after Melanoma Surgery." *New England Journal of Medicine* 348 (2003).
Wieczorek-Godlewska R., et al. "Dramatic Recurrence of Cancer in a Patient who Underwent Kidney Transplantation--Case Report." *Transplant Proceedings* 46, 8 (October 2014).

CHAPTER 23: MAN DIES OF OVARIAN CANCER

Dingli, D., and M. A. Nowak. "Cancer Biology: Infectious Tumour Cells." *Nature* 443, 7107 (2006).
Forbes, G. B., et al. "Accidental Transplantation of Bronchial Carcinoma from a Cadaver Donor to Two Recipients of Renal Allografts." *Journal of Clinical Pathology*, 34, 2 (February 1981).
Gugel, E. A. and M. E. Sanders. "Needle-Stick Transmission of Human Colonic Adenocarcinoma." *New England Journal of Medicine* 315, 23 (1986).
Lazebnik, Y. and G. E. Parris. "Comment on: 'Guidelines for the Use of Cell Lines in Biomedical Research': Human-to-Human Cancer Transmission as a Laboratory Safety Concern." *British Journal of Cancer* 112 (2015).
"Man gets woman's cancer after kidney transplant." *Associated Press/CBS News,* May 28, 2010. http://www.cbsnews.com/2100-204_162-6527303.html (accessed November 23, 2015).
Pasquier, B., et al. "Le Potentiel Métastatique des Tumeurs Primitives du Systeme Nerveux Central." *Revue Neurologique (Paris)*, 135, 3 (March 1979). (Article in French except for abstract)
Pasquier, B. "Accidental Transplantation of Tumour Cells." *Journal of Clinical Pathology* 34, 9 (September 1981).
Scanlon, E. F, et al. "Fatal Homotransplanted Melanoma: A Case Report." *Cancer* 18 (1965).
Strauss, D. C. and J. M. Thomas. "Transmission of Donor Melanoma by Organ Transplantation." *Lancet Oncology* 11 (2010).
Welsh, James S. "Contagious Cancer." *The Oncologist* 16, 1 (2011).

CHAPTER 24: MAN'S LIFE SAVED BY MOSQUITO BITE

Banfield, W. G., P. A. Woke, C. M. MacKay and H. L. Cooper. "Mosquito Transmission of a Reticulum Cell Sarcoma of Hamsters." *Science* 148, 3674 (May 1965). *Note: The original scientific paper on the weird transmission of cancer by mosquito bites. Quite frightening even if only in lab rodents!*

Fabrizio, A. M. "An Induced Transmissible Sarcoma in Hamsters: Eleven-Year Observation Through 288 Passages." *Cancer Research* 25 (1965). *Note: Description of the creation of a transmissible malignancy in laboratory golden Syrian hamsters.*

Gartner, H. V., et al. "Genetic Analysis of a Sarcoma Accidentally Transplanted from a Patient to a Surgeon." *New England Journal of Medicine* 335, 20 (1996).

Gugel E. A. and M. E. Sanders. "Needle-Stick Transmission of Human Colonic Adenocarcinoma." *New England Journal of Medicine* 315 (1986).

Lazebnik, Y. and G. E. Parris. "Comment on "Guidelines for the Use of Cell Lines in Biomedical Research." Human to Human Cancer Transmission and a Laboratory Safety Concern." *British Journal of Cancer* 112 (2015).

Mulford, R. D. "Experimentation on Human Beings." *Stanford Law Review* 20, 1 (1967).

Sepkowitz, Kent, M. D. "A Virus's Debut in a Doctor's Syringe." *New York Times*, August 24, 2009. http://www.nytimes.com/2009/08/25/health/25nile.html?mwrsm=Email&_r=0 (accessed November 23, 2013). *Note: A most interesting, well-written and informative article written by a prominent physician at Dr Southam's renowned institution.*

Southam, C. M. "Homotransplantation of Human Cell Lines." *Bulletin of the New York Academy of Medicine* 34 (1958).

Southam, Chester M., Alice E. Moore and Cornelius P. Rhoads. "Homotransplantation of Human Cell Lines." *Science* 125 (1957).

CHAPTER 25: MOLES, MOLES, AND MORE MOLES

Braun-Parvez L., et al. "Gestational Choriocarcinoma Transmission Following Multiorgan Donation." *American Journal of Transplantation* 10 (2010).

CHAPTER 27: THE IMPOSTER

Chen D. Y., et al. "Interspecies Implantation and Mitochondria Fate of Panda-Rabbit Clone Embryos." *Biology of Reproduction* 67, 2 (August 2002).

Fernandez-Arias, A., J. L. Alabart, J. Folch and J. F. Beckers. "Interspecies Pregnancy of Spanish Ibex Fetus in Domestic Goat Recipients Induces Abnormally High Plasmatic Levels of Pregnancy-Associated Glycoprotein." *Theriogenology* 51, 8 (1999).

Hufana-Duran, Danilda, et al. "Full-Term Delivery of River Buffalo Calves (2n = 50) from In Vitro-Derived Vitrified Embryos by Swamp Buffalo Recipients (2n = 48)" *Livestock Science* 107, 2–3 (2007).

CHAPTER 28: COMPETITION: THE CAUSE OF THE CELLULAR DISEASE

Becker, Wayne M., Lewis J. Kleinsmith, Jeff Hardin and Gregory Paul Bertoni. *The World of the Cell, 7th edition.* San Francisco, CA: Pearson Education Inc, Publishing as Pearson Benjamin Cummings, 2009. Chapter 24: "*Cancer Cells.*" *Note: An excellent chapter introducing the cellular biology of cancer as part of a more comprehensive undergraduate cell biology textbook. The concept of tumor initiation promotion and progression are clearly explained.*

Rozhok, A. I. and J. DeGregori. "Toward an Evolutionary Model of Cancer: Considering the Mechanisms that Govern the Fate of Somatic Mutations." *Proceedings of the National Academy of Science USA*, 112, 29 (July 2015).

CHAPTER 29: A STANDARD MODEL OF MOLECULAR ONCOLOGY

Cathcart, Brian. "*The Fly in the Cathedral: How a Small Group of 12.5 ptCambridge Scientists Won the Race to Split the Atom.*" New York: Farrar, Straus and Giroux, LLC, 2005. (originally published by Penguin Group, Great Britain, 2004).

Lederman, Leon M. with Dick Teresi. "*The God Particle: If the Universe Is the Answer, What Is the Question?*" New York, NY: Dell Publishing, 1993.

Lederman, Leon M. and Christopher T. Hill. "*Beyond the God Particle.*" Amherst, NY: Prometheus Books, 2013.

Pollock, Steven. "*Particle Physics For Non Physicists: A Tour Of The Microcosmos.*" Produced by The Great Courses®. Course No. 1247. *Note: An excellent overview of particle physics.*

Trefil, James S. "*From Atoms to Quarks.*" New York: Anchor Books, 1994.

Welsh, James S. "Quarks, Leptons, Fermions, Bosons: The Subatomic Pharmacology of Radiation Therapy." *Science & Medicine* 10, 2 (April 2005). https://www.sciandmed.com/sm/journalviewer.aspx?issue=-1&year=2005#issue1164 (accessed November 23, 2015)

Welsh, James S. "Quarks, Leptons, Fermions, and Bosons—The Subatomic World of Radiation Therapy." *The Oncologist* 11 (2006).
Available at: http://theoncologist.alphamedpress.org/content/11/1/62.full.pdf (accessed November 23, 2015).

CHAPTER 30: RUNAWAY TRAIN!

The Nobel Foundation. "Peyton Rous—Biographical." Nobelprize.org. http://www.nobelprize.org/nobel_prizes/medicine/laureates/1966/rous-bio.html (accessed Dec. 7, 2015).

Rous, Peyton. "A Transmissible Avian Neoplasm. (Sarcoma of The Common Fowl)." *The Journal of Experimental Medicine* 12, 5 (1910).

Wapner, Jessica. "The Philadelphia Chromosome: A Genetic Mystery, A Lethal Cancer, and the Improbable Invention of a Lifesaving Treatment." New York, NY: The Experiment, LLC, 2014.

Wapner, Jessica. "The Story of Peyton Rous and Chicken Cancer." *PLOS Blogs*, February 9, 2012. http://blogs.plos.org/workinprogress/2012/02/09/the-story-of-peyton-rous-and-chicken-cancer/ (accessed November 24, 2015).

CHAPTER 31: ORDER OUT OF CHAOS

Azevedo, P., M. S. Pedrosa, and V. Resende. "Gastroenterology: Hepatosplenic T Cell Lymphoma Associated with Schistosomiasis." *Journal of Gastroenterology and Hepatology* 30, 9 (September 2015).

Baden, H. P., et al. "Treatment of Ichthyosis with Isotretinoin." *Journal of the American Academy of Dermatology,* 6, 4, Pt 2 Suppl (April 1982).

Bizjak, M., et al. "Silicone Implants and Lymphoma: The Role of Inflammation." *Journal of Autoimmunity* (August 2015). [Epub ahead of print].

Chalifoux, L. V. and R. T. Bronson. "Colonic Adenocarcinoma Associated with Chronic Colitis in Cotton Top Marmosets, Saguinus oedipus." *Gastroenterology* 80, 5, prt 1 (May 1981).

Chang, L. M. and Reyes M. "A Case of Harlequin Ichthyosis Treated with Isotretinoin." *Dermatology Online Journal*, 20, 2 (February 2014).

"Definition of Acute Promyelocytic Leukemia." *Medicine.Net.com,* Last editorial review: June 14, 2012. http://www.medicinenet.com/script/main/art.asp?articlekey=19758 (accessed November 23, 2015). *Note: A very brief (one page) online medical article about acute promyelocytic leukemia for interested readers.*

Ferreri A. J., et al. "Bacteria Eradicating Therapy with Doxycycline in Ocular Adnexal

MALT Lymphoma: A Multi-Center Prospective Trial." *Journal National Cancer Institute* 98 (2006).

Hanahan D., and R. A. Weinberg. "The Hallmarks of Cancer." *Cell*, 100, 1 (January 2000). *Note: Among the most significant papers in the history of cancer biology, this article put all the known cellular and molecular features of cancer together into a coherent framework.*

Hvas, C. L, et al. "Celiac Disease: Diagnosis and Treatment." *Danish Medical Journal* 62, 4 (April 2015).

Kotiah, Sandy D. "Acute Promyelocytic Leukemia." *Medscape. Drugs and Diseases*, Updated May 7, 2015. http://emedicine.medscape.com/article/1495306-overview (accessed November 23, 2015).

Lushbaugh, C. C., et al. "Spontaneous Colonic Adenocarcinoma in Marmosets." *Primates in Medical Research* 10 (1987).

Vennervald, B. J. and K. Polman. "Helminths and Malignancy." *Parasite Immunology* 31, 11 (November 2009).

Two references on retinoids and their use as differentiating agents and cancer prevention drugs (which for the most part have not been successful):

Lippman, S. M., et al. "Randomized phase III intergroup trial of Isotretinoin to Prevent Second Primary Tumors in Stage I Non-Small-Cell Lung Cancer." *Journal National Cancer Institute* 93, 8 (April 2001).

Khuri, F. R., et al. "Randomized Phase III Trial of Low-Dose Isotretinoin for Prevention of Second Primary Tumors in Stage I and II Head and Neck Cancer Patients." *Journal National Cancer Institute* 98, 7 (April 2006).

CHAPTER 32: IMMUNE THEORY OF CANCER

Abegglen L. M., et al. "Potential Mechanisms for Cancer Resistance in Elephants and Comparative Cellular Response to DNA Damage in Humans." *JAMA* 314, 17 (2015). *Note: This is the actual scientific publication on elephants and cancer incidence. The reduced cancer incidence is postulated to be linked to an amplification of TP53 genes.*

Becker, Wayne M., Lewis J. Kleinsmith, Jeff Hardin, and Gregory Paul Bertoni. *The World of the Cell, 7th edition.* San Francisco, CA: Pearson Education Inc, Publishing as Pearson Benjamin Cummings, 2009. Chapter 24: "Cancer Cells." *Note: An excellent chapter introducing the cellular biology of cancer as part of a comprehensive undergraduate cell biology textbook. The concepts of tumor initiation, promotion and progression are clearly explained.*

Dunn, G. P., et al. "Cancer Immunoediting: From Immunosurveillance to Tumor Escape." *Nature Immunology* 3, 11 (November 2002).

Dunn, G. P., L. J. Old and R. D. Schreiber "The Three Es of Cancer Immunoediting." *Annual Review of Immunology* 22 (2004).

Greaves, Mel and Luca Ermini. "Evolutionary Adaptations to Risk of Cancer: Evidence from Cancer Resistance in Elephants." *JAMA* 314, 17 (2015). *Note: An interesting editorial to accompany the above JAMA article. It is co-authored by Mel Greaves, the author of the 2000 Oxford University Press book "Cancer: The Evolutionary Legacy."*

Rozhok, A. I. and J. DeGregori. "Toward an Evolutionary Model of Cancer: Considering the Mechanisms that Govern the Fate of Somatic Mutations." *Proceedings of the National Academy of Sciences of the United States of America* 112, 29 (July 2015).

Swann, J. B. and M. J. Smyth "Immune Surveillance of Tumors." *Journal of Clinical Investigation* 117, 5 (2007).

Waters, D. J. and K. Wildasin. "Cancer Clues from Pet Dogs." *Scientific American* 295 (November 2006). *Note: This is where I first read about very long-lived dogs and people (e.g. those over 100) having a reduced likelihood of ultimately succumbing to cancer.*

Zimmer, Carl. "Elephants: Large, Long-Living and Less Prone to Cancer." *New York Times,* October 8, 2015. http://www.nytimes.com/2015/10/13/science/why-elephants-get-less-cancer.html?mabReward=CTM&action=click&pgtype=Homepage®ion=CColumn&module=Recommendation&src=rechp&WT.nav=RecEngine&_r=2 (accessed November 24, 2015).

A print version was published on October 13, 2015, on page D3 of the New York edition with the headline: "Mighty Down to Its DNA." *Note: This recent NY Times piece nicely summarizes yet another clear and puzzling example of Peto's paradox, namely that elephants do NOT get cancer at the rate one would predict based on the mutation theory. In fact, they have surprisingly low rates of cancer. The recent research indicates that elephants use yet another mechanism to ward off cancer, namely multiple copies of the TP53 gene.*

CHAPTER 33: WHAT CAN COWS TEACH US ABOUT CONQUERING CANCER?

Herr, Harry W. and Alvaro Morales. "History of Bacillus Calmette-Guerin and Bladder Cancer: An Immunotherapy Success Story." *The Journal of Urology* 179, 1 (2008). *Note: A historical account written by researchers actually involved with the original discovery decades ago.*

Hopkins, Donald R. *The Greatest Killer: Smallpox in History, with a New Introduction*. Chicago, IL: University Of Chicago Press, 2002. *Note: An authoritative history of smallpox.*

Pearl, R. "Cancer and Tuberculosis" *American Journal of Hygiene* 9 (1929).

Wittes, Robert C. "Immunology of Bacille Calmette-Guerin and Related Topics." *Clinical Infectious Diseases* 31, Supl 3 (2000).

Zbar, B, I. D. Bernstein and H. J. Rapp. "Suppression of Tumor Growth at the Site of Infection with Living Bacillus Calmette-Guérin." *Journal of the National Cancer Institute* 46 (1971).

CHAPTER 35: AS CRAZY AS THE QUANTUM CAFÉ

Chavele, Konstantia-Maria and Michael R. Ehrenstein. "Regulatory T-cells in Systemic Lupus Erythematosus and Rheumatoid Arthritis." *FEBRUARY S Letters* 585, 23 (2011).

CHAPTER 36: CONNECTED DOTS: LOOKING BACK AND GLIMPSING THE FUTURE

Demaria S., et al. "Immune-Mediated Inhibition of Metastases after Treatment with Local Radiation and CTLA-4 Blockade in a Mouse Model of Breast Cancer." *Clinical Cancer Research*, 11 (2005).

Kantoff, P. W., et al. "Sipuleucel-T Immunotherapy for Castration-Resistant Prostate Cancer." *New England Journal of Medicine* 363 (2010). *Note: This controversial study did document an improvement in overall survival with the use of sipuleucel-T immunotherapy in prostate cancer patients. Part of the controversy surrounded the dissociation between clinical outcome and PSA levels, which was something not previously encountered.*

Lock, Michael, et al. "Abscopal Effects: Case Report and Emerging Opportunities." *Cureus* 7, 10 (2015). *Note: A recent case study describing an abscopal phenomenon in a patient with metastatic hepatocellular carcinoma with lung metastases. After stereotactic body radiation therapy to the liver lesion, the multiple lung tumors apparently regressed on CT imaging.*

Postow, Michael A., et al. "Nivolumab and Ipilimumab versus Ipilimumab in Untreated Melanoma." *New England Journal of Medicine* 372, 21 (2015).

Victor, Twyman-Saint C., et al. "Radiation and Dual Checkpoint Blockade Activate Non-Redundant Immune Mechanisms in Cancer." *Nature* 520, 7547 (April 2015). *Note: One of the numerous new studies suggesting that a* combination *of radiation plus checkpoint inhibition of CTLA4 and PD-L1 could prove more beneficial than any one alone.*

INDEX

Page references followed by *g* indicate glossary entries.